Basic Electrical
Installation Work

Basic Electrical Installation Work

Fourth Edition

TREVOR LINSLEY

Senior Lecturer
Blackpool and The Fylde College

AMSTERDAM • BOSTON • HEIDELBERG • LONDON • NEW YORK • OXFORD
PARIS • SAN DIEGO • SAN FRANCISCO • SINGAPORE • SYDNEY • TOKYO

Newnes is an imprint of Elsevier

ELSEVIER

Newnes

Newnes
An imprint of Elsevier
Linacre House, Jordan Hill, Oxford OX2 8DP
30 Corporate Drive, Burlington, MA 01803

First published by Arnold 1998
Reprinted by Butterworth-Heinemann 2001, 2002, 2003 (twice), 2004 (twice)
Fourth edition 2005

British Library Cataloguing in Publication Data
A catalogue record for this book is available from the British Library

For information on all Newnes publications
visit our web site at www.newnespress.com

ISBN 0 7506 66242

Typeset by Charon Tec Pvt. Ltd, Chennai, India
www.charontec.com
Printed and bound in Great Britain

Working together to grow
libraries in developing countries

www.elsevier.com | www.bookaid.org | www.sabre.org

ELSEVIER   BOOK AID
            International   Sabre Foundation

CONTENTS

—

PREFACE

—

The fourth edition of *Basic Electrical Installation Work* has been written as a complete textbook for the City and Guilds 2330 Level 2 Certificate in Electrotechnical Technology and the City and Guilds 2356 Level 2 NVQ in Installing Electrotechnical Systems. The book meets the combined requirements of these courses, that is the core units and the electrical installation occupational units and therefore students need purchase only this one textbook for all subjects in the Level 2 examinations.

The book will also assist students taking the SCOTVEC and BTEC Electrical and Utilization units at levels I and II and many taking engineering NVQ and Modern Apprentiship courses.

Although the text is based upon the City and Guilds syllabus, the book also provides a sound basic knowledge and comprehensive guide for other professionals in the construction and electrotechnical industry.

Modern regulations place a greater responsibility upon the installing electrician for safety and the design of an installation. The latest regulations governing electrical installations are the 16th edition of the IEE Wiring Regulations (BS 7671: 2001). The fourth edition of this book has been revised and updated to incorporate the requirements and amendments of the 16th edition of the IEE Wiring Regulations BS7671: 2001.

The City and Guilds examinations comprise assignments and multiple-choice written papers. For this reason multiple-choice questions can be found at the end of each chapter. More traditional questions are included as an aid to private study and to encourage a thorough knowledge of the subject.

I would like to acknowledge the assistance given by the following manufacturers and organizations in the preparation of this book:

Crabtree Electrical Industries Limited
Wylex Ltd
RS Components Ltd
The Institution of Electrical Engineers
The British Standards Institution
The City & Guilds of London Institute

I would also like to thank my colleagues and students at Blackpool and The Fylde College for their suggestions and assistance during the preparation of this book.

Finally, I would like to thank Joyce, Samantha and Victoria for their support and encouragement.

Trevor Linsley

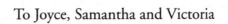
To Joyce, Samantha and Victoria

WORKING EFFECTIVELY AND SAFELY

—

Safety regulations and laws

At the beginning of the nineteenth century children formed a large part of the working population of Great Britain. They started work early in their lives and they worked long hours for unscrupulous employers or masters.

The Health and Morals of Apprentices Act of 1802 was introduced by Robert Peel in an attempt at reducing apprentice working hours to twelve per day and improving the conditions of their employment. The Factories Act of 1833 restricted the working week for children aged 13 to 18 years to sixty-nine hours in any working week.

With the introduction of the Factories Act of 1833, the first four full time Factory Inspectors were appointed. They were allowed to employ a small number of assistants and were given the responsibility of inspecting factories throughout England, Scotland, Ireland and Wales. This small, overworked band of men, were the forerunners of the modern HSE Inspectorate, enforcing the safety laws passed by Parliament. As the years progressed, new Acts of Parliament increased the powers of the Inspectorate and the growing strength of the Trade Unions meant that employers were increasingly being pressed to improve health, safety and welfare at work.

The most important recent piece of health and safety law was passed by Parliament in 1974 called the Health and Safety at Work Act. This Act gave added powers to the Inspectorate and is the basis of all modern statutory health and safety laws. This Law not only increased the employer's liability for safety measures but also put the responsibility for safety on employees too.

Health, safety and welfare legislation has increased the awareness of everyone to the risks involved in the workplace. All statutes within the Acts of Parliament must be obeyed and, therefore, we all need an understanding of the laws as they apply to our electrotechnical industry.

STATUTORY LAWS

Acts of Parliament are made up of Statutes. Statutory Regulations have been passed by Parliament and have, therefore, become laws. Non-compliance with the laws of this land may lead to prosecution by the Courts and possible imprisonment for offenders.

We shall now look at three Statutory Regulations as they apply to the electrotechnical industry. They are: The Health and Safety at Work Act 1974, The Electricity, Safety, Quality and Continuity Regulations 2002 (formerly Electricity Supply Regulations 1989), and the Electricity at Work Regulations 1989.

The Health and Safety at Work Act 1974

Many governments have passed laws aimed at improving safety at work but the most important recent legislation has been the Health and Safety at Work Act 1974. The purpose of the Act is to provide the legal framework for stimulating and encouraging high standards of health and safety at work; the Act puts the responsibility for safety at work on both workers and managers.

The employer has a duty to care for the health and safety of employees (Section 2 of the Act). To do this he must ensure that:

- the working conditions and standard of hygiene are appropriate;
- the plant, tools and equipment are properly maintained;
- the necessary safety equipment – such as personal protective equipment, dust and fume extractors and machine guards – is available and properly used;
- the workers are trained to use equipment and plant safely.

Employees have a duty to care for their own health and safety and that of others who may be affected by their actions (Section 7 of the Act). To do this they must

- take reasonable care to avoid injury to themselves or others as a result of their work activity;
- co-operate with their employer, helping him or her to comply with the requirements of the Act;
- not interfere with or misuse anything provided to protect their health and safety.

Failure to comply with the Health and Safety at Work Act is a criminal offence and any infringement of the law can result in heavy fines, a prison sentence or both.

ENFORCEMENT

Laws and rules must be enforced if they are to be effective. The system of control under the Health and Safety at Work Act comes from the Health and Safety Executive (HSE) which is charged with enforcing the law. The HSE is divided into a number of specialist inspectorates or sections which operate from local offices throughout the UK. From the local offices the inspectors visit individual places of work.

The HSE inspectors have been given wide-ranging powers to assist them in the enforcement of the law. They can:

1 enter premises unannounced and carry out investigations, take measurements or photographs;
2 take statements from individuals;
3 check the records and documents required by legislation;
4 give information and advice to an employee or employer about safety in the workplace;

5 demand the dismantling or destruction of any equipment, material or substance likely to cause immediate serious injury;
6 issue an improvement notice which will require an employer to put right, within a specified period of time, a minor infringement of the legislation;
7 issue a prohibition notice which will require an employer to stop immediately any activity likely to result in serious injury, and which will be enforced until the situation is corrected;
8 prosecute all persons who fail to comply with their safety duties, including employers, employees, designers, manufacturers, suppliers and the self-employed.

SAFETY DOCUMENTATION

Under the Health and Safety at Work Act, the employer is responsible for ensuring that adequate instruction and information is given to employees to make them safety-conscious. Part 1, Section 3 of the Act instructs all employers to prepare a written health and safety policy statement and to bring this to the notice of all employees. Figure 1.1 shows a typical Health and Safety Policy Statement of the type which will be available within your Company. Your employer must let you know who your safety representatives are and the new Health and Safety poster shown in Fig. 1.2 has a blank section into which the names and contact information of your specific representatives can be added. This is a large laminated poster, 595 × 415 mm suitable for wall or notice board display.

All workplaces employing five or more people must display the type of poster shown in Fig. 1.2 after 30th June 2000.

To promote adequate health and safety measures the employer must consult with the employees' safety representatives. In companies which employ more than 20 people this is normally undertaken by forming a safety committee which is made up of a safety officer and employee representatives, usually nominated by a trade union. The safety officer is usually employed full-time in that role. Small companies might employ a safety supervisor, who will have other duties within the company, or alternatively they could join a 'safety group'. The safety group then shares the cost of employing a safety adviser or safety officer, who visits each company in rotation. An employee who identifies a dangerous situation should initially report to his site

FLASH-BANG ELECTRICAL

Statement of Health and Safety at Work Policy in accordance with the Health and Safety at Work Act 1974

Company objective

The promotion of health and safety measures is a mutual objective for the Company and for its employees at all levels. It is the intention that all the Company's affairs will be conducted in a manner which will not cause risk to the health and safety of its members, employees or the general public. For this purpose it is the Company policy that the responsibility for health and safety at work will be divided between all the employees and the Company in the manner outlined below.

Company's responsibilities

The Company will, as a responsible employer, make every endeavour to meet its legal obligations under the Health and Safety at Work Act to ensure the health and safety of its employees and the general public. Particular attention will be paid to the provision of the following:

1 Plant equipment and systems of work that are safe.
2 Safe arrangements for the use, handling, storage and transport of articles, materials and substances.
3 Sufficient information, instruction, training and supervision to enable all employees to contribute positively to their own safety and health at work and to avoid hazards.
4 A safe place of work, and safe access to it.
5 A healthy working environment.
6 Adequate welfare services.

Note: Reference should be made to the appropriate safety etc. manuals.

Employees' responsibilities

Each employee is responsible for ensuring that the work which he/she undertakes is conducted in a manner which is safe to himself or herself, other members of the general public, and for obeying the advice and instructions on safety and health matters issued by his/her superior. If any employee considers that a hazard to health and safety exists it is his/her responsibility to report the matter to his/her supervisor or through his/her Union Representative or such other person as may be subsequently defined.

Management and Supervisors' responsibilities

Management and supervisors at all levels are expected to set an example in safe behaviour and maintain a constant and continuing interest in employee safety, in particular by:

1 acquiring the knowledge of health and safety regulations and codes of practice necessary to ensure the safety of employees in the workplace,
2 acquainting employees with these regulations on codes of practice and giving guidance on safety matters,
3 ensuring that employees act on instructions and advice given.

General Managers are ultimately responsible to the Company for the rectification or reporting of any safety hazard which is brought to their attention.

Joint consultations

Joint consultation on health and safety matters is important. The Company will agree with its staff, or their representatives, adequate arrangements for joint consultation on measures for promoting safety and health at work, and make and maintain satisfactory arrangements for the participation of their employees in the development and supervision of such measures. Trade Union representatives will initially be regarded as undertaking the role of Safety Representatives envisaged in the Health and Safety at Work Act. These representatives share a responsibility with management to ensure the health and safety of their members and are responsible for drawing the attention of management to any shortcomings in the Company's health and safety arrangements. The Company will in so far as is reasonably practicable provide representatives with facilities and training in order that they may carry out this task.

Review

A review, addition or modification of this statement may be made at any time and may be supplemented as appropriate by further statements relating to the work of particular departments and in accordance with any new regulations or codes of practice.
   This policy statement will be brought to the attention of all employees.

Fig. 1.1 Typical Health and Safety Policy Statement.

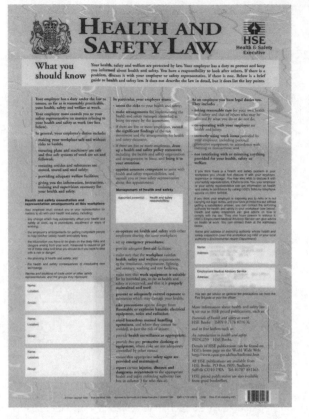

Fig. 1.2 New Health and Safety Law Poster. Source: HSE © Crown copyright material is reproduced with the permission of the Controller of HMSO and Her Majesty's Stationery Office, Norwich.

safety representative. The safety representative should then bring the dangerous situation to the notice of the safety committee for action which will remove the danger. This may mean changing company policy or procedures or making modifications to equipment. All actions of the safety committee should be documented and recorded as evidence that the company takes seriously its health and safety policy.

The Electricity Safety, Quality and Continuity Regulations 2002 (formerly Electricity Supply Regulations 1989)

The Electricity Safety, Quality and Continuity Regulations 2002 are issued by the Department of Trade and Industry. They are statutory regulations

which are enforceable by the laws of the land. They are designed to ensure a proper and safe supply of electrical energy up to the consumer's terminals.

These regulations impose requirements upon the regional electricity companies regarding the installation and use of electric lines and equipment. The regulations are administered by the Engineering Inspectorate of the Electricity Division of the Department of Energy and will not normally concern the electrical contractor except that it is these regulations which lay down the earthing requirement of the electrical supply at the meter position.

The regional electricity companies must declare the supply voltage and maintain its value between prescribed limits or tolerances.

The government agreed on 1 January 1995 that the electricity supplies in the United Kingdom would be harmonized with those of the rest of Europe. Thus the voltages used previously in low-voltage supply systems of 415 V and 240 V have become 400 V for three-phase supplies and 230 V for single-phase supplies. The permitted tolerances to the nominal voltage have also been changed from ±6% to +10% and −6%. This gives a voltage range of 216 V to 253 V for a nominal voltage of 230 V and 376 V to 440 V for a nominal supply voltage of 400 V.

The next change will come in 2005, when the tolerance levels will be adjusted to ±10% of the declared nominal voltage.

The frequency is maintained at an average value of 50 Hz over 24 hours so that electric clocks remain accurate.

Regulation 29 gives the area boards the power to refuse to connect a supply to an installation which in their opinion is not constructed, installed and protected to an appropriately high standard. This regulation would only be enforced if the installation did not meet the requirements of the IEE Regulations for Electrical Installations.

The Electricity at Work Regulations 1989 (EWR)

This legislation came into force in 1990 and replaced earlier regulations such as the Electricity (Factories Act) Special Regulations 1944. The Regulations are made under the Health and Safety at Work Act 1974, and enforced by the Health and Safety Executive. The purpose of the Regulations is to 'require precautions to be taken against the risk of death or personal injury from electricity in work activities'.

Section 4 of the EWR tells us that 'all systems must be constructed so as to prevent danger ..., and be properly maintained.... Every work activity shall be carried out in a manner which does not give rise to danger.... In the case of work of an electrical nature, it is preferable that the conductors be made dead before work commences'.

The EWR do not tell us specifically how to carry out our work activities and ensure compliance, but if proceedings were brought against an individual for breaking the EWR, the only acceptable defence would be 'to prove that all reasonable steps were taken and all diligence exercised to avoid the offence' (Regulation 29).

An electrical contractor could reasonably be expected to have 'exercised all diligence' if the installation was wired according to the IEE Wiring Regulations (see below). However, electrical contractors must become more 'legally aware' following the conviction of an electrician for manslaughter at Maidstone Crown Court in 1989. The Court accepted that an electrician had caused the death of another man as a result of his shoddy work in wiring up a central heating system. He received a 9 month suspended prison sentence. This case has set an important legal precedent, and in future any tradesman or professional who causes death through negligence or poor workmanship risks prosecution and possible imprisonment.

Non-statutory regulations

Statutory laws and regulations are written in a legal framework, they do not actually tell us how to comply with the laws at an everyday level.

Non-statutory regulations and codes of practice interpret the statutory regulations. They have been written for every specific section of industry, commerce and situation to enable everyone to comply with the Health and Safety laws.

When the Electricity at Work Regulations (EWR) tell us to 'ensure that all systems are constructed so as to prevent danger' they do not tell us how to actually do this in a specific situation. However, the IEE Regulations tell us precisely how to carry out our electrotechnical work safely in order to meet the statutory requirements of the EWR. If your electrotechnical work meets the requirements of the IEE Regulations you will also meet the requirements of the Electricity at Work Regulations.

Over the years non-statutory regulations and codes of practice have built upon previous good practice and responded to changes by bringing out new editions of the various regulations and new codes of practice to meet the changing needs of industry and commerce.

We will now look at six non-statutory regulations – the Management of Health and Safety Regulations 1992, the COSHH Regulations 2002, the Provision and Use of Work Equipment Regulations 1992, the Construction (Health, Safety and Welfare) Regulations 1996, Personal Protective Equipment Regulations 1992 and finally, at what is sometimes called 'the electricians' bible' the most important set of regulations for anyone working in the electrotechnical industry, the IEE Requirements for Electrical Installations (BS7671:2001).

The Management of Health and Safety at Work Regulations 1999

The Health and Safety at Work Act 1974 places responsibilities on employers to have robust health and safety systems and procedures in the workplace. Directors and managers of any company who employ more than five employees can be held personally responsible for failures to control health and safety.

The Management of Health and Safety at Work Regulations 1999 tell us that employers must systematically examine the workplace, the work activity and the management of safety in the establishment through a process of 'risk assessments.' A record of all significant risk assessment findings must be kept in a safe place and be available to an HSE inspector if required. Information based on these findings must be communicated to relevant staff and if changes in work behaviour patterns are recommended in the interests of safety, then they must be put in place. The process of risk assessment is considered in detail in Chapter 3 of this book.

Risks, which may require a formal assessment in the electrotechnical industry, might be:

■ working at heights;
■ using electrical power tools;
■ falling objects;
■ working in confined places;
■ electrocution and personal injury;
■ working with 'live' equipment;

- using hire equipment;
- manual handling – pushing – pulling – lifting;
- site conditions – falling objects – dust – weather – water – accidents and injuries.

And any other risks which are particular to a specific type of workplace or work activity.

The Control of Substances Hazardous to Health Regulations 1988

The original COSHH Regulations were published in 1988 and came into force in October 1989. They were re-enacted in 1994 with modifications and improvements, and the latest modifications and additions came into force in 2002.

The COSHH Regulations control people's exposure to hazardous substances in the workplace. Regulation 6 requires employers to assess the risks to health from working with hazardous substances, to train employees in techniques which will reduce the risk and provide personal protective equipment (PPE) so that employees will not endanger themselves or others through exposure to hazardous substances. Employees should also know what cleaning, storage and disposal procedures are required and what emergency procedures to follow. The necessary information must be available to anyone using hazardous substances as well as to visiting HSE inspectors.

Hazardous substances include:

1 any substance which gives off fumes causing headaches or respiratory irritation;
2 man-made fibres which might cause skin or eye irritation (e.g. loft insulation);
3 acids causing skin burns and breathing irritation (e.g. car batteries, which contain dilute sulphuric acid);
4 solvents causing skin and respiratory irritation (strong solvents are used to cement together PVC conduit fittings and tube);
5 fumes and gases causing asphyxiation (burning PVC gives off toxic fumes);
6 cement and wood dust causing breathing problems and eye irritation;
7 exposure to asbestos – although the supply and use of the most hazardous asbestos material is now prohibited, huge amounts were installed between

1950 and 1980 in the construction industry and much of it is still in place today. In their latest amendments the COSHH Regulations focus on giving advice and guidance to builders and contractors on the safe use and control of asbestos products. These can be found in Guidance Notes EH 71.

Where personal protective equipment is provided by an employer, employees have a duty to use it to safeguard themselves.

Provision and Use of Work Equipment Regulations 1998

These regulations tidy up a number of existing requirements already in place under other regulations such as the Health and Safety at Work Act 1974, the Factories Act 1961 and the Offices, Shops and Railway Premises Act 1963.

The Provision and Use of Work Equipment Regulations 1998 places a general duty on employers to ensure minimum requirements of plant and equipment. If an employer has purchased good quality plant and equipment, which is well maintained, there is little else to do. Some older equipment may require modifications to bring it in line with modern standards of dust extraction, fume extraction or noise, but no assessments are required by the regulations other than those generally required by the Management Regulations 1999 discussed previously.

The Construction (Health, Safety and Welfare) Regulations 1996

An electrical contractor is a part of the construction team, usually as a subcontractor, and therefore the regulations particularly aimed at the construction industry also influence the daily work procedures and environment of an electrician. The most important recent piece of legislation are the Construction Regulations.

The temporary nature of construction sites makes them one of the most dangerous places to work. These regulations are made under the Health and Safety at Work Act 1974 and are designed specifically to promote safety at work in the construction industry. Construction work is defined as any building or civil engineering work, including construction, assembly, alterations, conversions, repairs, upkeep, maintenance or dismantling of a structure.

The general provision sets out minimum standards to promote a good level of safety on site. Schedules specify the requirements for guardrails, working platforms, ladders, emergency procedures, lighting and welfare facilities. Welfare facilities set out minimum provisions for site accommodation: washing facilities, sanitary conveniences and protective clothing. There is now a duty for all those working on construction sites to wear head protection, and this includes electricians working on site as subcontractors.

Personal Protective Equipment (PPE) at Work Regulations 1998

PPE is defined as all equipment designed to be worn, or held, to protect against a risk to health and safety. This includes most types of protective clothing, and equipment such as eye, foot and head protection, safety harnesses, life jackets and high visibility clothing.

Under the Health and Safety at Work Act, employers must provide free of charge any personal protective equipment and employees must make full and proper use of it. Safety signs such as those shown at Fig. 1.3 are useful reminders of the type of PPE to be used in a particular area. The vulnerable parts of the body which may need protection are the head, eyes, ears, lungs, torso, hands and feet and, additionally, protection from falls may need to be considered. Objects falling from a height present the major hazard against which head protection is provided. Other hazards include striking the head against projections and hair becoming entangled in machinery. Typical methods of protection include helmets, light duty scalp protectors called 'bump caps' and hairnets.

The eyes are very vulnerable to liquid splashes, flying particles and light emissions such as ultraviolet light, electric arcs and lasers. Types of eye protectors include safety spectacles, safety goggles and face shields. Screen based workstations are being used increasingly in industrial and commercial locations by all types of personnel. Working with VDUs (visual display units) can cause eye strain and fatigue and, therefore, this hazard is the subject of a separate section at the beginning of Chapter 3 headed VDU operation hazards.

Noise is accepted as a problem in most industries and surprisingly there has been very little control legislation. The Health and Safety Executive have published a 'Code of Practice' and 'Guidance Notes' HSG 56 for reducing the exposure of employed persons to

Fig. 1.3 Safety signs showing type of PPE to be worn.

noise. A continuous exposure limit of below 85 dB for an eight hour working day is recommended by the code.

Noise may be defined as any disagreeable or undesirable sound or sounds, generally of a random nature, which do not have clearly defined frequencies. The usual basis for measuring noise or sound level is the decibel scale. Whether noise of a particular level is harmful or not also depends upon the length of exposure to it. This is the basis of the widely accepted limit of 85 dB of continuous exposure to noise for eight hours per day.

A peak sound pressure of above 200 pascals or about 120 dB is considered unacceptable and 130 dB is the threshold of pain for humans. If a person has to shout to be understood at two metres, the background noise is about 85 dB. If the distance is only one metre, the noise level is about 90 dB. Continuous noise at work causes deafness, makes people irritable, affects concentration, causes fatigue and accident proneness and may mask sounds which need to be heard in order to work efficiently and safely.

It may be possible to engineer out some of the noise, for example by placing a generator in a separate soundproofed building. Alternatively, it may be possible to provide job rotation, to rearrange work locations or provide acoustic refuges.

Where individuals must be subjected to some noise at work it may be reduced by ear protectors. These may be disposable ear plugs, re-usable ear plugs or ear muffs. The chosen ear protector must be suited to the user and suitable for the type of noise and individual personnel should be trained in its correct use.

Breathing reasonably clean air is the right of every individual, particularly at work. Some industrial processes produce dust which may present a potentially serious hazard. The lung disease asbestosis is caused by the inhalation of asbestos dust or particles and the coal dust disease pneumoconiosis, suffered by many coal miners, has made people aware of the dangers of breathing in contaminated air.

Some people may prove to be allergic to quite innocent products such as flour dust in the food industry or wood dust in the construction industry. The main effect of inhaling dust is a measurable impairment of lung function. This can be avoided by wearing an appropriate mask, respirator or breathing apparatus as recommended by the company's health and safety policy and indicated by local safety signs such as those shown in Fig. 1.4.

Fig. 1.4   Breathing protection signs.

A worker's body may need protection against heat or cold, bad weather, chemical or metal splash, impact or penetration and contaminated dust. Alternatively, there may be a risk of the worker's own clothes causing contamination of the product, as in the food industry. Appropriate clothing will be recommended in the company's health and safety policy. Ordinary working clothes and clothing provided for food hygiene purposes are not included in the Personal Protective Equipment at Work Regulations. Figure 1.5 shows typical safety signs to be found in the food industry.

Hands and feet may need protection from abrasion, temperature extremes, cuts and punctures, impact or skin infection. Gloves or gauntlets provide protection from most industrial processes but should not be worn when operating machinery because they may become entangled in it. Care in selecting the appropriate

Fig. 1.5   Safety signs to be found in the food industry.

protective device is required; for example, barrier creams provide only a limited protection against infection.

Boots or shoes with in-built toe caps can give protection against impact or falling objects and, when fitted with a mild steel sole plate, can also provide protection from sharp objects penetrating through the sole. Special slip resistant soles can also be provided for employees working in wet areas.

Whatever the hazard to health and safety at work, the employer must be able to demonstrate that he or she has carried out a risk analysis, made recommendations which will reduce that risk and communicated these recommendations to the workforce. Where there is a need for PPE to protect against personal injury and to create a safe working environment, the employer must provide that equipment and any necessary training which might be required and the employee must make full and proper use of such equipment and training.

The IEE Wiring Regulations to BS 7671: 2001 Requirements for Electrical Installations

The Institution of Electrical Engineers Requirements for Electrical Installations (the IEE Regulations) are non-statutory regulations. They relate principally to the design, selection, erection, inspection and testing of electrical installations, whether permanent or temporary, in and about buildings generally and to agricultural and horticultural premises, construction sites and caravans and their sites. Paragraph 7 of the introduction to the EWR says: 'the IEE Wiring Regulations is a code of practice which is widely recognised and accepted in the United Kingdom and compliance with them is likely to achieve compliance with all relevant aspects of the Electricity at Work Regulations'. The IEE Wiring Regulations only apply to installations operating at a voltage up to 1000 V a.c. They do not apply to electrical installations in mines and quarries, where special regulations apply because of the adverse conditions experienced there.

The current edition of the IEE Wiring Regulations is the 16th edition incorporating amendment number 1: 2002 and 2: 2004. The main reason for incorporating the IEE Wiring Regulations into British

Standard BS 7671 was to create harmonization with European standards.

To assist electricians in their understanding of the Regulations a number of guidance notes have been published. The guidance notes which I will frequently make reference to in this book are those contained in the *On Site Guide*. Seven other guidance notes booklets are also currently available. These are:

- *Selection and Erection*
- *Isolation and Switching*
- *Inspection and Testing*
- *Protection against Fire*
- *Protection against Electric Shock*
- *Protection against Overcurrent*
- *Special Locations.*

These guidance notes are intended to be read in conjunction with the Regulations.

The IEE Wiring Regulations are the electricians bible and provide the authoritative framework of information for anyone working in the electrotechnical industry.

Health and safety responsibilities

We have now looked at three statutory and six non-statutory regulations which influence working conditions in the electrotechnical industry today. So, who has *responsibility* for these workplace Health and Safety Regulations?

In 1970 a Royal Commission was set up to look at the health and safety of employees at work. The findings concluded that the main cause of accidents at work was apathy on the part of *both* employers and employees.

The Health and Safety at Work Act 1974 was passed as a result of recommendations made by the Royal Commission and, therefore, the Act puts legal responsibility for safety at work on *both* the employer and employee.

In general terms, the employer must put adequate health and safety systems in place at work and the employee must use all safety systems and procedures responsibly.

In specific terms the employer must:

- provide a Health and Safety policy statement if there are five or more employees such as that shown in Fig. 1.1;

- display a current employers liability insurance certificate as required by the Employers Liability (Compulsory Insurance) Act 1969;
- report certain injuries, diseases and dangerous occurrences to the enforcing authority (HSE Area Office – see Appendix for address);
- provide adequate first aid facilities (see Tables 1.1 and 1.2);
- provide personal protective equipment;
- provide information, training and supervision to ensure staffs' health and safety;
- provide adequate welfare facilities;
- put in place adequate precautions against fire, provide a means of escape and means of fighting fire;
- ensure plant and machinery are safe and that safe systems of operation are in place;
- ensure articles and substances are moved, stored and used safely;
- make the workplace safe and without risk to health by keeping dust, fumes and noise under control.

In specific terms the employee must:

- take reasonable care of his/her own health and safety and that of others who may be affected by what they do;
- co-operate with his/her employer on health and safety issues by not interfering or misusing anything provided for health, safety and welfare in the working environment;
- report any health and safety problem in the workplace to, in the first place, a supervisor, manager or employer.

Employment – rights and responsibilities

As a trainee in the electrotechnical industry you will be employed by a member company and receive a weekly or monthly wage, which will be dependent upon your age and grade as agreed by the appropriate trade union, probably Amicus.

We have seen in the beginning of this chapter that there are many rules and regulations which your employer must comply with in order to make your work-place healthy and safe. There are also responsibilities that apply to you, as an employee (or worker)

in the electrotechnical industry, in order to assist your employer to obey the law.

As an Employee you must:

- obey all lawful and reasonable requests;
- behave in a sensible and responsible way at work;
- work with care and reasonable skill.

Your Employer must:

- take care for your safety;
- not ask you to do anything unlawful or unreasonable;
- pay agreed wages;
- not change your contract of employment without your agreement.

Most of the other things that can be expected of you are things like honesty, punctuality, reliability and hard work. Really, just common sense things like politeness will help you to get on at work.

If you have problems relating to your employment rights you should talk it through with your supervisor or trade union representative at work.

WAGES AND TAX

When you start work you will be paid either weekly or monthly. It is quite common to work a week in hand if you are paid weekly, which means that you will be paid for the first week's work at the end of the second week. When you leave that employment, if you have worked a week in hand, you will have a week's wage to come. Money that you have worked for belongs to you and cannot be kept by your employer if you leave without giving notice.

Every employee is entitled to a payslip along with their wages, which should show how must you have earned (gross), how much has been taken off for tax and national insurance and what your take home pay (net) is.

If you are not given a payslip, ask for one, it is your legal right and you may be required to show payslips as proof of income. Always keep your payslips in a safe place.

We all pay tax on the money we earn (income tax). The Government uses tax to pay for services such as health, education, defence, social security and pensions.

We are all allowed to earn a small amount of money tax free each year and this is called the personal allowance. The personal allowance for the tax year in which I am writing this book 2004/2005 is £4745. So every pound that we earn above £4745 is taxed. The tax year starts on the 6th of April each year and finishes on the 5th of April the following year. Your personal tax code enables the personal allowance to be spread out throughout the year and you pay tax on each of your wages on a system called PAYE, pay as you earn.

At the end of the tax year your employers will give you a form called a P.60, which shows your tax code, how much you have earned and how much tax you have paid during a particular tax year. It is important to keep your P.60 somewhere safe, along with your payslips. If at some time you want to buy a house a building society will want proof of your earnings, which these documents show.

When leaving a particular employment you must obtain from your employer a form P.45. On starting new employment this form will be required by your new employer and will ensure that you do not initially pay too much tax.

WORKING HOURS

Employees cannot be forced to work more than forty-eight hours each week on average, and forty hours for 16 to 18 year old trainees. Trainees must also have twelve hours uninterrupted rest from work each day. Older workers, required to work for more than six hours continuously, are entitled to a twenty minute rest break, to be taken within the six hours, and must have eleven hours uninterrupted rest from work each day. If you think you are not getting the correct number of breaks, talk to your supervisor or trade union representative.

SICKNESS

If you are sick and unable to go to work you should contact your employer or supervisor as soon as you can on the first day of illness. When you go back to work, if you have been sick for up to seven days, you will have to fill in a self-certification form. After seven days you will need a medical certificate from your doctor and you must send it to work as soon as you can. If you are sick for four days or more your employer must pay you statutory sick pay (SSP), which can be paid for up to twenty-eight weeks. If you are sick after twenty-eight weeks you can claim incapacity benefit. To claim this you will need a form from your employer or Social Security Office. If you have a sickness problem, talk to your supervisor or someone at work who you trust, or telephone the local Social Security Office.

ACCIDENTS

It is the employer's duty to protect the health and safety and welfare of its employees, so if you do have an accident at work, however small, inform your supervisor, safety officer or first aid person. Make sure that the details are recorded in the accident/first aid book. Failure to do so may affect compensation if the accident proves to be more serious than you first thought.

Always be careful, use commonsense and follow instructions. If in doubt, ask someone. A simple accident might prevent you playing your favourite sport for a considerable period of time.

HOLIDAYS

Most employees are entitled to at least four weeks paid holiday each year. Your entitlement to paid holidays builds up each month, so a month after you start work you are entitled to one twelfth of the total holiday entitlement for the year. After two months it becomes two twelfths and so on. Ask your supervisor or the kind lady in the office who makes up the wages to explain your holiday entitlement to you.

PROBLEMS AT WORK

It is not unusual to find it hard to fit in when you start a new job. Give it a chance, give it time and things are likely to settle down. As a new person you might seem to get all the rotten jobs, but sometimes, being new, these are the only jobs that you can do for now.

In some companies there can be a culture of 'teasing', which may be O.K, if everyone is treated the same, but not so good if you are always the one being teased. If this happens, see if it stops after a while, if not, talk to someone about it. Don't give up your job without trying to get the problem sorted out.

If you feel that you are being discriminated against or harassed because of your race, sex or disability, then talk to your supervisor, trainer or someone you trust at work. There are laws about discrimination that are discussed in Advanced Electrical Installation Work.

You can join a Trade Union when you are 16 years of age or over. Trade Unions work toward fair deals for their members. If you join a Trade Union there will be subscriptions (subs) to pay. These are often reduced or suspended during the training period. As a member of a Trade Union you can get advice and support from them. If there is a problem of any kind at work, you can ask for Union support. However, you cannot get this support unless you are a member.

RESIGNATION/DISMISSAL

Most employers like to have your resignation or 'Notice' in writing. Your Contract of Employment will tell you how much Notice is expected. The minimum Notice you should give is one week if you have been employed for one month or more by that employer. However, if your Contract states a longer period, then that is what is expected.

If you have worked for one month or more, but less than two years, you are entitled to one week's Notice. If you have worked for two years you are entitled to two week's Notice and a further week's Notice for every additional continuous year of employment (with the same employer) up to twelve weeks for twelve years service.

If you are dismissed or 'sacked' you are entitled to the same periods of Notice. However, if you do something very serious, like stealing or hitting someone, your employer can dismiss you without Notice.

You can also be dismissed if you are often late or your behaviour is inappropriate to the type of work being done. You should have verbal or written warnings before you are dismissed.

If there are twenty or more employees at your place of employment then there should be a disciplinary procedure written down which must be followed. If you do get a warning, then you might like to see this as a second chance to start again.

If you have been working for the same employer for one year or more, you can complain to an Employment Tribunal if you think you have been unfairly dismissed. If you haven't worked for the same employer for this length of time, then you should talk to your training officer or Trade Union.

I do not want to finish this section in a negative way, talking about problems at work, so let me finally say

that each year over eight thousand young people are in apprenticeships in the electrical contracting industry and very few of them have problems. The small problems that may arise, because moving into full-time work is very different to school, can usually be resolved by your training officer or supervisor. Most of the trainees go on to qualify as craftsmen and enjoy a well paid and fulfilling career in the electrotechnical industry.

Safety signs

The rules and regulations of the working environment are communicated to employees by written instructions, signs and symbols. All signs in the working environment are intended to inform. They should give warning of possible dangers and must be obeyed. At first there were many different safety signs but British Standard BS 5378 Part 1 (1980) and the Health and Safety (Signs and Signals) Regulations 1996 have introduced a standard system which gives health and safety information with the minimum use of words. The purpose of the regulations is to establish an internationally understood system of safety signs and colours which draw attention to equipment and situations that do, or could, affect health and safety. Text-only safety signs became illegal from 24th December 1998. From that date, all safety signs have had to contain a pictogram or symbol such as those shown in Fig. 1.6. Signs fall into four categories: prohibited activities; warnings; mandatory instructions and safe conditions.

PROHIBITION SIGNS

These are circular white signs with a red border and red cross bar, and are given in Fig. 1.7. They indicate an activity which *must not* be done.

Fig. 1.6 Text only safety signs do not comply.

WARNING SIGNS

These are triangular yellow signs with a black border and symbol, and are given in Fig. 1.8. They *give warning* of a hazard or danger.

MANDATORY SIGNS

These are circular blue signs with a white symbol, and are given in Fig. 1.9. They *give instructions* which must be obeyed.

SAFE CONDITION SIGNS

These are square or rectangular green signs with a white symbol, and are given in Fig. 1.10. They *give information* about safety provision.

Accidents at work

Despite new legislation, improved information, education and training, accidents at work do still happen.

No entry   No smoking   Do not use ladders   No fork lift trucks

Fig. 1.7 Prohibition signs.

DANGER High voltage   DANGER Guard dogs   DANGER Radiation risk   DANGER Men working on machines

Fig. 1.8 Warning signs.

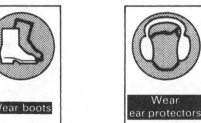

Wear Helmet   Wear gloves   Wear boots   Wear ear protectors

Fig. 1.9 Mandatory signs.

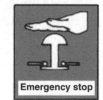

FIRE EXIT   Emergency stop   First aid post   Drinking water

Fig. 1.10 Safe condition signs.

An accident may be defined as an uncontrolled event causing injury or damage to an individual or property. An accident can nearly always be avoided if correct procedures and methods of working are followed. Any accident which results in an absence from work for more than 3 days, causes a major injury or death, is notifiable to the HSE. There are more than 40 000 accidents reported to the HSE each year which occur as a result of some building-related activity. To avoid having an accident you should:

1 follow all safety procedures (e.g. fit safety signs when isolating supplies and screen off work areas from the general public);
2 not misuse or interfere with equipment provided for health and safety;
3 dress appropriately and use personal protective equipment (PPE) when appropriate;
4 behave appropriately and with care;
5 avoid over-enthusiasm and foolishness;
6 stay alert and avoid fatigue;
7 not use alcohol or drugs at work;
8 work within your level of competence;
9 attend safety courses and read safety literature;
10 take a positive decision to act and work safely.

If you observe a hazardous situation at work, first make the hazard safe, using an appropriate method, or screen it off, but only if you can do so without putting yourself or others at risk, then report the situation to your safety representative or supervisor.

Fire control

A fire is a chemical reaction which will continue if fuel, oxygen and heat are present. To eliminate a fire *one* of these components must be removed. This is often expressed by means of the fire triangle shown in Fig. 1.11; all three corners of the triangle must be present for a fire to burn.

FUEL

Fuel is found in the construction industry in many forms: petrol and paraffin for portable generators and heaters; bottled gas for heating and soldering. Most solvents are flammable. Rubbish also represents a source of fuel: off-cuts of wood, roofing felt, rags,

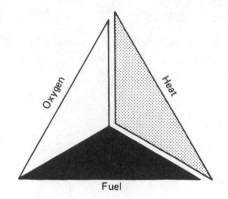

Fig. 1.11   The fire triangle.

empty solvent cans and discarded packaging will all provide fuel for a fire.

To eliminate fuel as a source of fire, all flammable liquids and gases should be stored correctly, usually in an outside locked store. The working environment should be kept clean by placing rags in a metal bin with a lid. Combustible waste material should be removed from the work site or burned outside under controlled conditions by a competent person.

OXYGEN

Oxygen is all around us in the air we breathe, but can be eliminated from a small fire by smothering with a fire blanket, sand or foam. Closing doors and windows but not locking them will limit the amount of oxygen available to a fire in a building and help to prevent it spreading.

Most substances will burn if they are at a high enough temperature and have a supply of oxygen. The minimum temperature at which a substance will burn is called the 'minimum ignition temperature' and for most materials this is considerably higher than the surrounding temperature. However, a danger does exist from portable heaters, blow torches and hot air guns which provide heat and can cause a fire by raising the temperature of materials placed in their path above the minimum ignition temperature. A safe distance must be maintained between heat sources and all flammable materials.

HEAT

Heat can be removed from a fire by dousing with water, but water must not be used on burning liquids since

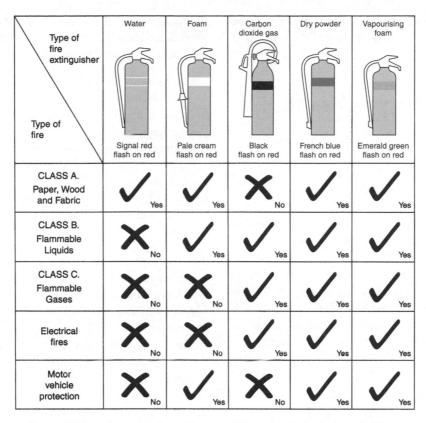

Type of fire extinguisher / Type of fire	Water	Foam	Carbon dioxide gas	Dry powder	Vapourising foam
	Signal red flash on red	Pale cream flash on red	Black flash on red	French blue flash on red	Emerald green flash on red
CLASS A. Paper, Wood and Fabric	✓ Yes	✓ Yes	✗ No	✓ Yes	✓ Yes
CLASS B. Flammable Liquids	✗ No	✓ Yes	✓ Yes	✓ Yes	✓ Yes
CLASS C. Flammable Gases	✗ No	✗ No	✓ Yes	✓ Yes	✓ Yes
Electrical fires	✗ No	✗ No	✓ Yes	✓ Yes	✓ Yes
Motor vehicle protection	✗ No	✓ Yes	✗ No	✓ Yes	✓ Yes

Fig. 1.12 Fire extinguishers and their applications (colour codes to BS EN3: 1996). The base colour of all fire extinguishers is red, with a different coloured flash to indicate the type.

the water will spread the liquid and the fire. Some fire extinguishers have a cooling action which removes heat from the fire.

Fires in industry damage property and materials, injure people and sometimes cause loss of life. Everyone should make an effort to prevent fires, but those which do break out should be extinguished as quickly as possible.

In the event of fire you should:

■ raise the alarm;
■ turn off machinery, gas and electricity supplies in the area of the fire;
■ close doors and windows but without locking or bolting them;
■ remove combustible materials and fuels away from the path of the fire, if the fire is small, and if this can be done safely;
■ attack small fires with the correct extinguisher.

Only attack the fire if you can do so without endangering your own safety in any way. Always leave your own exit from the danger zone clear. Those not involved

in fighting the fire should walk to a safe area or assembly point.

Fires are divided into four classes or categories:

■ Class A are wood, paper and textile fires.
■ Class B are liquid fires such as paint, petrol and oil.
■ Class C are fires involving gas or spilled liquefied gas.
■ Class D are very special types of fire involving burning metal.

Electrical fires do not have a special category because, once started, they can be identified as one of the four above types.

Fire extinguishers are for dealing with small fires, and different types of fire must be attacked with a different type of extinguisher. Using the wrong type of extinguisher could make matters worse. For example, water must not be used on a liquid or electrical fire. The normal procedure when dealing with electrical fires is to cut off the electrical supply and use an extinguisher which is appropriate to whatever is burning. Figure 1.12 shows the correct type of extinguisher to be used on the various categories of fire. The colour coding shown is in accordance with BS EN3: 1996.

Electrical safety and isolation

Electrical supplies at voltages above extra low voltages (ELV) – that is, above 50 V a.c. – can kill human beings and livestock and should therefore be treated with the greatest respect. As an electrician working on electrical installations and equipment, you should always make sure that the supply is first switched off. Every circuit must be provided with a means of isolation (Regulation 130–06–01) and you should isolate and lock off before work begins. In order to deter anyone from reconnecting the supply, a 'Danger Electrician at Work' sign should be displayed on the isolation switch. Where a test instrument or voltage indicator such as that shown in Fig. 1.13 is used to prove

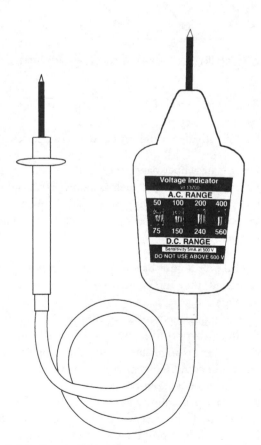

Fig. 1.13 Typical voltage indicator.

conductors dead, Regulation 4(3) of the Electricity at Work Regulations 1989 recommends that the following procedure be adopted so that the device itself is 'proved':

1 Connect the test device to the supply which is to be isolated; this should indicate mains voltage.
2 Isolate the supply and observe that the test device now reads 0 V.
3 Connect the test device to another source of supply to 'prove' that the device is still working correctly.
4 Lock off the supply and place warning notices. Only then should work commence on the 'dead' supply.

The test device must incorporate fused test leads to comply with HSE Guidance Note GS 38, *Electrical Test Equipment Used by Electricians*. Electrical isolation of supplies is further discussed in Chapter 3 of this book.

Temporary electrical supplies on construction sites can save many person-hours of labour by providing energy for fixed and portable tools and lighting. However, as stated previously in this chapter, construction sites are dangerous places and the temporary electrical supplies must be safe. IEE Regulation 110–01 tells us that the regulations apply to temporary electrical installations such as construction sites. The frequency of inspection of construction sites is increased to every 3 months because of the inherent dangers. Regulation 604–02–02 recommends the following voltages for distributing to plant and equipment on construction sites:

400 V – fixed plant such as cranes
230 V – site offices and fixed floodlighting robustly installed
110 V – portable tools and hand lamps
50 V or 25 V – portable lamps used in damp or confined places.

Portable tools must be fed from a 110 V socket outlet unit (see Fig. 1.14(a)) incorporating splash-proof sockets and plugs with a keyway which prevents a tool from one voltage being connected to the socket outlet of a different voltage.

Socket outlet and plugs are also colour-coded for voltage identification: 25 V violet, 50 V white, 110 V yellow, 230 V blue and 400 V red, as shown in Fig. 1.14(b).

ELECTRIC SHOCK

Electric shock occurs when a person becomes part of the electrical circuit, as shown in Fig. 1.15. The level or intensity of the shock will depend upon many factors, such as age, fitness and the circumstances in which the shock is received. The lethal level is approximately 50 mA, above which muscles contract, the heart flutters and breathing stops. A shock above the 50 mA level is therefore fatal unless the person is quickly separated from the supply. Below 50 mA only an unpleasant tingling sensation may be experienced or you may be thrown across a room, roof or ladder, but the resulting fall may lead to serious injury.

To prevent people receiving an electric shock accidentally, all circuits contain protective devices. All exposed metal is earthed, fuses and miniature circuit breakers (MCBs) are designed to trip under fault conditions and residual current devices (RCDs) are designed to trip below the fatal level as described in Chapter 4.

Construction workers and particularly electricians do receive electric shocks, usually as a result of

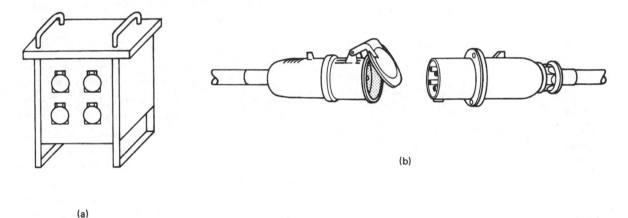

(b)

(a)

Fig. 1.14 110 V distribution unit and cable connector, suitable for construction site electrical supplies: (a) reduced-voltage distribution unit incorporating industrial sockets to BS 4343; (b) industrial plug and connector.

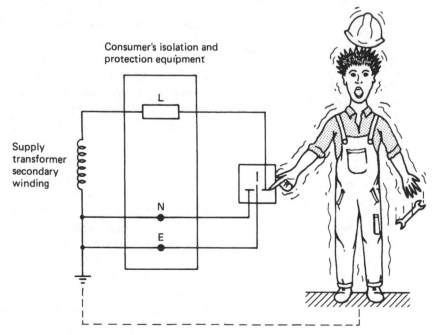

Fig. 1.15 Touching live and earth or live and neutral makes a person part of the electrical circuit and can lead to an electric shock.

carelessness or unforeseen circumstances. When this happens it is necessary to act quickly to prevent the electric shock becoming fatal. Actions to be taken upon finding a workmate receiving an electric shock are as follows:

- Switch off the supply if possible.
- Alternatively, remove the person from the supply *without touching him*, e.g. push him off with a piece of wood, pull him off with a scarf, dry towel or coat.
- If breathing or heart has stopped, immediately call professional help by dialling 999 or 112 and asking for the ambulance service. Give precise directions to the scene of the accident. The casualty stands the best chance of survival if the emergency services can get a rapid-response paramedic team quickly to the scene. They have extensive training and will have specialist equipment with them.
- Only then should you apply resuscitation or cardiac massage until the patient recovers, or help arrives.
- Treat for shock.

First aid

Despite all the safety precautions taken on construction sites to prevent injury to the workforce, accidents do happen and *you* may be the only other person able to take action to assist a workmate. If you are not a qualified first aider limit your help to obvious common-sense assistance and call for help *but* do remember that if a workmate's heart or breathing has stopped as a result of an accident he has only minutes to live unless you act quickly. The Health and Safety (First Aid) Regulations 1981 and relevant approved codes of practice and guidance notes place a duty of care on all employers to provide *adequate* first aid facilities appropriate to the type of work being undertaken. Adequate facilities will relate to a number of factors such as:

- How many employees are employed?
- What type of work is being carried out?
- Are there any special or unusual hazards?
- Are employees working in scattered and/or isolated locations?
- Is there shift work or 'out of hours' work being undertaken?
- Is the workplace remote from emergency medical services?

- Are there inexperienced workers on site?
- What were the risks of injury and ill health identified by the company's Hazard Risk Assessment?

The regulations state that:

> Employers are under a duty to provide such numbers of suitable persons as is *adequate and appropriate in the circumstances* for rendering first aid to his employees if they are injured or become ill at work. For this purpose a person shall not be suitable unless he or she has undergone such training and has such qualifications as the Health and Safety Executive may approve.

This is typical of the way in which the health and safety regulations are written. The regulations and codes of practice do not specify numbers, but set out guidelines in respect of the number of first aiders needed, dependent upon the type of company, the hazards present and the number of people employed.

Let us now consider the questions 'what is first aid?' and 'who might become a first aider?' The regulations give the following definitions of first aid. '*First aid* is the treatment of minor injuries which would otherwise receive no treatment or do not need treatment by a doctor or nurse' *or* 'In cases where a person will require help from a doctor or nurse, first aid is treatment for the purpose of preserving life and minimising the consequences of an injury or illness until such help is obtained.' A more generally accepted definition of first aid might be as follows: *first aid* is the initial assistance or treatment given to a casualty for any injury or sudden illness before the arrival of an ambulance, doctor or other medically qualified person.

Now having defined first aid, who might become a first aider? A *first aider* is someone who has undergone a training course to administer first aid at work and holds a current first aid certificate. The training course and certification must be approved by the HSE. The aims of a first aider are to preserve life, to limit the worsening of the injury or illness and to promote recovery.

A first aider may also undertake the duties of an *appointed person*. An *appointed person* is someone who is nominated to take charge when someone is injured or becomes ill, including calling an ambulance if required. The appointed person will also look after the first aid equipment, including re-stocking the first aid box.

Appointed persons should not attempt to give first aid for which they have not been trained but should limit their help to obvious common sense assistance and summon professional assistance as required. Suggested numbers of first aid personnel are given in Table 1.1. The actual number of first aid personnel must take into account any special circumstances such as remoteness from medical services, the use of several separate buildings and the company's hazard risk assessment. First aid personnel must be available at all times when people are at work, taking into account shift working patterns and providing cover for sickness absences.

Every company must have at least one first aid kit under the regulations. The size and contents of the kit will depend upon the nature of the risks involved in the particular working environment and the number of employees. Table 1.2 gives a list of the contents of any first aid box to comply with the HSE Regulations.

Table 1.1 Suggested numbers of first aid personnel

Category of risk	Numbers employed at any location	Suggested number of first aid personnel
Lower risk e.g. shops and offices, libraries	Fewer than 50 50–100 More than 100	At least one appointed person At least one first aider One additional first aider for every 100 employed
Medium risk e.g. light engineering and assembly work, food processing, warehousing	Fewer than 20 20–100 More than 100	At least one appointed person At least one first aider for every 50 employed (or part thereof) One additional first aider for every 100 employed
Higher risk e.g. most construction, slaughterhouses, chemical manufacture, extensive work with dangerous machinery or sharp instruments	Fewer than five 5–50 More than 50	At least one appointed person At least one first aider One additional first aider for every 50 employed

Table 1.2 Contents of first aid boxes

Item	Number of employees				
	1–5	6–10	11–50	51–100	101–150
Guidance card on general first aid	1	1	1	1	1
Individually wrapped sterile adhesive dressings	10	20	40	40	40
Sterile eye pads, with attachment (Standard Dressing No. 16 BPC)	1	2	4	6	8
Triangular bandages	1	2	4	6	8
Sterile covering for serious wounds (where applicable)	1	2	4	6	8
Safety pins	6	6	12	12	12
Medium sized sterile unmedicated dressings (Standard Dressings No. 9 and No. 14 and the Ambulance Dressing No. 1)	3	6	8	10	12
Large sterile unmedicated dressings (Standard Dressings No. 9 and No. 14 and the Ambulance Dressing No. 1)	1	2	4	6	10
Extra large sterile unmedicated dressings (Ambulance Dressing No. 3)	1	2	4	6	8

Where tap water is not available, sterile water or sterile normal saline in disposable containers (each holding a minimum of 300 ml) must be kept near the first aid box. The following minimum quantities should be kept:

Number of employees	1–10	11–50	51–100	101–150
Quantity of sterile water	1×300 ml	3×300 ml	6×300 ml	6×300 ml

There now follows a description of some first aid procedures which should be practised under expert guidance before they are required in an emergency.

Bleeding

If the wound is dirty, rinse it under clean running water. Clean the skin around the wound and apply a plaster, pulling the skin together.

If the bleeding is severe apply direct pressure to reduce the bleeding and raise the limb if possible. Apply a sterile dressing or pad and bandage firmly before obtaining professional advice.

To avoid possible contact with hepatitis or the AIDS virus, when dealing with open wounds, first aiders should avoid contact with fresh blood by wearing plastic or rubber protective gloves, or by allowing the casualty to apply pressure to the bleeding wound.

Burns

Remove heat from the burn to relieve the pain by placing the injured part under clean cold water. Do not remove burnt clothing sticking to the skin. Do not apply lotions or ointments. Do not break blisters or attempt to remove loose skin. Cover the injured area with a clean dry dressing.

Broken bones

Make the casualty as comfortable as possible by supporting the broken limb either by hand or with padding. Do not move the casualty unless by remaining in that position he is likely to suffer further injury. Obtain professional help as soon as possible.

Contact with chemicals

Wash the affected area very thoroughly with clean cold water. Remove any contaminated clothing. Cover the affected area with a clean sterile dressing and seek expert advice. It is a wise precaution to treat all chemical substances as possibly harmful; even commonly used substances can be dangerous if contamination is from concentrated solutions. When handling dangerous substances it is also good practice to have a neutralizing agent to hand.

Disposal of dangerous substances must not be into the main drains since this can give rise to an environmental hazard, but should be undertaken in accordance with local authority regulations.

Exposure to toxic fumes

Get the casualty into fresh air quickly and encourage deep breathing if conscious. Resuscitate if breathing has stopped. Obtain expert medical advice as fumes may cause irritation of the lungs.

Sprains and bruising

A cold compress can help to relieve swelling and pain. Soak a towel or cloth in cold water, squeeze it out and place it on the injured part. Renew the compress every few minutes.

Breathing stopped

Remove any restrictions from the face and any vomit, loose or false teeth from the mouth. Loosen tight clothing around the neck, chest and waist. To ensure a good airway, lay the casualty on his back and support the shoulders on some padding. Tilt the head backwards and open the mouth. If the casualty is faintly breathing, lifting the tongue clear of the airway may be all that is necessary to restore normal breathing. However, if the casualty does not begin to breathe, open your mouth wide and take a deep breath, close the casualty's nose by pinching with your fingers, and, sealing your lips around his mouth, blow into his lungs until the chest rises. Remove your mouth and watch the casualty's chest fall. Continue this procedure at your natural breathing rate. If the mouth is damaged or you have difficulty making a seal around the casualty's mouth, close his mouth and inflate the lungs through his nostrils. Give artificial respiration until natural breathing is restored or until professional help arrives.

Heart stopped beating

This sometimes happens following a severe electric shock. If the casualty's lips are blue, the pupils of his eyes widely dilated and the pulse in his neck cannot be felt, then he may have gone into cardiac arrest. Act quickly and lay the casualty on his back. Kneel down beside him and place the heel of one hand in the centre of his chest. Cover this hand with your other hand and interlace the fingers. Straighten your arms and press down on his chest sharply with the heel of your hands and then release the pressure. Continue to do this 15 times at the rate of one push per second. Check the casualty's pulse. If none is felt, give two breaths of artificial respiration and then a further 15 chest compressions. Continue this procedure until the heartbeat is restored and the artificial respiration until normal breathing returns. Pay close attention to the condition of the casualty while giving heart massage. When a pulse is restored the blueness around the mouth will quickly go away and you should stop the heart massage. Look carefully at the rate of breathing. When this is also normal, stop giving artificial respiration. Treat the casualty for shock, place him in the recovery position and obtain professional help.

Shock

Everyone suffers from shock following an accident. The severity of the shock depends upon the nature and extent of the injury. In cases of severe shock the casualty will become pale and his skin become clammy from sweating. He may feel faint, have blurred vision, feel sick and complain of thirst. Reassure the casualty that everything that needs to be done is being done. Loosen tight clothing and keep him warm and dry until help arrives. *Do not* move him unnecessarily or give him anything to drink.

Accident reports

Every accident must be reported to an employer and the details of the accident and treatment given suitably documented. A first aid log book or Accident book such as that shown in Fig. 1.16 containing first aid treatment record sheets could be used to effectively document accidents which occur in the workplace and the treatment given. Failure to do so may influence the payment of compensation at a later date if an injury leads to permanent disability. To comply with the Data Protection Regulations, from the 31 December 2003 all First Aid Treatment Logbooks or Accident Report books must contain perforated

Fig. 1.16 First Aid logbook/Accident book with data protection compliant removable sheets.

sheets which can be removed after completion and filed away for personal security.

Structures of Electrotechnical Organisations

The construction industry

An electrician working for an electrical contracting company works as a part of the broader construction industry. This is a multi-million-pound industry carrying out all types of building work, from basic housing to hotels, factories, schools, shops, offices and airports. The construction industry is one of the UK's biggest employers, and carries out contracts to the value of about 10% of the UK's gross national product.

Although a major employer, the construction industry is also very fragmented. Firms vary widely in size, from the local builder employing two or three people to the big national companies employing thousands. Of the total workforce of the construction industry, 92% are employed in small firms of less than 25 people.

The yearly turnover of the construction industry is about £35 billion. Of this total sum, about 60% is spent on new building projects and the remaining 40% on maintenance, renovation or restoration of mostly housing.

In all these various construction projects the electrical contractor plays an important role, supplying essential electrical services to meet the needs of those who will use the completed building.

The building team

The construction of a new building is a complex process which requires a team of professionals working together to produce the desired results. We can call this team of professionals the building team, and their interrelationship can be expressed by Fig. 1.17.

The client is the person or group of people with the actual need for the building, such as a new house, office or factory. The client is responsible for financing all the work and, therefore, in effect, employs the entire building team.

The architect is the client's agent and is considered to be the leader of the building team. The architect must interpret the client's requirements and produce working drawings. During the building process the architect will supervise all aspects of the work until the building is handed over to the client.

The quantity surveyor measures the quantities of labour and material necessary to complete the building work from drawings supplied by the architect.

Specialist engineers advise the architect during the design stage. They will prepare drawings and calculations on specialist areas of work.

The clerk of works is the architect's 'on-site' representative. He or she will make sure that the contractors carry out the work in accordance with the drawings and other contract documents. They can also agree general matters directly with the building contractor as the architect's representative.

The local authority will ensure that the proposed building conforms to the relevant planning and building legislation.

The health and safety inspectors will ensure that the government's legislation concerning health and safety is fully implemented by the building contractor.

The building contractor will enter into a contract with the client to carry out the construction work in accordance with contract documents. The building contractor is the main contractor and he or she, in turn, may engage subcontractors to carry out specialist services such as electrical installation, mechanical services, plastering and painting.

The electrical team

The electrical contractor is the subcontractor responsible for the installation of electrical equipment within the building. An electrical contracting firm is made up of a group of individuals with varying duties and responsibilities (see Fig. 1.18). There is often no

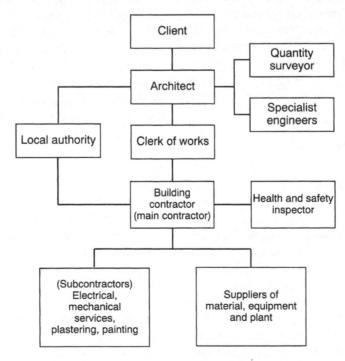

Fig. 1.17 The building team.

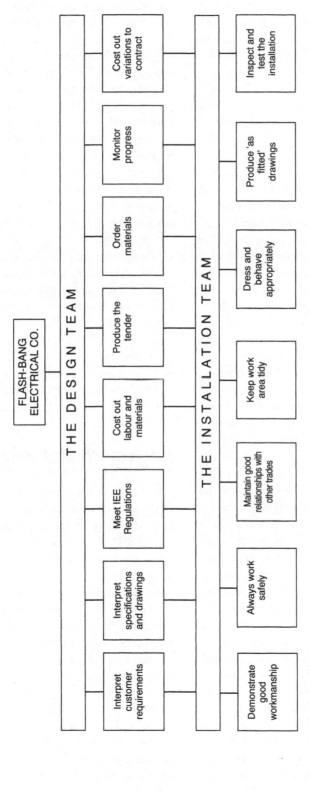

Fig. 1.18 The electrical team.

clear distinction between the duties of the individuals, and the responsibilities carried by an employee will vary from one employer to another. If the firm is to be successful, the individuals must work together to meet the requirements of their customers. Good customer relationships are important for the success of the firm and the continuing employment of the employee.

The customer or his representatives will probably see more of the electrician and the electrical trainee than the managing director of the firm and, therefore, the image presented by them is very important. They should always be polite and be seen to be capable and in command of the situation. This gives a customer confidence in the firm's ability to meet his or her needs. The electrician and his trainee should be appropriately dressed for the job in hand, which probably means an overall of some kind. Footwear is also important, but sometimes a difficult consideration for an electrician. For example, if working in a factory, the safety regulations insist that protective footwear be worn, but rubber boots may be most appropriate for a building site. However, neither of these would be the most suitable footwear for an electrician fixing a new light fitting in the home of the managing director!

The electrical installation in a building is often carried out alongside other trades. It makes sound sense to help other trades where possible and to develop good working relationships with other employees.

The employer has the responsibility of finding sufficient work for his employees, paying government taxes and meeting the requirements of the Health and Safety at Work Act described earlier. The rates of pay and conditions for electricians and trainees are determined by negotiation between the Joint Industry Board and the Amalgamated Engineering and Electrical Trades Union, Amicus, which will also represent their members in any disputes. Electricians are usually paid at a rate agreed for their grade as an electrician, approved electrician or technician electrician; movements through the grades are determined by a combination of academic achievement and practical experience.

One of the installation team will have special responsibility for the specific contract being carried out. He might be called the project manager or supervisor, and will be responsible to his electrical company to see that the design specification is carried out and will have overall responsibility on that site for the electrical installation. He will attend site meetings as the representative of the electrical contractor, supported by other members of the team, who will demonstrate a range of skills and responsibilities. The supervisor himself will probably be a mature electrician of 'technician' status. The trainee electrician will initially work alongside an electrician or 'approved electrician', who might have been given responsibility for a small part of a large installation by the supervisor on site.

The project manager or site supervisor will be supported by the design team. The design team might be made up of a contracts manager, who will oversee a number of individual electrical contracts at different sites, monitoring progress and costing out variations to the initial contractual agreement. He might also have responsibility for health and safety because he attends all sites and, therefore, has an overview of all company employees and projects that are being carried out.

The contracts manager will also be supported by the design engineer. The design engineer will meet with clients, architects and other trade professionals, to interpret the customer's requirements. He will produce the design specifications, which will set out the detailed design of the electrical installation and provide sufficient information to enable a competent person to carry out such installation. The design specifications will also enable a cost for the project to be estimated and included in the legal contracts between the client or main contractor and the electrical contractor.

Electrotechnical industry

The electrical team discussed above are working for an electrical contracting company, which I have called the Flash-Bang Electrical Company (as a joke between you and I). Any electrical contractor is part of the electroctechnical industry. The work of an electrical contractor is one of *installing* electrical equipment and systems, but a very similar role is also carried out by electrical teams working for local councils, the railways, the armed forces and hospitals.

White goods and electrical control panels are *manufactured* and assembled to meet specific specifications by electrical teams working in the manufacturing sector of the electrotechnical industry.

Whatever section of the electrotechnical industry *you* work for, the organisational structure of your company will be similar to the one described above for the electrical contractor. The electrical team in any section of the electrotechnical industry is made up of a dedicated team of 'electrical craftsmen' or 'operatives', carrying out their work to a high standard of competence and skill while complying with the requirements of the relevant regulations. The craftsmen are supported by a supervisor or a foreman, who pulls together the various parts of that specific job or product, thereby meeting the requirements of the client or customer. The supervisor is supported by the manager, who is responsible for designing the electrotechnical product within the requirements of all relevant regulations and specifications.

The craftsman, supervisor, foreman or manager, might be called something different in your electrotechnical company's organisation, but there will be a team of people *installing* equipment and systems or *maintaining* equipment and systems or *manufacturing* panels, equipment and machines or *rewinding* electrical machines and transformers. Each individual has a specific role to play within the team's discussed earlier for 'the electrical team'. Each individual is important to the success of the team and the success of the company.

The electrotechnical industry is made up of a variety of individual companies, all providing a service within their own specialism to a customer, client or user.

The electrical contracting industry provides lighting and power installations so that buildings and systems may be illuminated to an appropriate level, heated to a comfortable level and have the power circuits to drive electrical and electronic equipment. Emergency lighting and security systems are installed so that buildings are safe in unforeseen and adverse situations.

Building management and control systems provide a controlled environment for the people who use commercial buildings.

Instrumentation allows us to monitor industrial processes and systems.

Electrical maintenance allows us to maintain the efficiency of all installed systems.

Computer installations, fibre optic cables and data cabling provide speedy data processing and communications.

High voltage/low voltage (HV/LV) jointing provides a means of connecting new installations and services to live cables without the need to inconvenience existing supplies by electrical shutdown.

Highway electrical systems make our roads, pavements and alleyways safer for vehicle users and pedestrians.

Electrical panels provide electrical protection, isolation and monitoring for the electrical systems in commercial and industrial buildings.

Electrical machine drive installations drive everything that makes our modern life comfortable, from trains and trams to lifts and air conditioning units.

Finally, *consumer commercial electronics* allows us to live our modern life of rapid personal communication systems while listening to popular or classical music and watching wide screen television.

Designing an electrical installation

The designer of an electrical installation must ensure that the design meets the requirements of the IEE Wiring Regulations for electrical installations and any other regulations which may be relevant to a particular installation. The designer may be a professional technician or engineer whose only job is to design electrical installations for a large contracting firm as I have just described above. In a smaller firm, the designer may also be the electrician who will carry out the installation to the customer's requirements. The designer of any electrical installation is the person who interprets the electrical requirements of the customer within the regulations, identifies the appropriate types of installation, the most suitable methods of protection and control and the size of cables to be used.

A large electrical installation may require many meetings with the customer and his professional representatives in order to identify a specification of what is required. The designer can then identify the general characteristics of the electrical installation and its compatibility with other services and equipment, as indicated in Part 3 of the Regulations. The protection and safety of the installation, and of those who will use it, must be considered, with due regard to Part 4 of the Regulations. An assessment of the frequency and quality of the maintenance to be expected (Regulation 341–01–01) will give an indication of the type of installation which is most appropriate.

The size and quantity of all the materials, cables, control equipment and accessories can then be determined. This is called a 'bill of quantities'.

It is common practice to ask a number of electrical contractors to tender or submit a price for work specified by the bill of quantities. The contractor must cost all the materials, assess the labour cost required to install the materials and add on profit and overhead costs in order to arrive at a final estimate for the work. The contractor tendering the lowest cost is usually, but not always, awarded the contract.

To complete the contract in the specified time the electrical contractor must use the management skills required by any business to ensure that men and materials are on site as and when they are required. If alterations or modifications are made to the electrical installation as the work proceeds which are outside the original specification, then a variation order must be issued so that the electrical contractor can be paid for the additional work.

The specification for the chosen wiring system will be largely determined by the building construction and the activities to be carried out in the completed building.

An industrial building, for example, will require an electrical installation which incorporates flexibility and mechanical protection. This can be achieved by a conduit, tray or trunking installation.

In a block of purpose-built flats, all the electrical connections must be accessible from one flat without intruding upon the surrounding flats. A loop-in conduit system, in which the only connections are at the light switch and outlet positions, would meet this requirement.

For a domestic electrical installation an appropriate lighting scheme and multiple socket outlets for the connection of domestic appliances, all at a reasonable cost, are important factors which can usually be met by a PVC insulated and sheathed wiring system.

The final choice of a wiring system must rest with those designing the installation and those ordering the work, but whatever system is employed, good workmanship is essential for compliance with the regulations. The necessary skills can be acquired by an electrical trainee who has the correct attitude and dedication to his craft.

Legal contracts

Before work commences, some form of legal contract should be agreed between the two parties, that is, those providing the work (e.g. the subcontracting electrical company) and those asking for the work to be carried out (e.g. the main building company).

A contract is a formal document which sets out the terms of agreement between the two parties. A standard form of building contract typically contains four sections:

1 The articles of agreement – this names the parties, the proposed building and the date of the contract period.
2 The contractual conditions – this states the rights and obligations of the parties concerned, e.g. whether there will be interim payments for work or a penalty if work is not completed on time.
3 The appendix – this contains details of costings, e.g. the rate to be paid for extras as daywork, who will be responsible for defects, how much of the contract tender will be retained upon completion and for how long.
4 The supplementary agreement – this allows the electrical contractor to recoup any value-added tax paid on materials at interim periods.

In signing the contract, the electrical contractor has agreed to carry out the work to the appropriate standards in the time stated and for the agreed cost. The other party, say the main building contractor, is agreeing to pay the price stated for that work upon completion of the installation.

If a dispute arises the contract provides written evidence of what was agreed and will form the basis for a solution.

For smaller electrical jobs, a verbal contract may be agreed, but if a dispute arises there is no written evidence of what was agreed and it then becomes a matter of one person's word against another's.

On-site communications

Good communication is about transferring information from one person to another. Electricians and other professionals in the construction trades communicate with each other and the general public by means of drawings, sketches and symbols, in addition to what we say and do.

DRAWINGS AND DIAGRAMS

Many different types of electrical drawing and diagram can be identified: layout, schematic, block, wiring and circuit diagrams. The type of diagram to be used in any particular application is the one which most clearly communicates the desired information.

Layout drawings

These are scale drawings based upon the architect's site plan of the building and show the positions of the electrical equipment which is to be installed. The electrical equipment is identified by a graphical symbol.

The standard symbols used by the electrical contracting industry are those recommended by the British Standard EN 60617, *Graphical Symbols for Electrical Power, Telecommunications and Electronic Diagrams*. Some of the more common electrical installation symbols are given in Fig. 1.19.

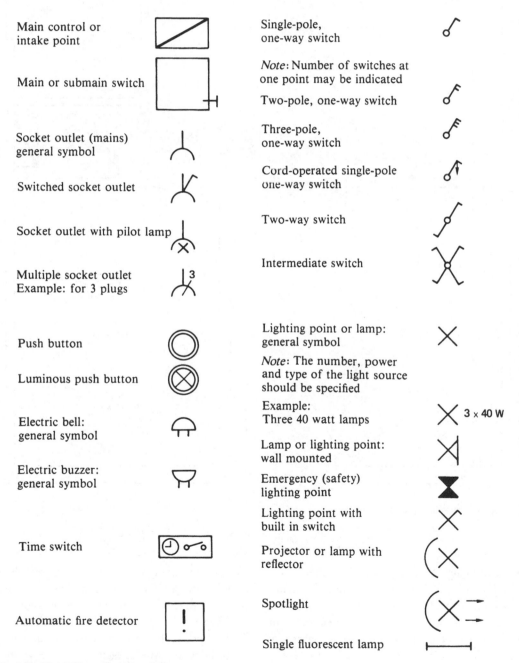

Fig. 1.19 Some EN 60617 installation symbols.

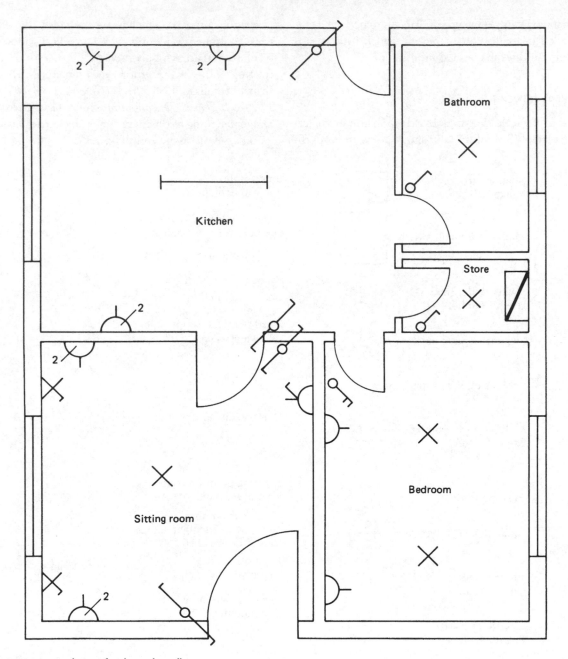

Fig 1.20 Layout drawing for electrical installation.

A layout drawing of a small domestic extension is shown in Fig. 1.20. It can be seen that the mains intake position, probably a consumer's unit, is situated in the store-room which also contains one light controlled by a switch at the door. The bathroom contains one lighting point controlled by a one-way switch at the door. The kitchen has two doors and a switch is installed at each door to control the fluorescent luminaire. There are also three double sockets situated around the kitchen. The sitting room has a two-way switch at each door controlling the centre lighting point. Two wall lights with built-in switches are to be wired, one at each side of the window. Two double sockets and one switched socket are also to be installed in the sitting room. The bedroom has two lighting points controlled independently by two one-way switches at the door.

The wiring diagrams and installation procedures for all these circuits can be found in the next chapter.

As-fitted drawings

When the installation is completed a set of drawings should be produced which indicate the final positions of all the electrical equipment. As the building and electrical installation progresses, it is sometimes necessary to modify the positions of equipment indicated on the layout drawing because, for example, the position of a doorway has been changed. The layout drawings indicate the original intentions for the positions of equipment, while the 'as-fitted' drawing indicates the actual positions of equipment upon completion of the job.

Detail drawings

These are additional drawings produced by the architect to clarify some point of detail. For example, a drawing might be produced to give a fuller description of the suspended ceiling arrangements.

Schematic diagrams

A schematic diagram is a diagram in outline of, for example, a motor starter circuit. It uses graphical symbols to indicate the interrelationship of the electrical elements in a circuit. These help us to understand the working operation of the circuit.

An electrical schematic diagram looks very like a circuit diagram. A mechanical schematic diagram gives a more complex description of the individual elements in the system, indicating, for example, acceleration, velocity, position, force sensing and viscous damping.

Block diagrams

A block diagram is a very simple diagram in which the various items or pieces of equipment are represented by a square or rectangular box. The purpose of the block diagram is to show how the components of the circuit relate to each other, and therefore the individual circuit connections are not shown. Figure 1.21 shows the block diagram of a space heating control system.

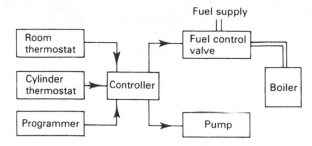

Fig. 1.21 Space heating control system.

Wiring diagrams

A wiring diagram or connection diagram shows the detailed connections between components or items of equipment. They do not indicate how a piece of equipment or circuit works. The purpose of a wiring diagram is to help someone with the actual wiring of the circuit. Figure 1.22 shows the wiring diagram for a space heating control system. Other wiring diagrams can be seen in Figs 4.8–4.11.

Circuit diagrams

A circuit diagram shows most clearly how a circuit works. All the essential parts and connections are represented by their graphical symbols. The purpose of a circuit diagram is to help our understanding of the circuit. It will be laid out as clearly as possible, without regard to the physical layout of the actual components, and therefore it may not indicate the most convenient way to wire the circuit. Figure 1.23 shows the circuit diagram of our same space heating control system. Most of the diagrams in Chapter 2 from Figs 2.2 to 2.24 are circuit diagrams.

Supplementary diagrams

A supplementary diagram conveys additional information in a way which is usually a mixture of the other categories of drawings. Figure 1.24 shows the supplementary diagram for our space heating control system and is probably the most useful diagram for initially setting out the wiring for the heating system.

Freehand working diagrams

Freehand working drawings or sketches are another important way in which we communicate our ideas. The drawings of the brackets in Chapter 4 in Fig. 4.35

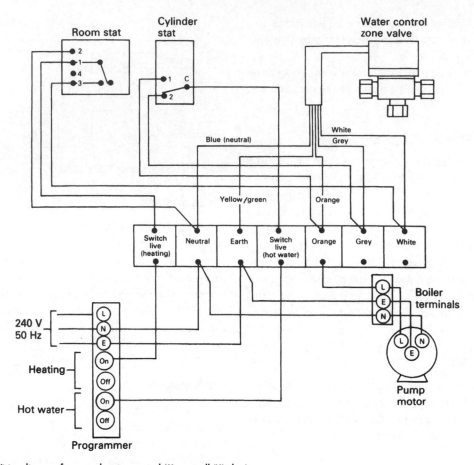

Fig. 1.22 Wiring diagram for space heating control (Honeywell 'Y' plan).

were done from freehand sketches. A freehand sketch may be done as an initial draft of an idea before a full working drawing is made or, in the case of Fig. 4.35, it may be used to enable someone actually to make the bracket. It is often much easier to produce a sketch of your ideas or intentions than to describe them or produce a list of instructions.

To convey the message or information clearly it is better to make your sketch large rather than too small. It should also contain all the dimensions necessary to indicate clearly the size of the finished object depicted by the sketch.

You could practise freehand sketching by drawing some of the tools and equipment used in our trade and shown in Figs 2.72 to 2.76.

TELEPHONE MESSAGES

Telephones today play one of the most important roles in enabling people to communicate with each other.

The advantage of a telephone message over a written message is its speed; the disadvantage is that no record is kept of an agreement made over the telephone. Therefore, business agreements made on the telephone are often followed up by written confirmation.

When *taking* a telephone call, remember that you cannot be seen and, therefore, gestures and facial expressions will not help to make you understood. Always be polite and helpful when answering your company's telephone – you are your company's most important representative at that moment. Speak clearly and loud enough to be heard without shouting, sound cheerful and write down messages if asked. Always read back what you have written down to make sure that you are passing on what the caller intended.

Many companies now use standard telephone message pads such as that shown in Fig. 1.25 because they prompt people to collect all the relevant information. In this case John Gall wants Dave Twem to pick up the Megger from Jim on Saturday and take it to the

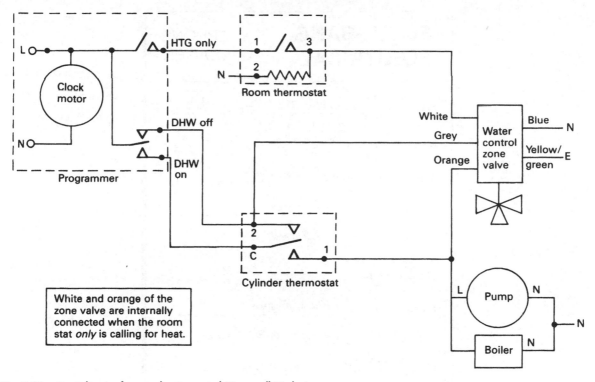

Fig. 1.23 Circuit diagram for space heating control (Honeywell 'Y' plan).

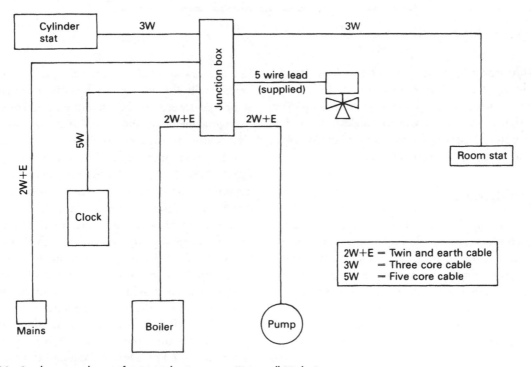

Fig. 1.24 Supplementary diagram for a space heating system (Honeywell 'Y' plan).

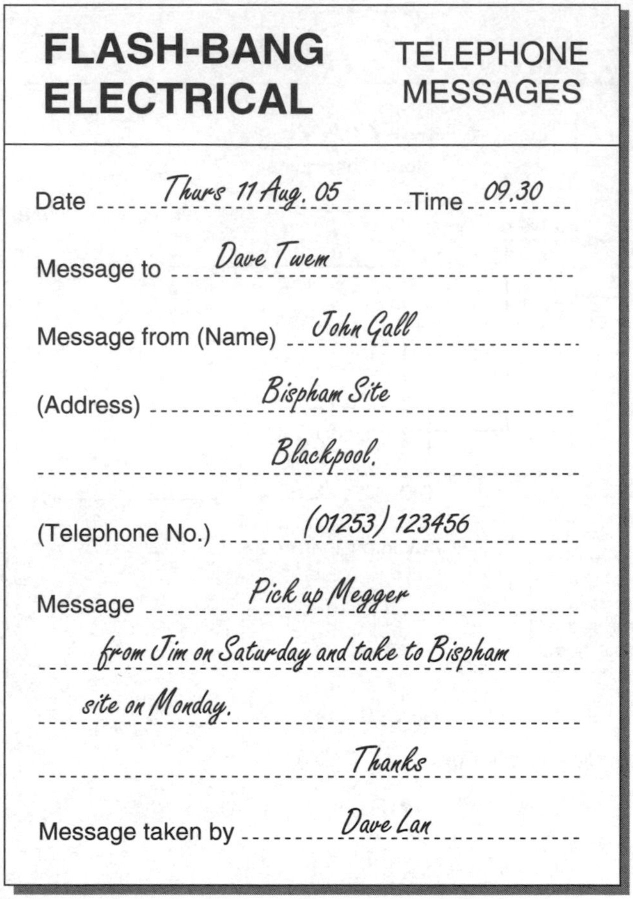

FLASH-BANG ELECTRICAL　　　TELEPHONE MESSAGES

Date .... *Thurs 11 Aug. 05* .... Time *09.30*

Message to .... *Dave Twem*

Message from (Name) .... *John Gall*

(Address) .... *Bispham Site*

.... *Blackpool.*

(Telephone No.) .... *(01253) 123456*

Message .... *Pick up Megger*

.... *from Jim on Saturday and take to Bispham*

.... *site on Monday.*

.... *Thanks*

Message taken by .... *Dave Lan*

Fig. 1.25 Typical standard telephone message pad.

Bispham site on Monday. The person taking the call and relaying the message is Dave Low.

When *making* a telephone call, make sure you know what you want to say or ask. Make notes so that you have times, dates and any other relevant information ready before you make the call.

WRITTEN MESSAGES

A lot of communications between and within larger organizations take place by completing standard forms or sending internal memos. Written messages have the advantage of being 'auditable'. An auditor can follow the paperwork trail to see, for example, who was responsible for ordering certain materials.

When completing standard forms, follow the instructions given and ensure that your writing is legible. Do not leave blank spaces on the form, always specifying 'not applicable' or 'N/A' whenever necessary.

Sign or give your name and the date as asked for on the form. Finally, read through the form again to make sure you have answered all the relevant sections correctly.

Internal memos are forms of written communication used within an organization; they are not normally used for communicating with customers or suppliers. Figure 1.26 shows the layout of a typical standard memo form used by Dave Twem to notify John Gall that he has ordered the hammer drill.

Letters provide a permanent record of communications between organizations and individuals. They may be handwritten, but formal business letters give a better impression of the organization if they are typewritten. A letter should be written using simple concise language, and the tone of the letter should always be polite even if it is one of complaint. Always include the date of the correspondence. The greeting on a formal letter should be 'Dear Sir/Madam' and concluded with 'Yours faithfully'. A less formal greeting would be

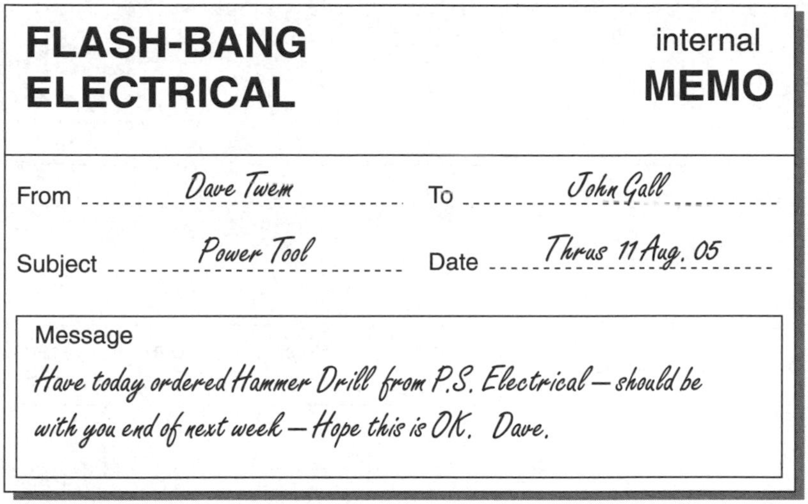

FLASH-BANG ELECTRICAL

internal **MEMO**

From _____ *Dave Twem* _____

To _____ *John Gall* _____

Subject _____ *Power Tool* _____

Date _____ *Thrus 11 Aug. 05* _____

Message

Have today ordered Hammer Drill from P.S. Electrical — should be with you end of next week — Hope this is OK. Dave.

Fig. 1.26 Typical standard memo form.

'Dear Mr Smith' and concluded 'Yours sincerely'. Your name and status should be typed below your signature.

DELIVERY NOTES

When materials are delivered to site, the person receiving the goods is required to sign the driver's 'delivery note'. This record is used to confirm that goods have been delivered by the supplier, who will then send out an invoice requesting payment, usually at the end of the month.

The person receiving the goods must carefully check that all the items stated on the delivery note have been delivered in good condition. Any missing or damaged items must be clearly indicated on the delivery note before signing, because, by signing the delivery note the person is saying 'yes, these items were delivered to me as my company's representative on that date and in good condition and I am now responsible for these goods'. Suppliers will replace materials damaged in transit provided that they are notified within a set time period, usually three days. The person receiving the goods should try to quickly determine their condition. Has the packaging been damaged, does the container 'sound' like it might contain broken items? It is best to check at the time of delivery if possible, or as soon as possible after delivery and within the notifiable period. Electrical goods delivered to site should be handled carefully and stored securely until they are

installed. Copies of delivery notes are sent to head office so that payment can be made for the goods received.

TIME SHEETS

A time sheet is a standard form completed by each employee to inform the employer of the actual time spent working on a particular contract or site. This helps the employer to bill the hours of work to an individual job. It is usually a weekly document and includes the number of hours worked, the name of the job and any travelling expenses claimed. Office personnel require time sheets such as that shown in Fig. 1.27 so that wages can be made up.

JOB SHEETS

A job sheet or job card such as that shown in Fig. 1.28 carries information about a job which needs to be done, usually a small job. It gives the name and address of the customer, contact telephone numbers, often a job reference number and a brief description of the work to be carried out. A typical job sheet work description might be:

■ Job 1 Upstairs lights not working.
■ Job 2 Funny fishy smell from kettle socket in kitchen.

TIME SHEET

FLASH-BANG ELECTRICAL

Employee's name (Print) _____

Week ending _____

Day	Job number and/or Address	Start time	Finish time	Total hours	Travel time	Expenses
Monday						
Tuesday						
Wednesday						
Thursday						
Friday						
Saturday						
Sunday						

Employee's signature _____ Date _____

Fig. 1.27 Typical time sheet.

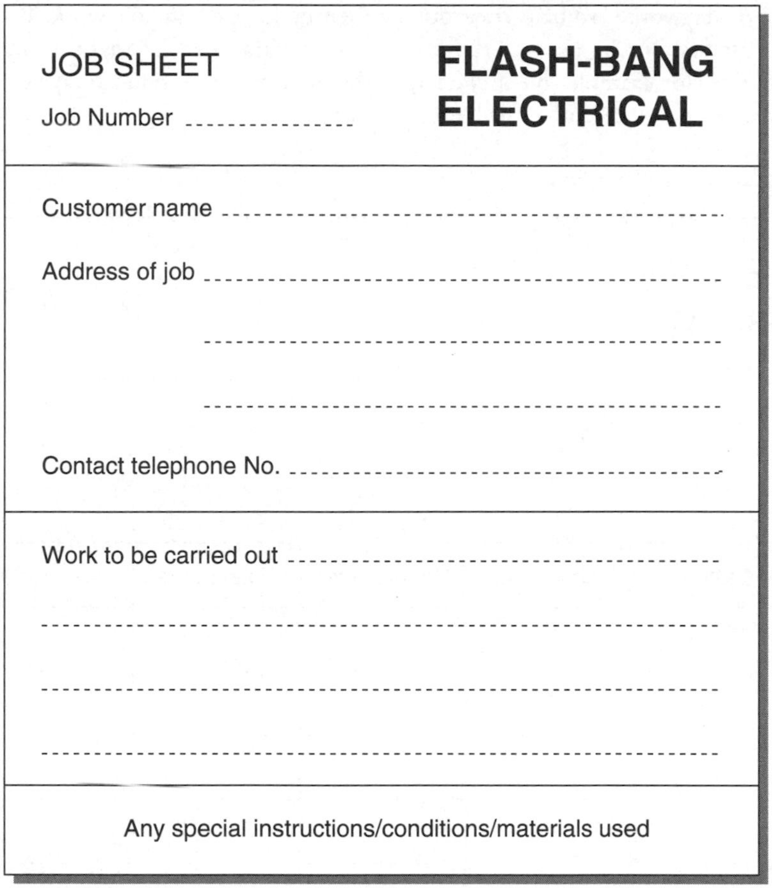

```
┌─────────────────────────────────────────────────┐
│  JOB SHEET            FLASH-BANG                  │
│  Job Number ............  ELECTRICAL              │
├───────────────────────────────────────────────── │
│                                                   │
│  Customer name ...............................    │
│                                                   │
│  Address of job ..............................    │
│                                                   │
│              ................................     │
│                                                   │
│              ................................     │
│                                                   │
│  Contact telephone No. .......................    │
├───────────────────────────────────────────────── │
│                                                   │
│  Work to be carried out ......................    │
│                                                   │
│         .......................................   │
│                                                   │
│         .......................................   │
│                                                   │
│         .......................................   │
│                                                   │
├───────────────────────────────────────────────── │
│     Any special instructions/conditions/materials used │
└─────────────────────────────────────────────────┘
```

Fig. 1.28 Typical job sheet.

An electrician might typically have a 'jobbing day' where he picks up a number of job sheets from the office and carries out the work specified.

Job 1, for example, might be the result of a blown fuse which is easily rectified, but the electrician must search a little further for the fault which caused the fuse to blow in the first place. The actual fault might, for example, be a decayed flex on a pendant drop which has become shorted out, blowing the fuse. The pendant drop would be re-flexed or replaced, along with any others in poor condition. The installation would then be tested for correct operation and the customer given an account of what has been done to correct the fault. General information and assurances about the condition of the installation as a whole might be requested and given before setting off to job 2.

The kettle socket outlet at job 2 is probably getting warm and, therefore, giving off that 'fishy' bakelite smell because loose connections are causing the bakelite socket to burn locally. A visual inspection would confirm the diagnosis. A typical solution would be to replace the socket and repair any damage to the conductors inside the socket box. Check the kettle plug top for damage and loose connections. Make sure all connections are tight before reassuring the customer that all is well; then, off to the next job or back to the office.

The time spent on each job and the materials used are sometimes recorded on the job sheet, but alternatively a daywork sheet can be used. This will depend upon what is normal practice for the particular electrical company. This information can then be used to 'bill' the customer for work carried out.

DAYWORK SHEETS

Daywork is one way of recording variations to a contract, that is, work done which is outside the scope of the original contract. If daywork is to be carried out, the site supervisor must first obtain a signature from the client's representative, for example, the architect, to authorize the extra work. A careful record must then be kept on the daywork sheets of all extra time and materials used so that the client can be billed for the extra work. A typical daywork sheet is shown in Fig. 1.29.

FLASH-BANG ELECTRICAL   DAYWORK SHEET

Client name ....

Job number/ref. ....

Date	Labour	Start time	Finish time	Total hours	Office use

Materials quantity	Description	Office use

Site supervisor or F.B. Electrical Representative responsible for carrying out work ....

Signature of person approving work and status e.g.

Client ☐   Architect ☐   Q.S. ☐   Main contractor ☐   Clerk of works ☐

Signature ....

Fig. 1.29 Typical daywork sheet.

REPORTS

On large jobs, the foreman or supervisor is often required to keep a report of the relevant events which happen on the site – for example, how many people from your company are working on site each day, what goods were delivered, whether there were any breakages or accidents, and records of site meetings attended. Some firms have two separate documents, a site diary to record daily events and a weekly report which is a summary of the week's events extracted from the site diary. The site diary remains on-site and the weekly report is sent to head office to keep managers informed of the work's progress.

PERSONAL COMMUNICATIONS

Remember that it is the customers who actually pay the wages of everyone employed in your company. You should always be polite and listen carefully to their wishes. They may be elderly or of a different religion or cultural background than you. In a domestic situation, the playing of loud music on a radio may not be approved of. Treat the property in which you are working with the utmost care. When working in houses, shops and offices use dust sheets to protect floor coverings and furnishings. Clean up periodically and made a special effort when the job is completed.

Dress appropriately: an unkempt or untidy appearance will encourage the customer to think that your work will be of poor quality.

The electrical installation in a building is often carried out alongside other trades. It makes good sense to help other trades where possible and to develop good working relationships with other employees. The customer will be most happy if the workers give an impression of working together as a team for the successful completion of the project.

Finally, remember that the customer will probably see more of the electrician and the electrical trainee than the managing director of your firm and, therefore, the image presented by you will be assumed to reflect the policy of the company. You are, therefore, your company's most important representative. Always give the impression of being capable and in command of the situation, because this gives customers confidence in the company's ability to meet their needs. However, if a problem does occur which is outside your previous experience and you do not feel confident to solve it successfully, then contact your supervisor for professional help and guidance. It is not unreasonable for a young member of the company's team to seek help and guidance from those employees with more experience. This approach would be preferred by most companies rather than having to meet the cost of an expensive blunder.

Professional organisations

There are two professional organisations supporting employers in the electrical contracting sector of the electrotechnical industries. These are the Electrical Contractors Association (ECA) and the National Inspection Council for Electrical Installation Contracting (NICEIC). They provide electrical companies with professional advice, maintain and encourage improved standards of workmanship and provide customers with a guarantee of the standard of the work carried out by member companies.

Amicus is the Trade Union representing all employees in the electrotechnical industries. It is involved in negotiating the pay and conditions of its members and in supporting them should a dispute with an employer arise.

Let us now look at these three professional organisations and their work in the electrotechnical industries.

THE ELECTRICAL CONTRACTORS ASSOCIATION (ECA)

The Electrical Contractors Association (ECA) was founded over 100 years ago and is the UK's largest Trade Association representing electrical, electronic, installation, engineering and building service companies. It is recognised by the Government, commerce and industry as the authoritative voice of the sector, working to improve industry standards and procedures.

ECA membership is made up of local electrical contractors with only a few operatives to multi-service companies employing thousands of people. The combined annual turnover of ECA member companies is over £4 billion, they employ more than 30 000 operatives and support 8000 apprentices in their craft training (2004 figures).

Customers employing an electrical contractor who has ECA membership means that the work undertaken is guaranteed by the ECA. If the work undertaken fails to meet the relevant standards, the ECA warranty ensures that the work will be rectified to comply with the relevant standards at no cost to the claimant. The ECA warranty is valid for six years after the completion of the contract. If a member company becomes insolvent while carrying out a contract, the ECA Bond will cover the cost of completing the contract by another company.

ECA membership gives customer guarantees, so how does the ECA ensure that its members meet the quality standards? It does this by:

- Undertaking periodic technical assessment of all its members to confirm that their work meets the industry standards, that is BS7671:2001 IEE Requirements for Electrical Installations and all other relevant standards, regulations and codes of practice.
- Insisting that member companies only employ properly qualified operatives.
- Providing an advice service to members on health and safety issues, technical interpretation of relevant regulations, employee relations and contractual and legal matters.
- Encouraging its members to join the ZAP Initiative, where ZAP means Zero Accident Potential. By 2004/5 the Government's ZAP Initiative plans to achieve zero fatal accidents by electrical contractors, a 30% reduction in reportable three day time lost injuries and a 40% reduction in major injuries.

The ECA is in the vanguard of training initiatives with the emphasis on constantly aiming for the highest standards. It played a key role in setting the National Vocational Standards (NVQs) across the electrical, heating/ventilation and plumbing sectors of the electrotechnical industry.

The ECA in partnership with Amicus, the electrotechnical industry's trade union, have negotiated the national working agreement, which sets levels of pay, holiday entitlement and the conditions of employment within the industry. This partnership has been in place since 1968 and its success is evident by the low levels of dispute and the high status of craft qualifications in the electrotechnical industry.

THE NATIONAL INSPECTION COUNCIL FOR ELECTRICAL INSTALLATION CONTRACTING (NICEIC)

The National Inspection Council for Electrical Installation Contracting (NICEIC) is an independent consumer safety organisation, set up to protect users of electricity against the hazards of unsafe and unsound electrical installations. It is the electrical industry's safety regulatory body and has been carrying out this work for over 40 years.

The NICEIC's independence and its thorough regular technical assessments of electrical contractors' work, provides an assurance that customers can put their trust in NICEIC Approved Contractors.

The NICEIC publishes lists of Approved Contractors and includes member companies in the Yellow Pages and Thompson local directories, the sources which most members of the public use when seeking an electrical contractor. Members of the NICEIC may display the NICEIC logo on stationery and vehicles, indicating to potential customers that this business is reputable, reliable and works to the industry's standards.

Electrical contractors enrolled with the NICEIC must comply with the Council's Rules. In particular they must:

- Ensure that all work carried out complies with BS 7671: 2001, the IEE Requirements for Electrical Installations and any other relevant regulations, standards or codes of practice.
- Issue appropriate forms of certification on completion of installation work in accordance with BS 7671.
- Hold up to date copies of BS 7671, IEE Guidance Notes 1 and 3 and the HSE Memorandum of Guidance of the Electricity at Work Regulations 1989.
- Have a written Health and Safety Policy statement and carry out risk assessments.
- Possess and demonstrate competence in the use of test instruments required to carry out the range of tests required by Section 7 of the IEE Regulations and detailed in Guidance Note 3 of the Regulations. (These are described in Chapter 4 of this book.)
- Hold public liability insurance of at least £2 million.
- Employ a suitably qualified and experienced 'Qualified Supervisor' who will be responsible for the competence of all operatives and the standard of their work.
- Appoint a 'Principal Duty Holder' who will have overall responsibility for Health and Safety. (Note – these two posts may be held by the same person.)

It is no use having rules if you do not enforce them, and in the case of the NICEIC, assessment of approved contractors is carried out by fifty-four local Area

Engineers. These engineers provide practical advice and guidance on electrical installation problems and inspect the work carried out by member companies in order to assess standards and quality. If the work does not meet the standards of BS 7671, the IEE Regulations, the company's name will be removed from the 'Approved Register' and the Council may, at its discretion, make public the removal of the 'Approved' contractors name from the Register and the rules that were broken – a very embarrassing situation for any respectable trading company.

The benefits of being an NICEIC Approved Contractor is that it is a statement to potential customers that this company takes its responsibilities seriously, that it is reputable and carries out work to the industry standards. The NICEIC also guarantees the standard of the work carried out by its Approved Contractors and provides an assurance that any departures from BS 7671, the IEE Regulations, will be corrected free of charge. Finally, membership of the NICEIC gives contractors access to additional work for local authorities. The majority of local authorities will only allow NICEIC Approved Contractors to carry out electrical work in schools and local government buildings.

TRADE UNIONS

Trade unions have a long history of representing workers in industry and commerce. Throughout the history of the trade union movement there has been a tendency for the smaller more specialised unions to come together and combine their resources in order to increase their influence and bargaining power, both with Government and employer organisations.

The Electrical Trade Union (ETU) was originally formed in 1868 and the Plumbing Trade Union (PTU) in 1865. These two Unions had a long separate history but subsequently amalgamated in 1968 to become the EETPU.

During the Second World War there was an increase in the number of people working in the engineering industry. This was chiefly due to the increased need for armaments, and these workers were represented by the Amalgamated Engineering Union (AEU). In 1992 the electricians, plumbers and engineering workers' Unions merged to form the AEEU.

Employees in the electrotechnical industry in the new millennium are represented by a trade union called Amicus. Since Amicus also represents workers in the sheet-metal industry, coppersmiths and scientific sector, the Union is known as 'Amicus' but each member will belong to an appropriate 'sector' of that Union. The 2003 rule book identifies twenty-one 'sectors', from aerospace through education, electrical engineering, finance, food, local authorities, servicing and transport.

Trade unions negotiate with the Governments and employers' representatives, levels of pay and conditions of employment for their members. Through a network of local area officers the Union can offer advice and support to its members, ranging from disputes with employers to more serious cases of, say, victimisation or wrongful dismissal. The Union will also provide legal advice and representation.

Because Unions have a large membership, they often also negotiate special membership discounts for insurance services, airport parking, car rental, eye care requirements, store discount cards and much more.

Trade union membership comes at a reasonable cost. The 2004 fees for full membership of Amicus are a little over £2 per week. Trainees and those serving an apprenticeship are allowed membership at a reduced or nil rate until their training is finished.

Exercises

1 For any fire to continue to burn three components must be present. These are:
   (a) fuel, wood, cardboard
   (b) petrol, oxygen, bottled gas
   (c) flames, fuel, heat
   (d) fuel, oxygen, heat.
2 The recommended voltage for portable hand tools on construction sites is:
   (a) 50 V
   (b) 110 V
   (c) 230 V
   (d) 400 V.
3 The person responsible for financing the building team is the:
   (a) main contractor
   (b) subcontractor
   (c) client
   (d) architect.
4 The person responsible for interpreting the client's requirements to the building team is the:
   (a) main contractor

   (b) subcontractor
   (c) client
   (d) architect.

5 The building contractor is also called the:
   (a) main contractor
   (b) subcontractor
   (c) client
   (d) architect.

6 The electrical contractor is also called the:
   (a) main contractor
   (b) subcontractor
   (c) client
   (d) architect.

7 The people responsible for interpreting the architect's electrical specifications and drawings are the:
   (a) building team
   (b) electrical design team
   (c) electrical installation team
   (d) construction industry.

8 The people responsible for demonstrating good workmanship and maintaining good relationships with other trades are the:
   (a) building team
   (b) electrical design team
   (c) electrical installation team
   (d) construction industry.

9 'A scale drawing showing the position of equipment by graphical symbols' is a description of a:
   (a) block diagram
   (b) layout diagram
   (c) wiring diagram
   (d) circuit diagram.

10 'A diagram which shows the detailed connections between individual items of equipment' is a description of a:
   (a) block diagram
   (b) layout diagram
   (c) wiring diagram
   (d) circuit diagram.

11 'A diagram which shows most clearly how a circuit works, with all items represented by graphical symbols' is a description of a:
   (a) block diagram
   (b) layout diagram
   (c) wiring diagram
   (d) circuit diagram.

12 A record of work done which is outside the scope of the original contract would be kept on a:
   (a) memo
   (b) daywork sheet
   (c) time sheet
   (d) delivery note.

13 A record of goods delivered to site is recorded on a:
   (a) memo
   (b) daywork sheet
   (c) time sheet
   (d) delivery note.

14 A small blow-torch burn to the arm of a workmate should be treated by:
   (a) immersing in cold water before applying a clean dry dressing
   (b) pricking blisters before applying a clean dry dressing
   (c) covering burned skin with cream or petroleum jelly to exclude the air before applying a clean dry dressing
   (d) applying direct pressure to the burned skin to remove the heat from the burn and relieve the pain.

15 Briefly describe the duties of each of the following people:
   (a) the clerk of works
   (b) the health and safety inspector
   (c) the electrician
   (d) the foreman electrician.

16 Describe the importance of a correct attitude towards the customer by an apprentice electrician and other members of the installation team.

17 State eight separate tasks carried out by the electrical design team.

18 State seven separate tasks carried out by the electrical installation team.

19 State the purpose of a 'variation' order.

20 State the advantages of a written legal contract as compared to a verbal contract.

21 Describe a suitable electrical distribution system for a construction site comprising
   (a) heavy current using fixed machines
   (b) site cabins
   (c) robustly installed site lighting
   (d) portable tools.
Identify suitable voltages and how the various voltages may be obtained from the mains input position which is at 400 V.

22 Describe the action to be taken upon finding a workmate apparently dead on the floor and connected to a live electrical supply.

23 Describe how you would arrange for electrical accessories, switchgear and cable to be stored on a large construction site.

24 State the responsibilities under the Health and Safety at Work Act of:
(a) an employer to his employees
(b) an employee to his employer and fellow workers.

25 Safety signs are used in the working environment to give information and warnings. Describe the purpose of the four categories of signs and state their colour code and shape. You may use sketches to illustrate your answer.

26 A trainee electrician discovers a large well-established fire in a store-room of the building in which he is working. The building is an office block which is under construction but almost complete. There are six offices on each of six floors and the store-room and fire are in an office on the fourth floor of the building. The trainee knows that there are between 10 and 20 other construction workers somewhere in the building and that the fire alarm system is not connected. Describe the actions which the trainee should take to prevent this emergency becoming a disaster.

27 State the name of three Statutory Regulations and three Non-Statutory Regulations which apply to the electrotechnical industry.

28 Briefly describe the difference between Statutory Regulations and Non-Statutory Regulations.

29 What is meant by PPE. State five pieces of PPE which an apprentice electrician might be expected to wear in the course of his work.

30 Explain how the Data Protection Act has changed the way in which we record and store Accident/First Aid information.

2

PRINCIPLES OF ELECTROTECHNOLOGY

—

Units

Very early units of measurement were based on the things easily available – the length of a stride, the distance from the nose to the outstretched hand, the weight of a stone and the time-lapse of one day. Over the years, new units were introduced and old ones were modified. Different branches of science and engineering were working in isolation, using their own units, and the result was an overwhelming variety of units.

In all branches of science and engineering there is a need for a practical system of units which everyone can use. In 1960 the General Conference of Weights and Measures agreed to an international system called the *Système International d'Unités* (abbreviated to SI units). SI units are based upon a small number of fundamental units from which all other units may be derived (see Table 2.1).

Table 2.1   SI units

SI unit	Measure of	Symbol
The fundamental units		
Metre	Length	m
Kilogram	Mass	kg
Second	Time	s
Ampere	Electric current	A
Kelvin	Thermodynamic temperature	K
Candela	Luminous intensity	cd
Some derived units		
Coulomb	Charge	C
Joule	Energy	J
Newton	Force	N
Ohm	Resistance	Ω
Volt	Potential difference	V
Watt	Power	W

Like all metric systems, SI units have the advantage that prefixes representing various multiples or submultiples may be used to increase or decrease the size of the unit by various powers of 10. Some of the more common prefixes and their symbols are shown in Table 2.2.

Table 2.2   Prefixes for use with SI units

Prefix	Symbol	Multiplication factor		
Mega	M	$\times 10^6$	or	$\times 1\,000\,000$
Kilo	k	$\times 10^3$	or	$\times 1000$
Hecto	h	$\times 10^2$	or	$\times 100$
Deca	da	$\times 10$	or	$\times 10$
Deci	d	$\times 10^{-1}$	or	$\div 10$
Centi	c	$\times 10^{-2}$	or	$\div 100$
Milli	m	$\times 10^{-3}$	or	$\div 1000$
Micro	μ	$\times 10^{-6}$	or	$\div 1\,000\,000$

Basic circuit theory

All matter is made up of atoms which arrange themselves in a regular framework within the material. The atom is made up of a central, positively charged nucleus, surrounded by negatively charged electrons. The electrical properties of a material depend largely upon how tightly these electrons are bound to the central nucleus.

A *conductor* is a material in which the electrons are loosely bound to the central nucleus and are, therefore, free to drift around the material at random from one atom to another, as shown in Fig. 2.1(a). Materials which are good conductors include copper, brass, aluminium and silver.

An *insulator* is a material in which the outer electrons are tightly bound to the nucleus and so there are

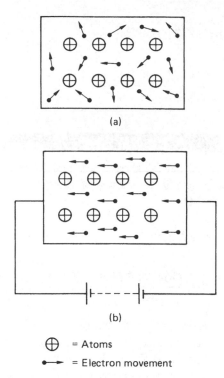

(a)

(b)

$\oplus$ = Atoms

•——→ = Electron movement

Fig. 2.1 Atoms and electrons on a material.

no free electrons to move around the material. Good insulating materials are PVC, rubber, glass and wood.

If a battery is attached to a conductor as shown in Fig. 2.1(b), the free electrons drift purposefully in one direction only. The free electrons close to the positive plate of the battery are attracted to it since unlike charges attract, and the free electrons near the negative plate will be repelled from it. For each electron entering the positive terminal of the battery, one will be ejected from the negative terminal, so the number of electrons in the conductor remains constant.

This drift of electrons within a conductor is known as an electric *current*, measured in amperes and given the symbol I. For a current to continue to flow, there must be a complete circuit for the electrons to move around. If the circuit is broken by opening a switch, for example, the electron flow and therefore the current will stop immediately.

To cause a current to flow continuously around a circuit, a driving force is required, just as a circulating pump is required to drive water around a central heating system. This driving force is the *electromotive force* (abbreviated to emf). Each time an electron passes through the source of emf, more energy is provided to send it on its way around the circuit.

An emf is always associated with energy conversion, such as chemical to electrical in batteries and mechanical to electrical in generators. The energy introduced into the circuit by the emf is transferred to the load terminals by the circuit conductors. The *potential difference* (abbreviated to p.d.) is the change in energy levels measured across the load terminals. This is also called the volt drop or terminal voltage, since emf and p.d. are both measured in volts. Every circuit offers some opposition to current flow, which we call the circuit *resistance*, measured in ohms (symbol Ω), to commemorate the famous German physicist Georg Simon Ohm, who was responsible for the analysis of electrical circuits.

OHM'S LAW

In 1826, Ohm published details of an experiment he had done to investigate the relationship between the current passing through and the potential difference between the ends of a wire. As a result of this experiment, he arrived at a law, now known as Ohm's law, which says that *the current passing through a conductor under constant temperature conditions is proportional to the potential difference across the conductor*. This may be expressed mathematically as

$$V = I \times R (\text{V})$$

Transposing this formula, we also have

$$I = \frac{V}{R} (\text{A}) \qquad \text{and} \qquad R = \frac{V}{I} (\Omega)$$

EXAMPLE 1

An electric heater, when connected to a 230 V supply, was found to take a current of 4 A. Calculate the element resistance.

$$R = \frac{V}{I}$$

$$\therefore R = \frac{230 \text{ V}}{4 \text{ A}} = 57.5 \, \Omega$$

EXAMPLE 2

The insulation resistance measured between phase conductors on a 400 V supply was found to be 2 MΩ. Calculate the leakage current.

$$I = \frac{V}{R}$$

$$\therefore I = \frac{400\,V}{2 \times 10^6\,\Omega} = 200 \times 10^{-6}\,A = 200\,\mu A$$

EXAMPLE 3

When a 4 Ω resistor was connected across the terminals of an unknown d.c. supply, a current of 3 A flowed. Calculate the supply voltage.

$$V = I \times R$$
$$\therefore V = 3A \times 4\,\Omega = 12\,V$$

RESISTIVITY

The resistance or opposition to current flow varies for different materials, each having a particular constant value. If we know the resistance of, say, 1 metre of a material, then the resistance of 5 metres will be five times the resistance of 1 metre.

The *resistivity* (symbol ρ – the Greek letter 'rho') of a material is defined as the resistance of a sample of unit length and unit cross-section. Typical values are given in Table 2.3. Using the constants for a particular material we can calculate the resistance of any length and thickness of that material from the equation.

$$R = \frac{\rho l}{a}\,(\Omega)$$

where

$\rho =$ the resistivity constant for the material (Ω m)
$l =$ the length of the material (m)
$a =$ the cross-sectional area of the material (m²).

Table 2.3 gives the resistivity of silver as 16.4×10^{-9} Ω m, which means that a sample of silver 1 metre long and 1 metre in cross-section will have a resistance of 16.4×10^{-9} Ω.

Table 2.3   Resistivity values

Material	Resistivity (Ω m)
Silver	16.4×10^{-9}
Copper	17.5×10^{-9}
Aluminium	28.5×10^{-9}
Brass	75.0×10^{-9}
Iron	100.0×10^{-9}

EXAMPLE 1

Calculate the resistance of 100 metres of copper cable of 1.5 mm² cross-sectional area if the resistivity of copper is taken as 17.5×10^{-9} Ω m.

$$R = \frac{\rho l}{a}\,(\Omega)$$

$$\therefore R = \frac{17.5 \times 10^{-9}\,\Omega\,m \times 100\,m}{1.5 \times 10^{-6}\,m^2} = 1.16\,\Omega$$

EXAMPLE 2

Calculate the resistance of 100 metres of aluminium cable of 1.5 mm² cross-sectional area if the resistivity of aluminium is taken as 28.5×10^{-9} Ω m.

$$R = \frac{\rho l}{a}\,(\Omega)$$

$$\therefore R = \frac{28.5 \times 10^{-9}\,\Omega\,m \times 100\,m}{1.5 \times 10^{-6}\,m^2} = 1.9\,\Omega$$

The above examples show that the resistance of an aluminium cable is some 60% greater than a copper conductor of the same length and cross-section. Therefore, if an aluminium cable is to replace a copper cable, the conductor size must be increased to carry the rated current as given by the tables in Appendix 4 of the IEE Regulations and Appendix 6 of the *On Site Guide*.

The other factor which affects the resistance of a material is the temperature, and we will consider this next.

TEMPERATURE COEFFICIENT

The resistance of most materials changes with temperature. In general, conductors increase their resistance as the temperature increases and insulators decrease their resistance with a temperature increase. Therefore, an increase in temperature has a bad effect upon the electrical properties of a material.

Each material responds to temperature change in a different way, and scientists have calculated constants for each material which are called the *temperature coefficient of resistance* (symbol α – the Greek letter 'alpha'). Table 2.4 gives some typical values.

Table 2.4   Temperature coefficient values

Material	Temperature coefficient (Ω/Ω°C)
Silver	0.004
Copper	0.004
Aluminium	0.004
Brass	0.001
Iron	0.006

Using the constants for a particular material and substituting values into the following formulae the resistance of a material at different temperatures may be calculated. For a temperature increase from 0°C:

$$R_t = R_0(1 + \alpha t) \ (\Omega)$$

where

R_t = the resistance at the new temperature t°C
R_0 = the resistance at 0°C
α = the temperature coefficient for the particular material.

For a temperature increase between two intermediate temperatures above 0°C:

$$\frac{R_1}{R_2} = \frac{(1 + \alpha t_1)}{(1 + \alpha t_2)}$$

where

R_t = the resistance at the original temperature
R_2 = the resistance at the final temperature
α = the temperature coefficient for the particular material.

If we take a 1 Ω resistor of, say, copper, and raise its temperature by 1°C, the resistance will increase by 0.004 Ω to 1.004 Ω. This increase of 0.004 Ω is the temperature coefficient of the material.

The field winding of a d.c. motor has a resistance of 100 Ω at 0°C. Determine the resistance of the coil at 20°C if the temperature coefficient is 0.004 Ω/Ω°C.

$$R_t = R_0(1 + \alpha t) \ (\Omega)$$

$$\therefore R_t = 100\,\Omega(1 + 0.004\,\Omega/\Omega°C \times 20°C)$$

$$R_t = 100\,\Omega(1 + 0.08)$$

$$R_t = 108\,\Omega$$

The field winding of a shunt generator has a resistance of 150 Ω at an ambient temperature of 20°C. After running for some time the mean temperature of the generator rises to 45°C. Calculate the resistance of the winding at the higher temperature if the temperature coefficient of resistance is 0.004 Ω/Ω°C.

$$\frac{R_1}{R_2} = \frac{(1 + \alpha t_1)}{(1 + \alpha t_2)}$$

$$\frac{150\,\Omega}{R_2} = \frac{1 + 0.004\,\Omega/\Omega°C \times 20°C}{1 + 0.004\,\Omega/\Omega°C \times 45°C}$$

$$\frac{150\,\Omega}{R_2} = \frac{1.08}{1.18}$$

$$\therefore R_2 = \frac{150\,\Omega \times 1.18}{1.08} = 164\,\Omega.$$

It is clear from the last two sections that the resistance of a cable is affected by length, thickness, temperature and type of material. Since Ohm's law tells us that current is inversely proportional to resistance, these factors must also influence the current carrying capacity of a cable. The tables of current ratings in Appendix 4 of the IEE Regulations and Appendix 6 of the *On Site Guide* contain correction factors so that current ratings may be accurately determined under defined installation conditions. Cable selection is considered in Chapter 6.

Resistors

In an electrical circuit resistors may be connected in series, in parallel, or in various combinations of series and parallel connections.

SERIES-CONNECTED RESISTORS

In any series circuit a current I will flow through all parts of the circuit as a result of the potential difference supplied by a battery V_T. Therefore, we say that in a series circuit the current is common throughout that circuit.

When the current flows through each resistor in the circuit, R_1, R_2 and R_3 for example in Fig. 2.2, there will be a voltage drop across that resistor whose value will be determined by the values of I and R, since from

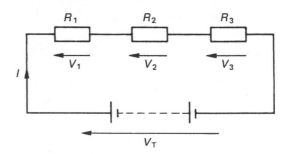

Fig. 2.2 A series circuit.

Ohm's law $V = I \times R$. The sum of the individual voltage drops, V_1, V_2 and V_3 for example in Fig. 2.2, will be equal to the total voltage V_T.

We can summarize these statements as follows. For any series circuit, I is common throughout the circuit and

$$V_T = V_1 + V_2 + V_3 \tag{1}$$

Let us call the total circuit resistance R_T. From Ohm's law we know that $V = I \times R$ and therefore

$$
\begin{aligned}
&\text{Total voltage } V_T = I \times R_T \\
&\text{Voltage drop across } R_1 \text{ is } V_1 = I \times R_1 \\
&\text{Voltage drop across } R_2 \text{ is } V_2 = I \times R_2 \\
&\text{Voltage drop across } R_3 \text{ is } V_3 = I \times R_3
\end{aligned}
\tag{2}
$$

We are looking for an expression for the total resistance in any series circuit and, if we substitute equations (2) into equation (1) we have:

$$V_T = V_1 + V_2 + V_3$$

$$\therefore I \times R_T = I \times R_1 + I \times R_2 + I \times R_3$$

Now, since I is common to all terms in the equation, we can divide both sides of the equation by I. This will cancel out I to leave us with an expression for the circuit resistance:

$$R_T = R_1 + R_2 + R_3$$

Note that the derivation of this formula is given for information only. Craft students need only state the expression $R_T = R_1 + R_2 + R_3$ for series connections.

PARALLEL-CONNECTED RESISTORS

In any parallel circuit, as shown in Fig. 2.3, the same voltage acts across all branches of the circuit. The total current will divide when it reaches a resistor junction, part of it flowing in each resistor. The sum of the individual currents, I_1, I_2 and I_3 for example in Fig. 2.3, will be equal to the total current I_T.

We can summarize these statements as follows. For any parallel circuit, V is common to all branches of the circuit and

$$I_T = I_1 + I_2 + I_3 \tag{3}$$

Let us call the total resistance R_T.

From Ohm's law we know, that $I = \dfrac{V}{R}$, and therefore

the total current $I_T = \dfrac{V}{R_T}$

the current through R_1 is $I_1 = \dfrac{V}{R_1}$

the current through R_2 is $I_2 = \dfrac{V}{R_2}$ $\qquad (4)$

the current through R_3 is $I_3 = \dfrac{V}{R_3}$

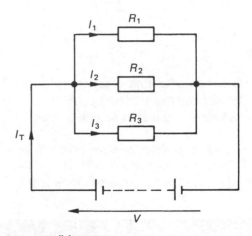

Fig. 2.3   A parallel circuit.

We are looking for an expression for the equivalent resistance R_T in any *parallel* circuit and, if we substitute equations (4) into equation (3) we have:

$$I_T = I_1 + I_2 + I_3$$

$$\therefore \frac{V}{R_T} = \frac{V}{R_1} + \frac{V}{R_2} + \frac{V}{R_3}$$

Now, since V is common to all terms in the equation, we can divide both sides by V, leaving us with an expression for the circuit resistance:

$$\frac{1}{R_T} = \frac{1}{R_1} + \frac{1}{R_2} + \frac{1}{R_3}$$

Note that the derivation of this formula is given for information only. Craft students need only state the expression $1/R_T = 1/R_1 + 1/R_2 + 1/R_3$ for parallel connections.

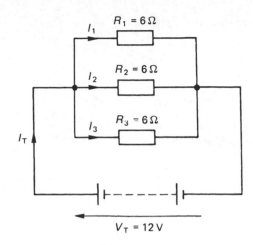

EXAMPLE

Three $6\,\Omega$ resistors are connected (n) in series (see Fig. 2.4), and (b) in parallel (see Fig. 2.5), across a $12\,V$ battery. For each method of connection, find the total resistance and the values of all currents and voltages.

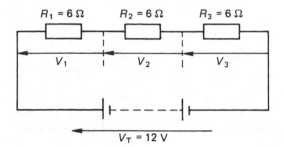

Fig. 2.5 Resistors in parallel.

Fig. 2.4 Resistors in series.

For any series connection

$$R_T = R_1 + R_2 + R_3$$
$$\therefore R_T = 6\,\Omega + 6\,\Omega + 6\,\Omega = 18\,\Omega$$

Total current $I_T = \dfrac{V_T}{R_T}$

$$\therefore I_T = \dfrac{12\,V}{18\,\Omega} = 0.67\,A$$

The voltage drop across R_1 is

$$V_1 = I_T \times R_1$$
$$\therefore V_1 = 0.67\,A \times 6\,\Omega = 4\,V$$

The voltage drop across R_2 is

$$V_2 = I_T \times R_2$$
$$\therefore V_2 = 0.67\,A \times 6\,\Omega = 4\,V$$

The voltage drop across R_3 is

$$V_3 = I_T \times R_3$$
$$\therefore V_3 = 0.67\,A \times 6\,\Omega = 4\,V$$

For any parallel connection,

$$\frac{1}{R_T} = \frac{1}{R_1} + \frac{1}{R_2} + \frac{1}{R_3}$$

$$\therefore \frac{1}{R_T} = \frac{1}{6\,\Omega} + \frac{1}{6\,\Omega} + \frac{1}{6\,\Omega}$$

$$\frac{1}{R_T} = \frac{1+1+1}{6\,\Omega} = \frac{3}{6\,\Omega}$$

$$R_T = \frac{6\,\Omega}{3} = 2\,\Omega$$

Total current $I_T = \dfrac{V_T}{R_T}$

$$\therefore I_T = \dfrac{12\,V}{2\,\Omega} = 6\,A$$

The current flowing through R_1 is

$$I_1 = \frac{V_T}{R_1}$$

$$\therefore I_1 = \frac{12\,V}{6\,\Omega} = 2\,A$$

The current flowing through R_2 is

$$I_2 = \frac{V_T}{R_2}$$

$$\therefore I_2 = \frac{12\,V}{6\,\Omega} = 2\,A$$

The current flowing through R_3 is

$$I_3 = \frac{V_T}{R_3}$$

$$\therefore I_3 = \frac{12\,V}{6\,\Omega} = 2\,A$$

SERIES AND PARALLEL COMBINATIONS

The most complex arrangement of series and parallel resistors can be simplified into a single equivalent resistor by combining the separate rules for series and parallel resistors.

EXAMPLE 1

Resolve the circuit shown in Fig. 2.6 into a single resistor and calculate the potential difference across each resistor.

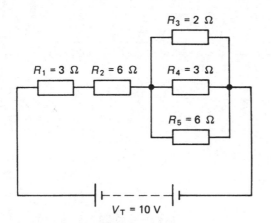

Fig. 2.6 A series/parallel circuit.

By inspection, the circuit contains a parallel group consisting of R_3, R_4 and R_5 and a series group consisting of R_1 and R_2 in series with the equivalent resistor for the parallel branch.

Consider the parallel group. We will label this group R_P. Then

$$\frac{1}{R_P} = \frac{1}{R_3} + \frac{1}{R_4} + \frac{1}{R_5}$$

$$\frac{1}{R_P} = \frac{1}{2\,\Omega} + \frac{1}{3\,\Omega} + \frac{1}{6\,\Omega}$$

$$\frac{1}{R_P} = \frac{3+2+1}{6\,\Omega} = \frac{6}{6\,\Omega}$$

$$R_P = \frac{6\,\Omega}{6} = 1\,\Omega$$

Figure 2.6 may now be represented by the more simple equivalent shown in Fig. 2.7.

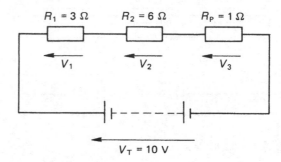

Fig. 2.7 Equivalent series circuit.

Since all resistors are now in series,

$$R_T = R_1 + R_2 + R_P$$

$$\therefore R_T = 3\,\Omega + 6\,\Omega + 1\,\Omega = 10\,\Omega$$

Thus, the circuit may be represented by a single equivalent resistor of value $10\,\Omega$ as shown in Fig. 2.8. The total current flowing in the circuit may be found by using Ohm's law:

$$I_T = \frac{V_T}{R_T} = \frac{10\,V}{10\,\Omega} = 1\,A$$

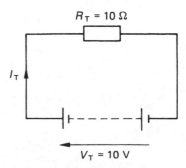

Fig. 2.8 Single equivalent resistor for Fig. 8.6.

The potential differences across the individual resistors are

$$V_1 = I_T \times R_1 = 1\,A \times 3\,\Omega = 3\,V$$
$$V_2 = I_T \times R_2 = 1\,A \times 6\,\Omega = 6\,V$$
$$V_P = I_T \times R_P = 1\,A \times 1\,\Omega = 1\,V$$

Since the same voltage acts across all branches of a parallel circuit the same p.d. of 1 V will exist across each resistor in the parallel branch R_3, R_4 and R_5.

EXAMPLE 2

Determine the total resistance and the current flowing through each resistor for the circuit shown in Fig. 2.9.

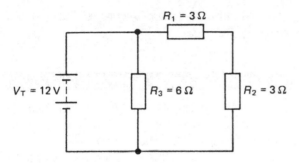

Fig. 2.9   A series/parallel circuit for Example 2.

By inspection, it can be seen that R_1 and R_2 are connected in series while R_3 is connected in parallel across R_1 and R_2. The circuit may be more easily understood if we redraw it as in Fig. 2.10.

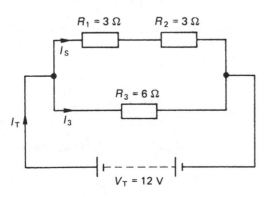

Fig. 2.10   Equivalent circuit for Example 2.

For the series branch, the equivalent resistor can be found from

$$R_S = R_1 + R_2$$
$$\therefore R_S = 3\,\Omega + 3\,\Omega = 6\,\Omega$$

Figure 2.10 may now be represented by a more simple equivalent circuit, as in Fig. 2.11.

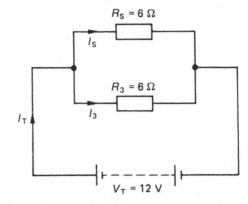

Fig. 2.11   Simplified equivalent circuit for Example 2.

Since the resistors are now in parallel, the equivalent resistance may be found from

$$\frac{1}{R_T} = \frac{1}{R_S} + \frac{1}{R_3}$$
$$\therefore \frac{1}{R_T} = \frac{1}{6\,\Omega} + \frac{1}{6\,\Omega}$$
$$\frac{1}{R_T} = \frac{1+1}{6\,\Omega} = \frac{2}{6\,\Omega}$$
$$R_T = \frac{6\,\Omega}{2} = 3\,\Omega$$

The total current is

$$I_T = \frac{V_T}{R_T} = \frac{12\,V}{3\,\Omega} = 4\,A$$

Let us call the current flowing through resistor R_3 I_3.

$$\therefore I_3 = \frac{V_T}{R_3} = \frac{12\,V}{6\,\Omega} = 2\,A$$

Let us call the current flowing through both resistors R_1 and R_2, as shown in Fig. 2.10, I_S.

$$\therefore I_S = \frac{V_T}{R_S} = \frac{12\,V}{6\,\Omega} = 2\,A$$

Power and energy

POWER

Power is the rate of doing work and is measured in watts:

$$\text{Power} = \frac{\text{Work done}}{\text{Time taken}} (\text{W})$$

In an electrical circuit,

$$\text{Power} = \text{Voltage} \times \text{Current (W)} \qquad (5)$$

Now from Ohm's law

$$\text{Voltage} = I \times R (\text{V}) \qquad (6)$$

$$\text{Current} = \frac{V}{R} (\text{A}) \qquad (7)$$

Substituting equation (6) into equation (5), we have

$$\text{Power} = (I \times R) \times \text{Current} = I^2 \times R (\text{W})$$

and substituting equation (7) into equation (5) we have

$$\text{Power} = \text{Voltage} \times \frac{V}{R} = \frac{V^2}{R} (\text{W})$$

We can find the power of a circuit by using any of the three formulae

$$P = V \times I, \quad P = I^2 \times R, \quad P = \frac{V^2}{R}$$

ENERGY

Energy is a concept which engineers and scientists use to describe the ability to do work in a circuit or system:

$$\text{Energy} = \text{Power} \times \text{Time}$$

but, since

$$\text{Power} = \text{Voltage} \times \text{Current}$$

then

$$\text{Energy} = \text{Voltage} \times \text{Current} \times \text{Time}$$

The SI unit of energy is the joule, where time is measured in seconds. For practical electrical installation circuits this unit is very small and therefore the kilowatt-hour (kWh) is used for domestic and commercial installations. Electricity Board meters measure 'units' of electrical energy, where each 'unit' is 1 kWh. So,

Energy in joules

$$= \text{Voltage} \times \text{Current} \times \text{Time in seconds}$$

$$\text{Energy in kWh} = \text{kW} \times \text{Time in hours}$$

EXAMPLE 1

A domestic immersion heater is switched on for 40 minutes and takes 15 A from a 200 V supply. Calculate the energy used during this time.

$$\text{Power} = \text{Voltage} \times \text{Current}$$
$$\text{Power} = 200\,\text{V} \times 15\,\text{A} = 3000\text{W or 3 kW}$$
$$\text{Energy} = \text{kW} \times \text{Time in hours}$$
$$\text{Energy} = 3\,\text{kW} \times \frac{40\,\text{min}}{60\,\text{min/h}} = 2\,\text{kWh}$$

This immersion heater uses 2 kWh in 40 minutes, or 2 'units' of electrical energy every 40 minutes.

EXAMPLE 2

Two 50 Ω resistors may be connected to a 200 V supply. Determine the power dissipated by the resistors when they are connected (a) in series, (b) each resistor separately connected and (c) in parallel.

For (a), the equivalent resistance when resistors are connected in series is given by

$$R_T = R_1 + R_2$$
$$\therefore R_T = 50\,\Omega + 50\,\Omega = 100\,\Omega$$
$$\text{Power} = \frac{V^2}{R_T} (\text{W})$$
$$\therefore \text{Power} = \frac{200\,\text{V} \times 200\,\text{V}}{100\,\Omega} = 400\,\text{W}$$

For (b), each resistor separately connected has a resistance of 50 Ω.

$$\text{Power} = \frac{V^2}{R} (\text{W})$$
$$\therefore \text{Power} = \frac{200\,\text{V} \times 200\,\text{V}}{50\,\Omega} = 800\,\text{W}$$

For (c), the equivalent resistance when resistors are connected in parallel is given by

$$\frac{1}{R_T} = \frac{1}{R_1} + \frac{1}{R_2}$$

$$\therefore \frac{1}{R_T} = \frac{1}{50\ \Omega} + \frac{1}{50\ \Omega}$$

$$\frac{1}{R_T} = \frac{1+1}{50\ \Omega} = \frac{2}{50\ \Omega}$$

$$R_T = \frac{50\ \Omega}{2} = 25\ \Omega$$

$$\text{Power} = \frac{V^2}{R_T}\ (W)$$

$$\therefore \text{Power} = \frac{200\ V \times 200\ V}{25\ \Omega} = 1600\ W$$

This example shows that by connecting resistors together in different combinations of series and parallel connections, we can obtain various power outputs: in this example, 400, 800 and 1600 W. This theory finds a practical application in the three heat switch used to control a boiling ring.

Alternating current theory

The supply which we obtain from a car battery is a unidirectional or d.c. supply, whereas the mains electricity supply is alternating or a.c. (see Fig. 2.12).

One of the reasons for using alternating supplies for the electricity mains supply, is because we can very easily change the voltage levels by using a transformer which will only work on an a.c. supply.

The generated alternating supply at the power station is transformed up to 132 000 V, or more, for efficient transmission along the National Grid conductors. It is then transformed down to 11 000 V for local underground distribution and finally, down to 400 V for commercial and industrial consumers and 230 V for domestic consumers. We will discuss transformers and supply systems later in this chapter and in Chapter 3.

Most electrical equipment makes use of alternating current supplies, and for this reason a knowledge of alternating waveforms and their effect upon resistive, capacitive and inductive loads is necessary for all practising electricians.

When a coil of wire is rotated inside a magnetic field a voltage is induced in the coil. The induced voltage follows a mathematical law known as the sinusoidal law and, therefore, we can say that a sine wave has been generated. Such a waveform has the characteristics displayed in Fig. 2.13.

In the UK we generate electricity at a frequency of 50 Hz and the time taken to complete each cycle is given by

$$T = \frac{1}{f}$$

$$\therefore T = \frac{1}{50\ Hz} = 0.02\ s$$

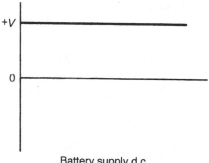

Battery supply d.c.

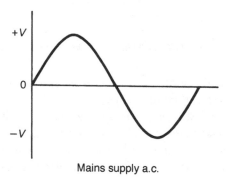

Mains supply a.c.

Fig. 2.12 Unidirectional and alternating supply.

An alternating waveform is constantly changing from zero to a maximum, first in one direction, then in the opposite direction, and so the instantaneous values of the generated voltage are always changing. A useful description of the electrical effects of an a.c. waveform can be given by the maximum, average and rms values of the waveform.

The maximum or peak value is the greatest instantaneous value reached by the generated waveform.

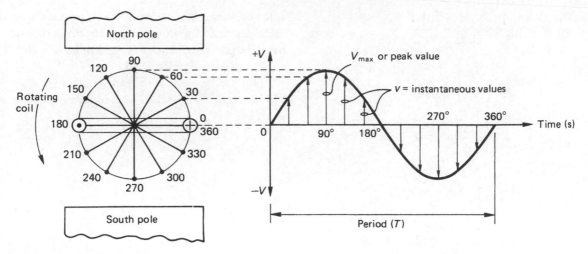

Fig. 2.13 Characteristics of a sine wave.

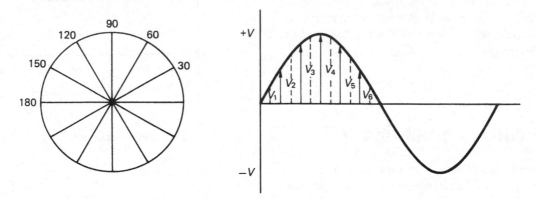

Fig. 2.14 Sinusoidal waveform showing instantaneous values of voltage.

Cable and equipment insulation levels must be equal to or greater than this value.

The average value is the average over one half-cycle of the instantaneous values as they change from zero to a maximum and can be found from the following formula applied to the sinusoidal waveform shown in Fig. 2.14:

$$V_{av} = \frac{V_1 + V_2 + V_3 + V_4 + V_5 + V_6}{6} = 0.637\,V_{max}$$

For any sinusoidal waveform the average value is equal to 0.637 of the maximum value.

The rms value is the square root of the mean of the individual squared values and is the value of an a.c. voltage which produces the same heating effect as a d.c. voltage. The value can be found from the following formula applied to the sinusoidal waveform shown in Fig. 2.14.

$$V_{rms} = \sqrt{\frac{V_1^2 + V_2^2 + V_3^2 + V_4^2 + V_5^2 + V_6^2}{6}}$$
$$= 0.7071\,V_{max}$$

For any sinusoidal waveform the rms value is equal to 0.7071 of the maximum value.

EXAMPLE

The sinusoidal waveform applied to a particular circuit has a maximum value of 325.3 V. Calculate the average and rms value of the waveform.

$$\text{average value } V_{av} = 0.637 \times V_{max}$$
$$\therefore V_{av} = 0.637 \times 325.3 = 207.2\,V$$
$$\text{rms value } V_{rms} = 0.7071 \times V_{max}$$
$$V_{rms} = 0.7071 \times 325.3 = 230\,V$$

When we say that the main supply to a domestic property is 230 V we really mean 230 V$_{rms}$. Such a waveform has an average value of about 207.2 V and a maximum value of almost 325.3 V but because the rms value gives the d.c. equivalent value we almost always give the rms value without identifying it as such.

THE THREE EFFECTS OF AN ELECTRIC CURRENT

When an electric current flows in a circuit it can have one or more of the following three effects: *heating, magnetic* or *chemical*.

Heating effect

The movement of electrons within a conductor, which is the flow of an electric current, causes an increase in the temperature of the conductor. The amount of heat generated by this current flow depends upon the type and dimensions of the conductor and the quantity of current flowing. By changing these variables, a conductor may be operated hot and used as the heating element of a fire, or be operated cool and used as an electrical installation conductor.

The heating effect of an electric current is also the principle upon which a fuse gives protection to a circuit. The fuse element is made of a metal with a low melting point and forms a part of the electrical circuit. If an excessive current flows, the fuse element overheats and melts, breaking the circuit.

Magnetic effect

Whenever a current flows in a conductor a magnetic field is set up around the conductor like an extension of the insulation. The magnetic field increases with the current and collapses if the current is switched off. A conductor carrying current and wound into a solenoid produces a magnetic field very similar to a permanent magnet, but has the advantage of being switched on and off by any switch which controls the circuit current.

The magnetic effect of an electric current is the principle upon which electric bells, relays, instruments, motors and generators work.

Chemical effect

When an electric current flows through a conducting liquid, the liquid is separated into its chemical parts. The conductors which make contact with the liquid are called the anode and cathode. The liquid itself is called the electrolyte, and the process is called *electrolysis*.

Electrolysis is an industrial process used in the refining of metals and electroplating. It was one of the earliest industrial applications of electric current. Most of the aluminium produced today is extracted from its ore by electrochemical methods. Electroplating serves a double purpose by protecting a base metal from atmospheric erosion and also giving it a more expensive and attractive appearance. Silver and nickel plating has long been used to enhance the appearance of cutlery, candlesticks and sporting trophies.

An anode and cathode of dissimilar metal placed in an electrolyte can react chemically and produce an emf. When a load is connected across the anode and cathode, a current is drawn from this arrangement, which is called a cell. A battery is made up of a number of cells. It has many useful applications in providing portable electrical power, but electrochemical action can also be undesirable since it is the basis of electrochemical corrosion which rots our motor cars, industrial containers and bridges.

Magnetism

The Greeks knew as early as 600 BC that a certain form of iron ore, now known as magnetite or lodestone, had the property of attracting small pieces of iron. Later, during the Middle Ages, navigational compasses were made using the magnetic properties of lodestone. Small pieces of lodestone attached to wooden splints floating in a bowl of water always came to rest pointing in a north–south direction. The word lodestone is derived from an old English word meaning 'the way', and the word magnetism is derived from Magnesia, the place where magnetic ore was first discovered.

Iron, nickel and cobalt are the only elements which are attracted strongly by a magnet. These materials are said to be *ferromagnetic*. Copper, brass, wood, PVC and glass are not attracted by a magnet and are, therefore, described as *non-magnetic*.

SOME BASIC RULES OF MAGNETISM

1 Lines of magnetic flux have no physical existence, but they were introduced by Michael Faraday (1791–1867) as a way of explaining the magnetic

energy existing in space or in a material. They help us to visualize and explain the magnetic effects. The symbol used for magnetic flux is the Greek letter Φ (phi) and the unit of magnetic flux is the weber (symbol Wb), pronounced 'veber', to commemorate the work of the German physicist Wilhelm Weber (1804–91).

2 Lines of magnetic flux always form closed loops.

3 Lines of magnetic flux behave like stretched elastic bands, always trying to shorten themselves.

4 Lines of magnetic flux never cross over each other.

5 Lines of magnetic flux travel along a magnetic material and always emerge out of the 'north pole' end of the magnet.

6 Lines of magnetic flux pass through space and non-magnetic materials undisturbed.

7 The region of space through which the influence of a magnet can be detected is called the *magnetic field* of that magnet.

8 The number of lines of magnetic flux within a magnetic field is a measure of the flux density. Strong magnetic fields have a high flux density. The symbol used for flux density is B, and the unit of flux density is the tesla (symbol T), to commemorate the work of the Croatian-born American physicist Nikola Tesla (1857–1943).

9 The places on a magnetic material where the lines of flux are concentrated are called the magnetic poles.

10 Like poles repel; unlike poles attract. These two statements are sometimes called the 'first laws of magnetism' and are shown in Fig. 2.16.

EXAMPLE

The magnetizing coil of a radio speaker induces a magnetic flux of 360 μWb in an iron core of cross-sectional area 300 mm^2. Calculate the flux density in the core.

$$\text{Flux density } B = \frac{\Phi}{\text{area}} \text{ (tesla)}$$

$$B = \frac{360 \times 10^{-6}(\text{Wb})}{300 \times 10^{-6}(\text{m}^2)}$$

$$B = 1.2\,\text{T}$$

MAGNETIC FIELDS

If a permanent magnet is placed on a surface and covered by a piece of paper, iron filings can be shaken

on to the paper from a dispenser. Gently tapping the paper then causes the filings to take up the shape of the magnetic field surrounding the permanent magnet. The magnetic fields around a permanent magnet are shown in Figs 2.15 and 2.16.

Bar magnet

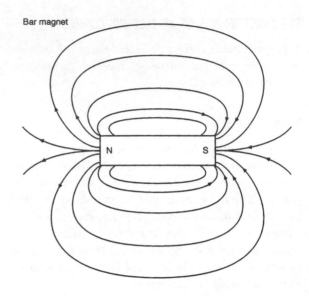

Horse shoe magnet

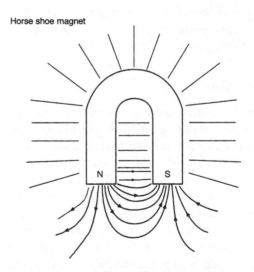

Fig. 2.15   Magnetic fields around a permanent magnet.

Electromagnetism

Electricity and magnetism have been inseparably connected since the experiments by Oersted and Faraday

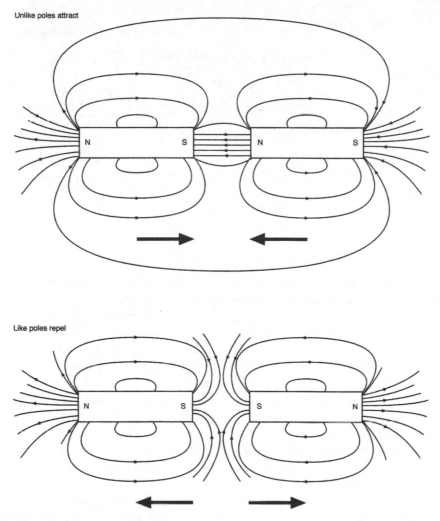

Unlike poles attract

Like poles repel

Fig. 2.16 The first laws of magnetism.

in the early nineteenth century. An electric current flowing in a conductor produces a magnetic field 'around' the conductor which is proportional to the current. Thus a small current produces a weak magnetic field, while a large current will produce a strong magnetic field. The magnetic field 'spirals' around the conductor, as shown in Fig. 2.17 and its direction can be determined by the 'dot' or 'cross' notation and the 'screw rule'. To do this, we think of the current as being represented by a dart or arrow inside the conductor. The dot represents current coming towards us when we would see the point of the arrow or dart inside the conductor. The cross represents current going away from us when we would see the flights of the dart or arrow. Imagine a corkscrew or screw being turned so that it will move in the direction of the

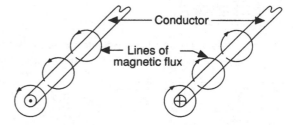

(a) The dot indicates current flowing towards our viewing position

(b) The cross indicates current flowing away from our viewing position

Fig. 2.17 Magnetic fields around a current carrying conductor.

current. Therefore, if the current was coming out of the paper, as shown in Fig. 2.17(a), the magnetic field would be spiralling anticlockwise around the conductor.

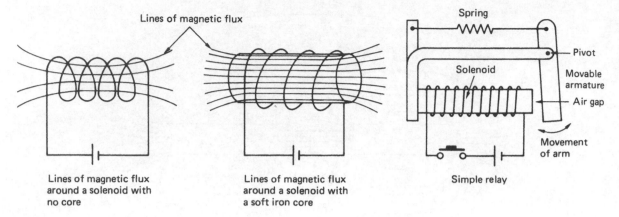

Fig. 2.18 The solenoid and one practical application: the relay.

If the current was going into the paper, as shown by Fig. 2.17(b), the magnetic field would spiral clockwise around the conductor.

A current flowing in a *coil* of wire or solenoid establishes a magnetic field which is very similar to that of a bar magnet. Winding the coil around a soft iron core increases the flux density because the lines of magnetic flux concentrate on the magnetic material. The advantage of the electromagnet when compared with the permanent magnet is that the magnetism of the electromagnet can be switched on and off by a functional switch controlling the coil current. This effect is put to practical use in the electrical relay as used in a motor starter or alarm circuit. Figure 2.18 shows the structure and one application of the solenoid.

Inductance

If a coil of wire is wound on to an iron core as shown in Fig. 2.19, a magnetic field will become established

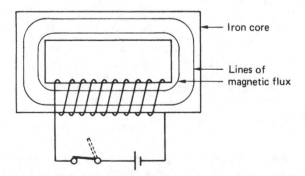

Fig. 2.19 An inductive coil or choke.

in the core when a current flows in the coil due to the switch being closed.

When the switch is opened the current stops flowing and, therefore, the magnetic flux collapses. The collapsing magnetic flux cuts the electrical conductors of the coil and induces an emf into it. This voltage appears across the switch contacts. The effect is known as *inductance* and is one property of any coil. The unit of inductance is the henry (symbol H), to commemorate the work of the American physicist Joseph Henry (1797–1878) who, quite independently, discovered electromagnetic induction just one year after Michael Faraday in 1831.

Faraday's law states that when a conductor cuts or is cut by a magnetic field, an emf is induced in that conductor. The amount of induced emf is proportional to the rate or speed at which the magnetic field cuts the conductor. This is the principle of operation of the simple generator shown in Fig. 2.40. Rotating a coil of wire in a magnetic field induces an emf or voltage in the coil. Increasing the speed of rotation will increase the generated voltage.

Modern power station generators work on this principle and, indeed, the foundations of all our modern knowledge of electricity were laid down by Michael Faraday. He is one of the most famous English scientists, and when Robert Peel, the Prime Minister of the day, asked Faraday 'What use will electricity be?', Faraday replied 'I know not, sir, but I'll wager that one day you will tax it'!

Any circuit in which a change of magnetic flux induces an emf is said to be 'inductive' or to possess 'inductance'.

Fluorescent light fittings contain a choke or inductive coil in series with the tube and starter lamp. The starter lamp switches on and off very quickly, causing rapid current changes which induce a large voltage across the tube electrodes sufficient to strike an arc in the tube.

Further information on fluorescent lighting circuits, and the regulations associated with inductive circuits, are given in Chapter 5 of this book. Inductance is also the principle upon which a transformer works and we will look at transformers a little later in this chapter.

Electrostatics

If a battery is connected between two insulated plates, the emf of the battery forces electrons from one plate to another until the p.d. between the plates is equal to the battery emf.

The electrons flowing through the battery constitute a current, I (in amperes), which flows for a time, t (in seconds). The plates are then said to be charged.

The amount of charge transferred is given by

$$Q = It \text{ (coulomb [Symbol C])}$$

Figure 2.20 shows the charges on a capacitor's plates.

When the voltage is removed the charge Q is trapped on the plates, but if the plates are joined together, the same quantity of electricity, $Q + It$, will flow back from one plate to the other, so discharging them. The property of a pair of plates to store an electric charge is called its *capacitance*.

By definition, a capacitor has a capacitance (C) of one farad (symbol F) when a p.d. of one volt maintains a charge of one coulomb on that capacitor, or

$$C = \frac{Q}{V} \text{(F)}$$

Collecting these important formulae together, we have

$$Q = It = CV$$

CAPACITORS

A capacitor consists of two metal plates, separated by an insulating layer called the dielectric. It has the ability of storing a quantity of electricity as an excess of electrons on one plate and a deficiency on the other.

EXAMPLE

A 100 μF capacitor is charged by a steady current of 2 mA flowing for 5 seconds. Calculate the total charge stored by the capacitor and the p.d. between the plates.

$$Q = It \text{ (C)}$$
$$\therefore Q = 2 \times 10^{-3} \text{A} \times 5\text{s} = 10\,\text{mC}$$
$$Q = CV$$
$$\therefore V = \frac{Q}{C} \text{(V)}$$

$$V = \frac{10 \times 10^{-3}\,\text{C}}{100 \times 10^{-6}\,\text{F}} = 100\,\text{V}$$

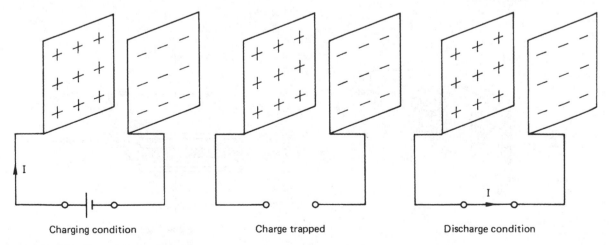

Charging condition   Charge trapped   Discharge condition

Fig. 2.20 The charge on a capacitor's plates.

The p.d. which may be maintained across the plates of a capacitor is determined by the type and thickness of the dielectric medium. Capacitor manufacturers usually indicate the maximum safe working voltage for their products.

Capacitors are classified by the type of dielectric material used in their construction. Figure 2.21 shows the general construction and appearance of some capacitor types to be found in installation work.

Air-dielectric capacitors

Air-dielectric capacitors are usually constructed of multiple aluminium vanes of which one section moves to make the capacitance variable. They are often used for radio tuning circuits.

Mica-dielectric capacitors

Mica-dielectric capacitors are constructed of thin aluminium foils separated by a layer of mica. They are expensive, but this dielectric is very stable and has low dielectric loss. They are often used in high-frequency electronic circuits.

Paper-dielectric capacitors

Paper-dielectric capacitors usually consist of thin aluminium foils separated by a layer of waxed paper. This paper–foil sandwich is rolled into a cylinder and usually contained in a metal cylinder. These capacitors are used in fluorescent lighting fittings and motor circuits.

Electrolytic capacitors

The construction of these is similar to that of the paper-dielectric capacitors, but the dielectric material in this case is an oxide skin formed electrolytically by the manufacturers. Since the oxide skin is very thin, a large capacitance is achieved for a small physical size, but if a voltage of the wrong polarity is applied, the oxide skin is damaged and the gas inside the sealed

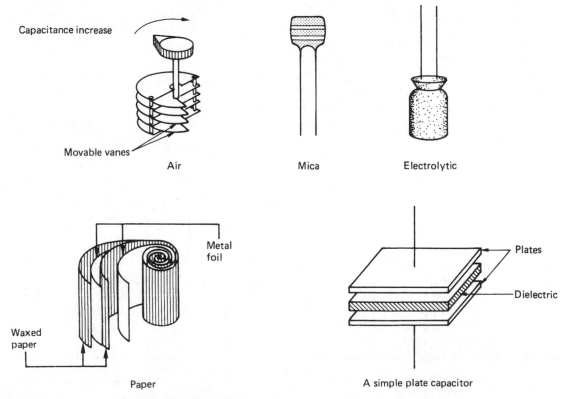

Fig. 2.21 Construction and appearance of capacitors.

container explodes. For this reason electrolytic capacitors must be connected to the correct voltage polarity. They are used where a large capacitance is required from a small physical size and where the terminal voltage never reverses polarity.

CAPACITORS IN COMBINATION

Capacitors, like resistors, may be joined together in various combinations of series or parallel connections (see Fig. 2.22). The equivalent capacitance, C_T, of a number of capacitors is found by the application of similar formulae to those used for resistors and discussed earlier in this chapter. *Note* that the form of the formulae is the opposite way round to that used for series and parallel resistors.

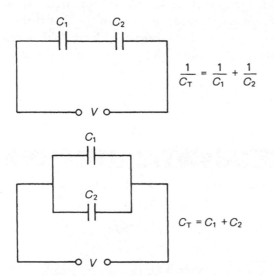

$$\frac{1}{C_T} = \frac{1}{C_1} + \frac{1}{C_2}$$

$$C_T = C_1 + C_2$$

Fig. 2.22 Connection of and formulae for series and parallel capacitors.

The most complex arrangement of capacitors may be simplified into a single equivalent capacitor by applying the separate rules for series or parallel capacitors in a similar way to the simplification of resistive circuits.

EXAMPLE

Capacitors of 10 μF and 20 μF are connected first in series, and then in parallel, as shown in Figs 2.23 and 2.24. Calculate the effective capacitance for each connection. For connection in series,

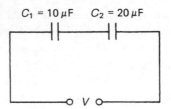

Fig. 2.23 Series capacitors.

$$\frac{1}{C_T} = \frac{1}{C_1} + \frac{1}{C_2}$$

$$\frac{1}{C_T} = \frac{1}{10\,\mu F} + \frac{1}{20\,\mu F}$$

$$\frac{1}{C_T} = \frac{2+1}{20\,\mu F} = \frac{3}{20\,\mu F}$$

$$\therefore C_T = \frac{20\,\mu F}{3} = 6.66\,\mu F$$

For connection in parallel,

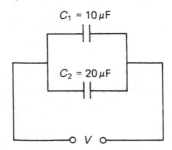

Fig. 2.24 Parallel capacitors.

$$C_T = C_1 + C_2$$

$$C_T = 10\,\mu F + 20\,\mu F = 30\,\mu F$$

Therefore, when capacitors of 10 μF and 20 μF are connected in series their combined effect is equivalent to a capacitor of 6.66 μF. But, when the same capacitors are connected in parallel their combined effect is equal to a capacitor of 30 μF.

Mechanics

Mechanics is the scientific study of 'machines', where a machine is defined as a device which transmits motion or force from one place to another. An engine is one particular type of machine, an energy-transforming machine, converting fuel energy into a more directly useful form of work.

Most modern machines can be traced back to the five basic machines described by the Greek inventor Hero of Alexandria who lived about the time of Christ. The machines described by him were the wedge, the screw, the wheel and axle, the pulley, and the lever. Originally they were used for simple purposes, to raise water and move objects which man alone could not lift, but today their principles are of fundamental importance to our scientific understanding of mechanics. Let us now consider some fundamental mechanical principles and calculations.

MASS

This is a measure of the amount of material in a substance, such as metal, plastic, wood, brick or tissue, which is collectively known as a body. The mass of a body remains constant and can easily be found by comparing it on a set of balance scales with a set of standard masses. The SI unit of mass is the kilogram (kg).

WEIGHT

This is a measure of the force which a body exerts on anything which supports it. Normally it exerts this force because it is being attracted toward the earth by the force of gravity.

For scientific purposes the weight of a body is *not* constant, because gravitational force varies from the equator to the poles; in space a body would be 'weightless' but here on earth under the influence of gravity a 1 kg mass would have a weight of approximately 9.81 newtons (see also the definition of 'force').

SPEED

The feeling of speed is something with which we are all familiar. If we travel in a motor vehicle we know that an increase in speed would, excluding accidents, allow us to arrive at our destination more quickly. Therefore, speed is concerned with distance travelled and time taken. Suppose we were to travel a distance of 30 miles in one hour; our speed would be an average of 30 miles per hour:

$$\text{Speed} = \frac{\text{Distance(m)}}{\text{Time(s)}}$$

VELOCITY

In everyday conversation we often use the word velocity to mean the same as speed, and indeed the units are the same. However, for scientific purposes this is not acceptable since velocity is also concerned with direction. Velocity is speed in a given direction. For example, the speed of an aircraft might be 200 miles per hour, but its velocity would be 200 miles per hour in, say, a westerly direction. Speed is a scalar quantity, while velocity is a vector quantity.

$$\text{Velocity} = \frac{\text{Distance(m)}}{\text{Time(s)}}$$

ACCELERATION

When an aircraft takes off, it starts from rest and increases its velocity until it can fly. This change in velocity is called its acceleration. By definition, acceleration is the rate of change in velocity with time.

$$\text{Acceleration} = \frac{\text{Velocity}}{\text{Time}} = (\text{m/s}^2)$$

EXAMPLE

If an aircraft accelerates from a velocity of 15 m/s to 35 m/s in 4 s, calculate its average acceleration.

$$\text{Average velocity} = 35\,\text{m/s} - 15\,\text{m/s} = 20\,\text{m/s}$$

$$\text{Average acceleration} = \frac{\text{Velocity}}{\text{Time}} = \frac{20}{4} = 5\,\text{m/s}^2$$

Thus, the average acceleration is 5 metres per second per second.

FORCE

The presence of a force can only be detected by its effect on a body. A force may cause a stationary object to move or bring a moving body to rest. For example, a number of people pushing a broken-down motor car exert a force which propels it forward, but applying the motor car brakes applies a force on the brake drums which slows down or stops the vehicle. Gravitational force causes objects to fall to the ground. The apple fell from the tree on to Isaac Newton's head as a result of gravitational force. The standard rate of acceleration due to gravity is accepted as 9.81 m/s^2. Therefore, an apple weighing 1 kg will exert a force of 9.81 N since

$$\text{Force} = \text{Mass} \times \text{Acceleration (N)}$$

The SI unit of force is the newton, symbol N, to commemorate the great English scientist Sir Isaac Newton (1642–1727).

EXAMPLE

A 50 kg bag of cement falls from a forklift truck while being lifted to a storage shelf. Determine the force with which the bag will strike the ground:

$$\text{Force} = \text{Mass} \times \text{Acceleration (N)}$$
$$\text{Force} = 50 \text{ kg} \times 9.81 \text{ m/s}^2 = 490.5 \text{ N}$$

A force can manifest itself in many different ways. Let us consider a few examples:

- 'Inertial force' is the force required to get things moving, to change direction or stop, like the motor car discussed above.
- 'Cohesive or adhesive force' is the force required to hold things together.
- 'Tensile force' is the force pulling things apart.
- Compressive force' is the force pushing things together.
- 'Friction force' is the force which resists or prevents the movement of two surfaces in contact.
- 'Shearing force' is the force which moves one face of a material over another.
- 'Centripetal force' is the force acting towards the centre when a mass attached to a string is rotated in a circular path.
- 'Centrifugal force' is the force acting away from the centre, the opposite to centripetal force.
- 'Gravitational force' is the force acting towards the centre of the earth due to the effect of gravity.
- 'Magnetic force' is the force created by a magnetic field.
- 'Electrical force' is the force created by an electrical field.

PRESSURE OR STRESS

To move a broken-down motor car I might exert a force on the back of the car to propel it forward.

My hands would apply a pressure on the body panel at the point of contact with the car. Pressure or stress is a measure of the force per unit area.

$$\text{Pressure or stress} = \frac{\text{Force}}{\text{Area}} (\text{N/m}^2)$$

EXAMPLE 1

A young woman of mass 60 kg puts all her weight on to the heel of one shoe which has an area of 1 cm^2. Calculate the pressure exerted by the shoe on the floor (assuming the acceleration due to gravity to be 9.81 m/s^2).

$$\text{Pressure} = \frac{\text{Force}}{\text{Area}} (\text{N/m}^2)$$

$$\text{Pressure} = \frac{60 \text{ kg} \times 9.81 \text{ m/s}^2}{1 \times 10^{-4} \text{m}^2} = 5886 \text{ kN/m}^2$$

EXAMPLE 2

A small circus elephant of mass 1 tonne (1000 kg) puts all its weight on to one foot which has a surface area of 400 cm^2. Calculate the pressure exerted by the elephant's foot on the floor, assuming the acceleration due to gravity to be 9.81 m/s^2.

$$\text{Pressure} = \frac{\text{Force}}{\text{Area}} (\text{N/m}^2)$$

$$\text{Pressure} = \frac{1000 \text{ kg} \times 9.81 \text{ m/s}^2}{400 \times 10^{-4} \text{m}^2} = 245.3 \text{ kN/m}^2$$

These two examples show that the young woman exerts 24 times more pressure on the ground than the elephant. This is because her mass exerts a force over a much smaller area than the elephant's foot, and is the reason why many wooden dance floors are damaged by high-heeled shoes.

WORK DONE

Suppose a broken-down motor car was to be pushed along a road; work would be done on the car by applying the force necessary to move it along the road. Heavy breathing and perspiration would be evidence of the work done:

$$\text{Work done} = \text{Force} \times \text{Distance moved in the direction of the force (J)}$$

The SI unit of work done is the newton metre or joule (symbol J). The joule is the preferred unit and it commemorates an English physicist, James Prescot Joule (1818–89).

EXAMPLE

A building hoist lifts ten 50 kg bags of cement through a vertical distance of 30 m to the top of a high rise building. Calculate the work done by the hoist, assuming the acceleration due to gravity to be 9.81 m/s^2.

$$\text{Work done} = \text{Force} \times \text{Distance moved (J)}$$

but $\quad$ Force = Mass × Acceleration (N)

∴ $\quad$ Work done = Mass × Acceleration × Distance moved (J)

$\quad$ Work done = 10 × 50 kg × 9.81 m/s^2 × 30 m

$\quad$ Work done = 147.15 kJ.

POWER

If one motor car can cover the distance between two points more quickly than another car, we say that the faster car is more powerful. It can do a given amount of work more quickly. By definition, power is the rate of doing work.

$$\text{Power} = \frac{\text{Work done}}{\text{Time taken}} (\text{W})$$

The SI unit of power, both electrical and mechanical, is the watt (symbol W). This commemorates the name of James Watt (1736–1819), the inventor of the steam engine.

EXAMPLE 1

A building hoist lifts ten 50 kg bags of cement to the top of a 30 m high building. Calculate the rating (power) of the motor to perform this task in 60 seconds if the acceleration due to gravity is taken as 9.81 m/s^2.

$$\text{Power} = \frac{\text{Work done}}{\text{Time taken}} (\text{W}).$$

but Work done = Force × Distance moved (J)

and Force = Mass × Acceleration (N)

By substitution,

$$\text{Power} = \frac{\text{Mass} \times \text{Acceleration} \times \text{Distance moved}}{\text{Time taken}} (\text{W})$$

$$\text{Power} = \frac{10 \times 50 \text{ kg} \times 9.81 \text{ m/s}^2 \times 30 \text{ m}}{60 \text{ s}}$$

$$\text{Power} = 2452.5 \text{ W}$$

The rating of the building hoist motor will be 2.45 kW.

EXAMPLE 2

A hydroelectric power station pump motor working continuously during a 7 hour period raises 856 tonnes of water through a vertical distance of 60 m. Determine the rating (power) of the motor, assuming the acceleration due to gravity is 9.81 m/s^2.

From Example 1,

$$\text{Power} = \frac{\text{Mass} \times \text{Acceleration} \times \text{Distance moved}}{\text{Time taken}} (\text{W})$$

$$\text{Power} = \frac{856 \times 1000 \text{ kg} \times 9.81 \text{ m/s}^2 \times 60 \text{ m}}{7 \times 60 \times 60 \text{ s}}$$

$$\text{Power} = 20\,000 \text{ W}$$

The rating of the pump motor is 20 kW.

EXAMPLE 3

An electric hoist motor raises a load of 500 kg at a velocity of 2 m/s. Calculate the rating (power) of the motor if the acceleration due to gravity is 9.81 m/s^2.

$$\text{Power} = \frac{\text{Mass} \times \text{Acceleration} \times \text{Distance moved}}{\text{Time taken}} (\text{W})$$

but Velocity $= \dfrac{\text{Distance}}{\text{Time}}$ (m/s)

∴ Power = Mass × Acceleration × Velocity

$\quad$ Power = 500 kg × 9.81 m/s^2 × 2 m/s

$\quad$ Power = 9810 W.

The rating of the hoist motor is 9.81 kW.

EFFICIENCY

In any machine the power available at the output is less than that which is put in because losses occur in the machine. The losses may result from friction in the bearings, wind resistance to moving parts, heat, noise or vibration.

The ratio of the output power to the input power is known as the *efficiency* of the machine. The symbol for efficiency is the Greek letter 'eta' (η). In general,

$$\eta = \frac{\text{Power output}}{\text{Power input}}$$

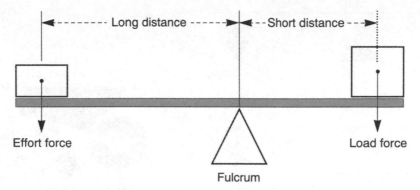

Fig. 2.25 Turning forces of a simple lever.

Since efficiency is usually expressed as a percentage we modify the general formula as follows.

$$\eta = \frac{\text{Power output}}{\text{Power input}} \times 100$$

EXAMPLE

A transformer feeds the 9.81 kW motor driving the mechanical hoist of the previous example. The input power to the transformer was found to be 10.9 kW. Find the efficiency of the transformer.

$$\eta = \frac{\text{Power output}}{\text{Power input}} \times 100$$

$$\eta = \frac{9.81\,\text{kW}}{10.9\,\text{kW}} \times 100 = 90\%$$

Thus the transformer is 90% efficient. Note that efficiency has no units, but is simply expressed as a percentage.

LEVERS

Every time we open a door, turn on a tap or tighten a nut with a spanner, we exert a lever-action turning force. A lever is any rigid body which pivots or rotates about a fixed axis or fulcrum. The simplest form of lever is the crowbar, which is useful because it enables a person to lift a load at one end which is greater than the effort applied through his or her arm muscles at the other end. In this way the crowbar is said to provide a 'mechanical advantage'. A washbasin tap and a spanner both provide a mechanical advantage through the simple lever action. The mechanical advantage of a simple lever is dependent upon the length of lever on either side of the fulcrum. Applying the principle of turning forces to a lever, we obtain the formula:

Load force × Distance from fulcrum

= Effort force × Distance from fulcrum

This formula can perhaps better be understood by referring to Fig. 2.25. A small effort at a long distance from the fulcrum can balance a large load at a short distance from the fulcrum. Thus a 'turning force' or 'turning moment' depends upon the distance from the fulcrum and the magnitude of the force.

EXAMPLE

Calculate the effort required to raise a load of 500 kg when the effort is applied at a distance of five times the load distance from the fulcrum (assume the acceleration due to gravity to be 10 m/s^2).

Load force = Mass × Acceleration (N)
Load force = 500 kg × 10 m/s^2 = 5000 N

Load force × Distance from fulcrum
 = Effort force × Distance from fulcrum

5000 N × 1 m = Effort force × 5 m

$$\therefore \text{Effort force} = \frac{5000\,\text{N} \times 1\,\text{m}}{5\,\text{m}} = 1000\,\text{N}$$

Thus an effort force of 1000 N can overcome a load force of 5000 N using the mechanical advantage of this simple lever.

Every lever has one pivot point, the fulcrum, and is acted upon by two forces, the load and the effort. There are three classes or types of lever, according to the position of the load, effort and fulcrum (Fig. 2.26).

In a first-class lever the load is applied to one side of the fulcrum and the effort to the other, as shown in

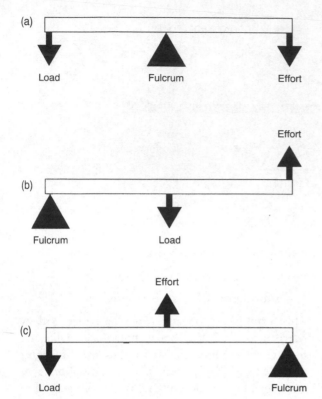

Fig. 2.26   The three classes of lever: (a) first class; (b) second class; (c) third class.

Fig. 2.26(a). A typical example of a first-class lever is a crowbar or a sack truck. A crocodile clip, a pair of pliers or side cutters, are examples of two first-class levers acting together as shown in Fig. 2.27.

A second-class lever has the load and effort applied to one side of the fulcrum, as shown in Fig. 2.26(b). A typical example of a second class lever is a beer bottle opener or nutcracker, a conduit bending machine or wheelbarrow, as shown in Fig. 2.28. The general principle is just the same, that a small effort applied at

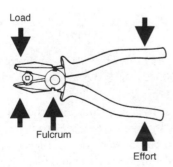

Fig. 2.27   A pair of pliers, an example of two first-class levers acting together.

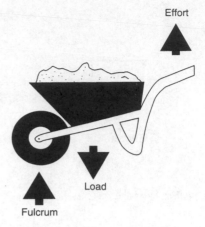

Fig. 2.28   A wheelbarrow, an example of a second-class lever.

the end of a long arm can be used to overcome a much larger load at the end of a short arm.

The third class of lever gives us the least mechanical advantage because the effort is closer to the fulcrum than the load, as shown in Fig. 2.26(c). However, it does have many useful applications. The forearm operates as a third-class lever with the fulcrum at the elbow. A load in the hand is raised by the effort exerted by the biceps acting on the forearm close to the elbow. Other applications are a builders shovel, sugar or coal tongs, or the heat shunt used when soldering electronic components, as shown in Fig. 2.29.

Simple machines

Our physical abilities in the field of lifting and moving heavy objects are limited. However, over the centuries we have used our superior intelligence to design tools, mechanisms and machines which have overcome this physical inadequacy. This concept is shown in Fig. 2.30.

By definition, a machine is an assembly of parts, some fixed, others movable, by which motion and force are transmitted. With the aid of a machine we are able to magnify the effort exerted at the input and lift or move large loads at the output.

MECHANICAL ADVANTAGE (*MA*)

This is the advantage given by the machine and is defined as the ratio of the load to the effort.

$$MA = \frac{\text{Load}}{\text{Effort}} \text{(no units)}$$

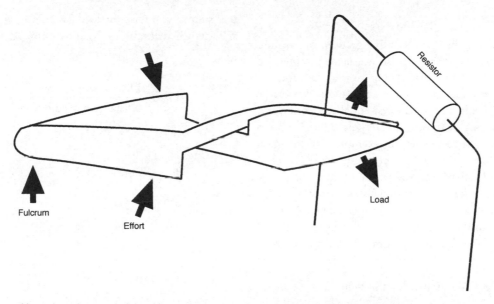

Fig. 2.29   A soldering heat shunt used when soldering electronic components, an example of two third-class levers acting together.

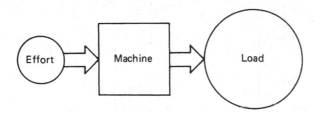

Fig. 2.30   Simple machine concept.

VELOCITY RATIO (*VR*)

This is the ratio of the distance moved by the effort to the distance moved by the load. Because the distance moved by the load and effort are dependent upon the construction of a particular machine, the velocity ratio is usually a constant for that machine.

$$VR = \frac{\text{Distance moved by effort}}{\text{Distance moved by load}} \text{(no units)}$$

EFFICIENCY

In all machines the power available at the output is less than that which is put in because losses occur in the machine. These losses may result from friction in the bearings, wind resistance to moving parts, heat, noise or vibrations.

The ratio of the output power to the input power is known as the *efficiency* of the machine. The symbol for efficiency is the Greek letter 'eta' (η). In general,

$$\text{Efficiency} = \frac{\text{Work output}}{\text{Work input}}$$

But

Work Output

  = Load × Distance moved by the load

and

Work Input

  = Effort × Distance moved by the effort

If we divide these two equations we see that

$$\frac{\text{Load}}{\text{Effort}} = MA$$

$$\frac{\text{Distance moved by load}}{\text{Distance moved by effort}} = \frac{1}{VR}$$

and therefore

$$\text{Efficiency} = MA \times \frac{1}{VR}$$

or

$$\eta = \frac{MA}{VR}$$

PULLEYS

A pulley is simply a wheel with a grooved rim around which a rope is passed. A load is attached to one end of the rope and an effort applied to the other end.

A single pulley offers no mechanical advantage because the effort applied must equal the load, but it does have uses in raising small loads. They are often used on construction sites, for example, to lift buckets of mortar up to the top of a building scaffold.

Pulley blocks which contain more than one pulley do offer a mechanical advantage since the total load is shared equally by each vertical rope; see Fig. 2.31.

If the load is raised by 1 cm, each length of rope in the system will shorten by 1 cm. Therefore, the effort applied at

- system A will raise the load by 1 cm

- system B will raise the load by 2 cm
- system C will raise the load by 4 cm.

Now

$$VR = \frac{\text{Distance moved by effort}}{\text{Distance moved by load}}$$

so

For system A $\quad VR = \dfrac{1}{1} = 1$

For system B $\quad VR = \dfrac{2}{1} = 2$

For system C $\quad VR = \dfrac{4}{1} = 4$

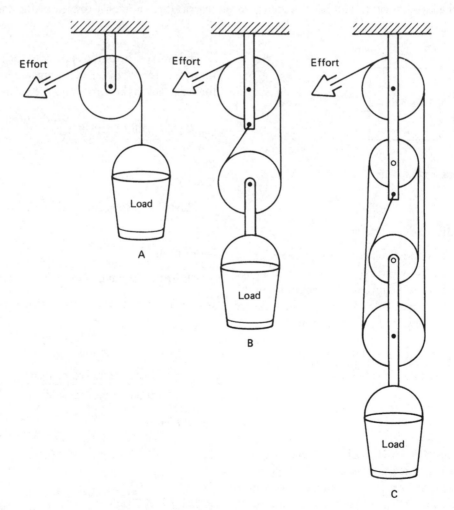

Fig. 2.31 Pulley systems.

The velocity ratio of any pulley system, such as a block and tackle, can be determined by counting the number of pulley wheels.

Pulley blocks are used for lifting heavy loads such as car engines from the body of the vehicle, and passenger lifts and cranes operate using this principle. The limitations of this machine are that the upper pulley must be secured at a higher level than the height through which the load will travel.

The following 'good practice' should be borne in mind when using any pulley or block and tackle system:

1 Ensure that the lifting equipment is suspended from a 'secure' point.
2 Do not exceed the maximum safe working load (SWL) indicated on the lifting equipment.
3 Use the lifting 'eye' fitted to the load. Heavy loads often have a lifting point or points cast into the outer casing.
4 If the load does not have a lifting eye, place slings under the load and adjust the lifting point to above the centre of gravity. Remember, the total weight acts through the centre of gravity.
5 Avoid 'shock loading' the lifting equipment.
6 Prevent the load from swinging and twisting while being lifted.
7 Finally, always look at the lifting problems first before taking any action. Ask yourself these questions:
   (a) Do I have the appropriate equipment for the job?
   (b) Do I have the necessary skill and experience for moving the load?

If you do not feel confident to tackle the job, seek help and guidance from your supervisor.

$$VR = \text{Total number of pulley wheels} = 5$$

$$\eta = \frac{MA}{VR} = \frac{5}{6} = 0.833$$

or Percentage efficiency $= 0.833 \times 100 = 83.3\%$

THE WEDGE

The wedge provides a means of converting motion in one direction to motion in another at right angles. Driving the wedge under the load causes the load to move up an inclined plane, thus changing the direction of the force by 90°. This is shown in Fig. 2.32.

The wedge is used to prevent horizontal motion, for example when placing a wedge under a door. The early Egyptians used the inclined plane of the wedge to raise the huge blocks of stone with which they built the pyramids.

An electrician might use the mechanical advantage of an inclined plane by pulling a heavy object up the sloping surface formed by a plank into, say, the loading bay of his or her van.

This uses the lever principle discussed earlier. A small effort exerted over a greater distance (the length of the plank) can overcome a heavy load travelling a shorter distance. (The load only travels through a vertical distance equal to the height of the van loading bay.)

THE SCREW JACK

The screw jack is a simple machine which makes use of a screw thread to lift a large load with a small effort.

EXAMPLE

A transformer having a mass of 100 kg is lifted on to a lorry by means of a pulley block containing three pulleys in the upper block and three in the lower block. If the effort required is 200 N, calculate the efficiency of the system. (Take g as $10\,\text{m/s}^2$.)

Force exerted by transformer $= \text{Mass} \times \text{Acceleration (N)}$

$$\text{Force} = 100\,\text{kg} \times 10\,\text{m/s}^2 = 1000\,\text{N}$$

$$MA = \frac{\text{Load}}{\text{Effort}} = \frac{1000\,\text{N}}{200\,\text{N}} = 6$$

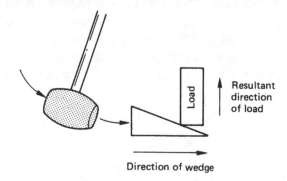

Fig. 2.32 A wedge.

The principle is similar to that of a wedge in that the load moves up an inclined plane, in this case, the screw thread. Figure 2.33 shows the screw jack.

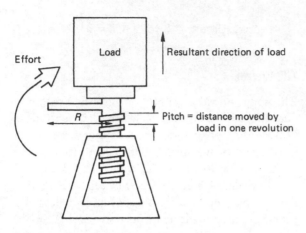

Fig. 2.33   A screw jack.

For this machine consider *one* revolution of the jack arm:

$$VR = \frac{\text{Distance moved by effort}}{\text{Distance moved by load}}$$

$$= \frac{\text{Circumference of jack arm}}{\text{Pitch}}$$

$$VR = \frac{2\pi R}{\text{Pitch}}$$

The screw jack is used extensively as a motor car jack, but the principle is also used for roof supports, an engineer's vice and the tipping mechanism of trucks.

EXAMPLE

A screw jack is used to raise a motor car of mass 1 tonne (1000 kg). The length of the jack arm is 250 mm and the pitch of the thread 5 mm. If an effort of 250 N is required to raise the arm, calculate the efficiency of this machine. (Take g as 10 m/s^2.) We must first change the mass into a force:

$$\text{Force} = \text{Mass} \times \text{Acceleration (N)}$$

$$\therefore \text{Force} = 1000\,\text{kg} \times 10\,\text{m/s}^2 = 10\,000\,\text{N}$$

$$MA = \frac{\text{Load}}{\text{Effort}} = \frac{10\,000\,\text{N}}{250\,\text{N}} = 40$$

$$VR = \frac{2\pi R}{\text{Pitch}} = \frac{2 \times 3.142 \times 0.25\,\text{m}}{0.005\,\text{m}} = 314.2$$

$$\eta = \frac{MA}{VR} = \frac{40}{314.2} = 0.127$$

or Percentage efficiency = $0.127 \times 100 = 12.7\%$

We usually expect a machine to have a high efficiency (better than about 80%), but the efficiency of the screw jack is very poor, as you can see from the example. This is because friction on the moving surfaces, the inclined plane of the screw thread, is very high. A mechanically more efficient car jack would use the liquid properties of hydraulic fluid in a hydraulic jack.

Transporting loads

If the load is to be moved after being lifted, first make sure it is secure.

Some quite simple machines can ease the burden of moving very heavy loads.

A sack truck is an effective means for one person to move very heavy loads. The mechanical advantage of the sack truck, which allows a little human effort to overcome a large load, is based on the principle of levers discussed earlier.

A flat bed truck has a sturdy horizontal base of approximately 1.5 m × 0.75 m supported in an angle iron frame with four heavy castors on each corner. This is ideal for moving heavy loads on hard smooth surfaces. You can often see them being used in supermarkets to move boxes of canned products and in DIY stores for moving building materials.

A fork-lift truck can be powered by batteries or bottled gas for inside use or by combustion engine for outside use. The operator will need training and must hold a certificate of competence. If the machine is to be used on the public highway the operator will also need a driving licence.

Moving loads safely is discussed further in Chapter 3 under the heading 'Manual Handling'.

CENTRE OF GRAVITY

We have discussed earlier in this chapter that a body is attracted to the centre of the earth by the force of gravity. This statement, however, says nothing about where the force is acting, so let us assume that any body is made up of a very large number of tiny particles. The force of gravity will act equally on every particle, thus creating a large number of parallel forces. These forces will have a resultant force equal to the total force of gravity and will act through a point called the *centre of gravity*. By definition, the centre of gravity of a body is the point through which the total weight of the body appears to act. The centre of gravity of a 30 cm rule will act through the 15 cm point. If you place the rule on your finger at this point, it will balance because equal forces are acting on both sides of your finger.

The centre of gravity of a disc is at the centre, as shown in Fig. 2.34. The centre of gravity of a ring is also at the centre, even though there is no material at the centre. This is the point through which the resultant forces act. The centre of gravity of a square, a rectangle or a triangle is at the point where the intersecting diagonal lines meet, as shown in Fig. 2.34.

The point at which the centre of gravity acts on an object is important to the 'stability' of that object. If a cone is placed on its point it will have a high centre of gravity and the slightest movement will cause the cone to topple over. This is called 'unstable equilibrium'. Placing the cone on its base gives it a lower centre of gravity and a broader base. The cone is now more difficult to knock over and is, therefore, said to be in 'stable equilibrium'. Laying the cone on its side gives it a lower centre of gravity, a very broad base and it becomes impossible to topple. This position is called 'neutral equilibrium'. In general, if the centre of gravity acts within the base width, the object will be stable. If a small displacement brings the centre of gravity outside the base width it becomes unstable. Figure 2.35 shows these three effects. The risk of unstable equilibrium is increased as the height of the centre of gravity is increased. The nearer the centre of gravity is to the ground, the more stable the equilibrium is likely to be.

In designing a car or a ship the engineer must take into account its stability under normal operating conditions. A car turning a sharp corner at high speed

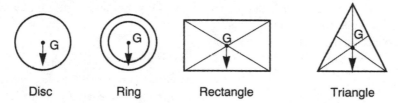

Disc   Ring   Rectangle   Triangle

Fig. 2.34 The centre of gravity (point G) of some regular shapes.

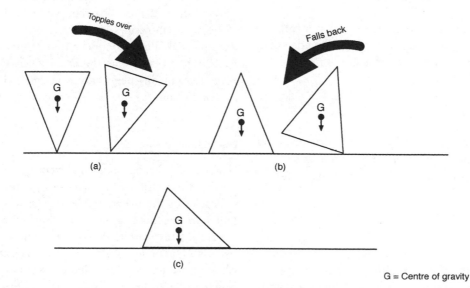

G = Centre of gravity

Fig. 2.35 Equilibrium of a cone: (a) unstable equilibrium; (b) stable equilibrium; (c) neutral equilibrium.

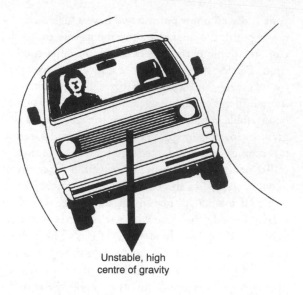

Unstable, high
centre of gravity

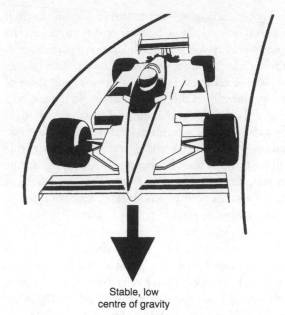

Stable, low
centre of gravity

Fig. 2.36　Unstable and stable vehicles.

may become unstable if its centre of gravity is high. Racing cars have a very low centre of gravity, as shown in Fig. 2.36. A ship is liable to considerable rolling and lurching in heavy seas and must be designed to be in stable equilibrium even in rough weather.

Temperature and heat

Heat has the capacity to do work and is a form of energy. Temperature is not an energy unit but describes the hotness or coldness of a substance or material.

HEAT TRANSFER

Heat energy is transferred by three separate processes which can occur individually or in combination. The processes are convection, radiation and conduction.

Convection

Air which passes over a heated surface expands, becomes lighter and warmer and rises, being replaced by descending cooler air. These circulating currents of air are called *convection currents*. In circulating, the warm air gives up some of its heat to the surfaces over which it passes and so warms a room and its contents.

Radiation

Molecules on a metal surface vibrating with thermal energies generate electromagnetic waves. The waves travel away from the surface at the speed of light, taking energy with them and leaving the surface cooler. If the waves meet another material they produce a disturbance of the surface molecules which raises the temperature of that material. Radiated heat requires no intervening medium between the transmitter and receiver, obeys the same laws as light energy and is the method by which the energy from the sun reaches the earth.

Conduction

Heat transfer through a material by conduction occurs because there is direct contact between the vibrating molecules of the material. The application of a heat source to the end of a metal bar causes the atoms to vibrate rapidly within the lattice framework of the material. This violent shaking causes adjacent atoms to vibrate and liberates any loosely bound electrons, which also pass on the heat energy. Thus the heat energy travels *through* the material by conduction.

TEMPERATURE SCALES

When planning any scale of measurement a set of reference points must first be established. Two obvious reference points on any temperature scale are the temperatures at which ice melts and water boils. Between these points the scale is divided into a convenient number of divisions. In the case of the Celsius or Centigrade scale, the lower fixed point is zero degrees and the upper fixed point 100 degrees Celsius. The Kelvin scale takes as its lower fixed point the lowest possible temperature which has a value of $-273°C$, called absolute zero. A temperature change of one kelvin is exactly the same as one degree Celsius and so we can say that

$$0°C = 273\,K \quad \text{or} \quad 0\,K = -273°C$$

EXAMPLE 1

Convert the following Celsius temperatures into kelvin: $-20°C$, $20°C$ and $200°C$.

$$-20°C = -20 + 273 = 253\,K$$
$$20°C = 20 + 273 = 293\,K$$
$$200°C = 200 + 273 = 473\,K$$

EXAMPLE 2

Convert into degrees Celsius: $250\,K$, $300\,K$ and $500\,K$.

$$250\,K = 250 - 273 = -23°C$$
$$300\,K = 300 - 273 = 27°C$$
$$500\,K = 500 - 273 = 227°C$$

Temperature measurement

One instrument which measures temperature is a thermometer. This uses the properties of an expanding liquid in a glass tube to indicate a temperature level. Most materials change their dimensions when heated and this property is often used to give a measure of temperature. Many materials expand with an increase in temperature and the *rate* of expansion varies with different materials. A bimetal strip is formed from two dissimilar metals joined together. As the temperature increases the metals expand at different rates and the bimetal strip bends or bows.

THERMOSTATS

A thermostat is a device for maintaining a constant temperature at some predetermined value. The operation of a thermostat is often based upon the principle of differential expansion between dissimilar metals which causes a contact to make or break at a chosen temperature. Figure 2.37 shows the principle of a rod-type thermostat which is often used with water heaters. An Invar rod, which has minimal expansion when heated, is housed within a copper tube and the two metals are brazed together at one end. The other end of the copper tube is secured to one end of the switch mechanism. As the copper expands and contracts under the influence of a varying temperature, the switch contacts are activated. In this case the contact breaks the electrical circuit when the temperature setting is reached.

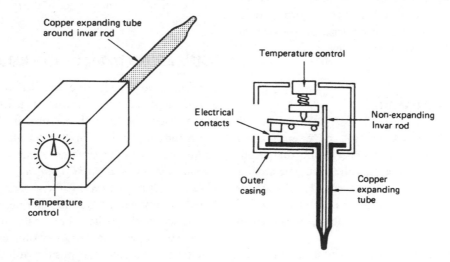

Fig. 2.37   A rod-type thermostat.

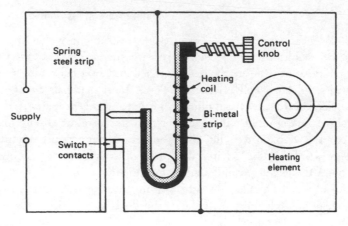

Fig. 2.38 Simmerstat control arrangement.

SIMMERSTATS

A simmerstat is a device used to control the temperature of an electrical element, typically the boiling ring
of a cooker. A snap-action switch is opened and closed
at time intervals by passing current through a heater
wrapped around a bimetal strip, as shown in Fig. 2.38.

With the switch contact made, current flows
through the load and the heating coil. The heater
warms the bimetal strip which expands and opens out,
pushing against the spring steel strip and the control
knob, so opening the switch contacts. The load and
the heating coil are then switched off and the bimetal
strip cools to its original shape, which allows the contacts to close, and the process repeats. The load and
heating coil are switched on and off frequently if the
control knob is arranged to allow little movement of
the bimetal strip. Alternatively, the heater will remain
switched on for longer periods if the control knob is
adjusted to allow a larger movement. In this way the
temperature of the load is controlled.

Electrical machines

Electrical machines are energy converters. If the
machine input is mechanical energy and the output
electrical energy then that machine is a generator, as
shown in Fig. 2.39(a). Alternatively, if the machine
input is electrical energy and the output mechanical
energy then the machine is a motor, as shown in
Fig. 2.39(b).

An electrical machine may be used as a motor or a
generator, although in practice the machine will operate

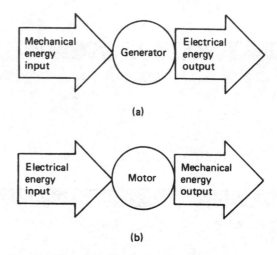

Fig. 2.39 Electrical machines as energy converters.

more efficiently when operated in the mode for which it
was designed.

SIMPLE A.C. GENERATOR OR ALTERNATOR

If a simple loop of wire is rotated between the poles of
a permanent magnet, as shown in Fig. 2.40 the loop of
wire will cut the lines of magnetic flux between the
north and south poles. This flux cutting will induce an
emf in the wire by Faraday's law which states that *when
a conductor cuts or is cut by a magnetic field, an emf is
induced in that conductor*. If the generated emf is col-
lected by carbon brushes at the slip rings and displayed
on the screen of a cathode ray oscilloscope, the waveform will be seen to be approximately sinusoidal.
Alternately changing, first positive and then negative,
then positive again, giving an alternating output.

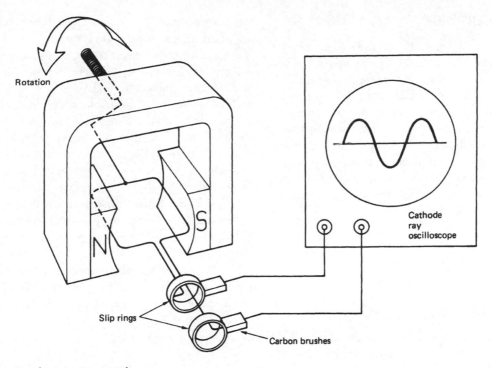

Fig. 2.40 Simple a.c. generator or alternator.

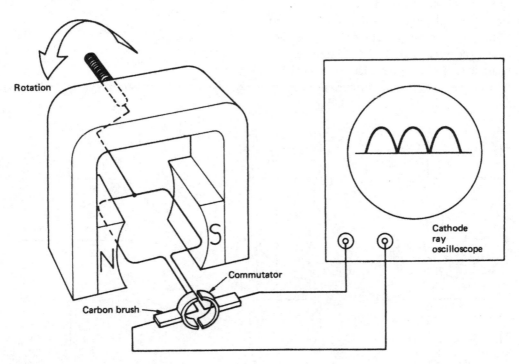

Fig. 2.41 Simple d.c. generator.

SIMPLE D.C. GENERATOR

If the slip rings of Fig. 2.40 are replaced by a single split ring, called a commutator, the generated emf will be seen to be in one direction, as shown in Fig. 2.41. The action of the commutator is to reverse the generated emf every half cycle, rather like an automatic change-over switch. However, this simple arrangement produces a very bumpy d.c. output. In a practical machine the commutator would contain many segments and many windings to produce a smoother d.c. output. Similar to the unidirectional battery supply shown earlier in Fig. 2.12.

Motors

THE D.C. MOTOR

If a current carrying conductor is placed into the field of a permanent magnet, as shown in Fig. 2.42(c), a force will be exerted on the conductor to push it out of the magnetic field. To understand the force, let us consider each magnetic field acting alone. Figure 2.42(a) shows the magnetic field due to the current carrying conductor only, shown as a cross-section. Figure 2.42(b) shows the magnetic field due to the permanent magnet, in which is placed the conductor carrying no current. Figure 2.42(c) shows the effect of the two magnetic fields and the force exerted on the conductor.

The magnetic field of the permanent magnet is distorted by the magnetic field from the current carrying conductor. Since lines of magnetic flux behave like stretched elastic bands, always trying to find the shorter distance between the north and south poles, a force is exerted on the conductor, pushing it out of the permanent magnetic field.

This is the basic principle which produces the rotation in a d.c. machine and a moving coil instrument. Current fed into the single coil winding of Fig. 2.41, through the commutator, will set up magnetic fluxes as shown in Fig. 2.43. The resultant forces will cause the coil to rotate – in the case of Fig. 2.43 in an anticlockwise direction. Reversing the current in the coil, or the polarity of the permanent field, would cause the coil to rotate in the opposite direction.

Direct current machines work on this basic principle but the permanent field magnets of Fig. 2.43 are usually replaced by electromagnets. This arrangement gives greater control of the magnetic field strength and the motor performance.

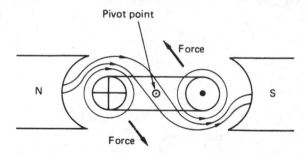

Fig. 2.43 Forces exerted on a current carrying coil in a magnetic field.

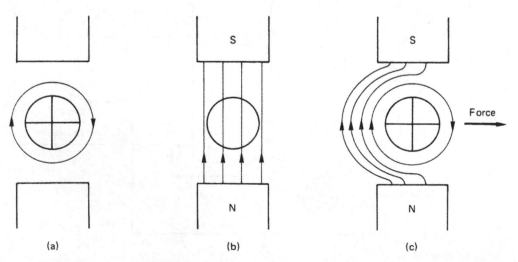

Fig. 2.42 Force on a conductor in a magnetic field.

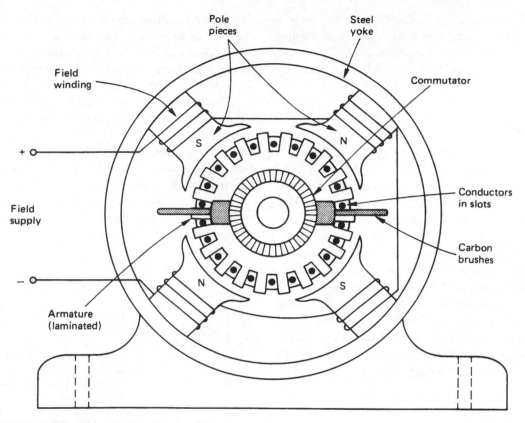

Fig. 2.44   Direct current machine construction.

PRACTICAL D.C. MOTORS

Practical motors are constructed as shown in Fig. 2.44. All d.c. motors contain a field winding wound on pole pieces attached to a steel yoke. The armature winding rotates between the poles and is connected to the commutator. Contact with the external circuit is made through carbon brushes rubbing on the commutator segments. Direct current motors are classified by the way in which the field and armature windings are connected, which may be in series or in parallel.

Series motor

The field and armature windings are connected in series and consequently share the same current. The series motor has the characteristics of a high starting torque but a speed which varies with load. Theoretically the motor would speed up to self-destruction, limited only by the windage of the rotating armature and friction, if the load were completely removed. Figure 2.45 shows series motor connections and characteristics. For this reason the motor is only suitable for

direct coupling to a load, except in very small motors, such as vacuum cleaners and hand drills, and is ideally suited for applications where the machine must start on load, such as electric trains, cranes and hoists.

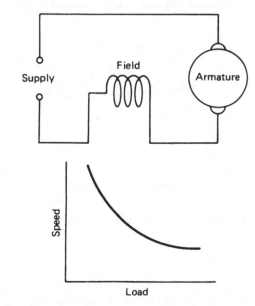

Fig. 2.45   Series motor connections and characteristics.

Reversal of rotation may be achieved by reversing the connections of either the field or armature windings but not both. This characteristic means that the machine will run on both a.c. or d.c. and is, therefore, sometimes referred to as a 'universal' motor.

Shunt motor

The field and armature windings are connected in parallel (see Fig. 2.46). Since the field winding is across the supply, the flux and motor speed are considered constant under normal conditions. In practice, however, as the load increases the field flux distorts and there is a small drop in speed of about 5% at full load, as shown in Fig. 2.46. The machine has a low starting torque and it is advisable to start with the load disconnected. The shunt motor is a very desirable d.c. motor because of its constant speed characteristics. It is used for driving power tools, such as lathes and drills. Reversal of rotation may be achieved by reversing the connections to either the field or armature winding but not both.

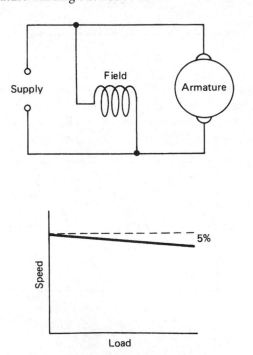

Fig. 2.46   Shunt motor connections and characteristics.

Compound motor

The compound motor has two field windings – one in series with the armature and the other in parallel.

If the field windings are connected so that the field flux acts in opposition, the machine is known as a *short shunt* and has the characteristics of a series motor. If the fields are connected so that the field flux is strengthened, the machine is known as a *long shunt* and has constant speed characteristics similar to a shunt motor. The arrangement of compound motor connections is given in Fig. 2.47. The compound motor may be designed to possess the best characteristics of both series and shunt motors, that is, good starting torque together with almost constant speed. Typical applications are for electric motors in steel rolling mills, where a constant speed is required under varying load conditions.

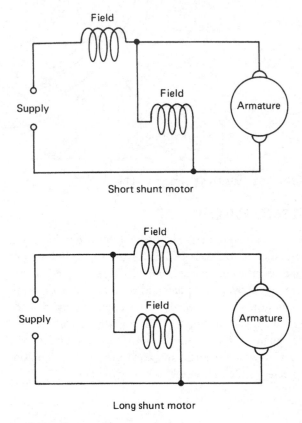

Fig. 2.47   Compound motor connections.

Simple transformers

A transformer is an electrical machine which is used to change the value of an alternating voltage. They vary in size from miniature units used in electronics to huge power transformers used in power stations. A transformer will only work when an alternating voltage is

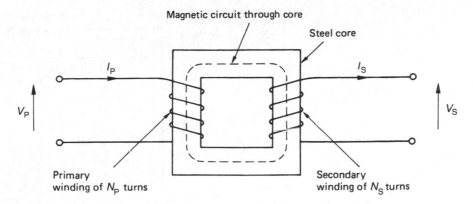

Fig. 2.48 A simple transformer.

connected. It will not normally work from a d.c. supply such as a battery.

A transformer, as shown in Fig. 2.48, consists of two coils, called the primary and secondary coils, or windings, which are insulated from each other and wound on to the same steel or iron core.

An alternating voltage applied to the primary winding produces an alternating current, which sets up an alternating magnetic flux throughout the core. This magnetic flux induces an emf in the secondary winding, as described by Faraday's law, which says that when a conductor is cut by a magnetic field, an emf is induced in that conductor. Since both windings are linked by the same magnetic flux, the induced emf per turn will be the same for both windings. Therefore, the emf in both windings is proportional to the number of turns. In symbols,

$$\frac{V_P}{N_P} = \frac{V_S}{N_S} \qquad (1)$$

Most practical power transformers have a very high efficiency, and for an ideal transformer having 100% efficiency the primary power is equal to the secondary power:

$$\text{Primary power} = \text{Secondary power}$$

and, since

$$\text{Power} = \text{Voltage} \times \text{Current}$$

then

$$V_P \times I_P = V_S \times I_S \qquad (2)$$

Combining equations (1) and (2), we have

$$\frac{V_P}{V_S} = \frac{N_P}{N_S} = \frac{I_S}{I_P}$$

EXAMPLE

A 230 V to 12 V bell transformer is constructed with 800 turns on the primary winding. Calculate the number of secondary turns and the primary and secondary currents when the transformer supplies a 12 V 12 W alarm bell.

Collecting the information given in the question into a usable form, we have

$$V_P = 230\,V$$
$$V_S = 12\,V$$
$$N_P = 800$$
$$\text{Power} = 12\,W$$

Information required: N_S, I_S and I_P

Secondary turns

$$N_S = \frac{N_P V_S}{V_P}$$

$$\therefore N_S = \frac{800 \times 12\,V}{230\,V} = 42\ \text{turns}$$

Secondary current

$$I_S = \frac{\text{Power}}{V_S}$$

$$\therefore I_S = \frac{12\,W}{12\,V} = 1\,A$$

Primary current

$$I_P = \frac{I_S \times V_S}{V_P}$$

$$\therefore I_P = \frac{1\,A \times 12\,V}{230\,V} = 0.052\,A$$

TRANSFORMER LOSSES

As they have no moving parts causing frictional losses, most transformers have a very high efficiency, usually better than 90%. However, the losses which do occur in a transformer can be grouped under two general headings: copper losses and iron losses.

Copper losses occur because of the small internal resistance of the windings. They are proportional to the load, increasing as the load increases because copper loss is an 'I^2R' loss.

Iron losses are made up of *hysteresis loss* and *eddy current loss*. The hysteresis loss depends upon the type of iron used to construct the core and consequently core materials are carefully chosen. Transformers will only operate on an alternating supply. Thus, the current which establishes the core flux is constantly changing from positive to negative. Each time there is a current reversal, the magnetic flux reverses and it is this build-up and collapse of magnetic flux in the core material which accounts for the hysteresis loss.

Eddy currents are circulating currents created in the core material by the changing magnetic flux. These are reduced by building up the core of thin slices or laminations of iron and insulating the separate laminations from each other. The iron loss is a constant loss consuming the same power from no load to full load.

TRANSFORMER CONSTRUCTION

Transformers are constructed in a way which reduces the losses to a minimum. The core is usually made of silicon-iron laminations, because at fixed low frequencies silicon-iron has a small hysteresis loss and the laminations reduce the eddy current loss. The primary and secondary windings are wound close to each other on the same limb. If the windings are spread over two limbs there will usually be half of each winding on each limb, as shown in Fig. 2.49.

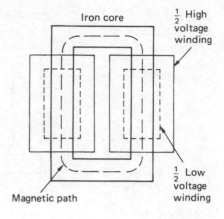

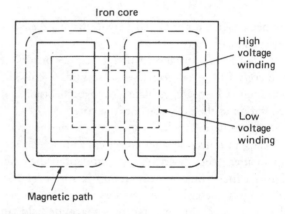

Fig. 2.49 Transformer construction.

AUTO-TRANSFORMERS

Transformers having a separate primary and secondary winding, as shown in Fig. 2.49 are called double-wound transformers, but it is possible to construct a transformer which has only one winding which is common to the primary and secondary circuits. The secondary voltage is supplied by means of a 'tapping' on the primary winding. An arrangement such as this is called an auto-transformer.

The auto-transformer is much cheaper and lighter than a double-wound transformer because less copper and iron are used in its construction. However, the primary and secondary windings are not electrically separate and a short circuit on the upper part of the winding shown in Fig. 2.50 would result in the primary voltage appearing across the secondary terminals. For this reason auto-transformers are mostly used where only a small difference is required between the primary and secondary voltages. When installing

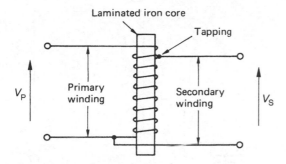

Fig. 2.50 An auto-transformer.

transformers, the regulations of Section 555 must be complied with, in addition to any other regulations relevant to the particular installation.

Measuring volts and amps

The type of instrument to be purchased for general use in the electrotechnical industries is a difficult choice because there are so many different types on the market and every manufacturer's representative is convinced that his company's product is the best. However, most instruments can be broadly grouped under two general headings: those having *analogue* and those with *digital* displays.

ANALOGUE METERS

These meters have a pointer moving across a calibrated scale. They are the only choice when a general trend or variation in value is to be observed. Hi-fi equipment often uses analogue displays to indicate how power levels vary with time, which is more informative than a specific value. Red or danger zones can be indicated on industrial instruments. The fuel gauge on a motor car often indicates full, half full or danger on an analogue display which is much more informative than an indication of the exact number of litres of petrol remaining in the tank.

These meters are only accurate when used in the calibrated position – usually horizontally.

Most meters using an analogue scale incorporate a mirror to eliminate parallax error. The user must look straight at the pointer on the scale when taking readings and the correct position is indicated when the pointer image in the mirror is hidden behind the actual pointer. That is the point at which a reading should be taken from the appropriate scale of the instrument.

DIGITAL METERS

These provide the same functions as analogue meters but they display the indicated value using a seven-segment LED to give a numerical value of the measurement. Modern digital meters use semiconductor technology to give the instrument a very high input impedance, typically about 10 MΩ and, therefore, they are ideal for testing most electrical or electronic circuits.

The choice between an analogue and a digital display is a difficult one and must be dictated by specific circumstances. However, if you are an electrician or service engineer intending to purchase a new instrument, I think on balance that a good-quality digital multimeter such as that shown in Fig. 2.51 would be best. Having no moving parts, digital meters tend to be more rugged and, having a very high input impedance, they are ideally suited to testing all circuits that an electrician might work on in his daily work.

Fig. 2.51 Digital multimeter suitable for testing electrical and electronic circuits.

The multimeter

Multimeters are designed to measure voltage, current or resistance. Before taking measurements the appropriate volt, ampere or ohm scale should be selected. To avoid damaging the instrument it is good practice first to switch to the highest value on a particular scale range. For example, if the 10 A scale is first selected and a reading of 2.5 A is displayed, we then know that a more appropriate scale would be the 3 A or 5 A range. This will give a more accurate reading which might be,

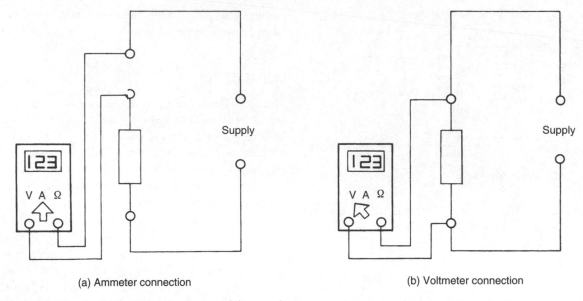

(a) Ammeter connection   (b) Voltmeter connection

Fig. 2.52   Using a multimeter (a) as an ammeter and (b) as a voltmeter.

say, 2.49 A. When the multimeter is used as an ammeter to measure current it must be connected in series with the test circuit, as shown in Fig. 2.52 (a). When used as a voltmeter the multimeter must be connected in parallel with the component, as shown in Fig. 2.52 (b).

When using a commercial multirange meter as an ohmmeter for testing electronic components, care must be exercised in identifying the positive terminal. The red terminal of the meter, identifying the positive input for testing voltage and current, usually becomes the negative terminal when the meter is used as an ohmmeter because of the way the internal battery is connected to the meter movement. Check the meter manufacturers handbook before using a multimeter to test electronic components.

Electrical cables

Let us first of all define some technical terms and discuss the properties of materials used in electrical installation work.

Conductor A material (usually a metal) which allows heat and electricity to pass easily through it.

Insulator A material (usually a non-metal) which will *not* allow heat and electricity to pass easily through it.

Ferrous A word used to describe all metals in which the main constituent is iron. The word 'ferrous' comes

from the Latin word *ferrum* meaning iron. Ferrous metals have magnetic properties. Cast iron, wrought iron and steel are all ferrous metals.

Non-ferrous Metals which *do not* contain iron are called non-ferrous. They are non-magnetic and resist rusting. Copper, aluminium, tin, lead, zinc and brass are examples of non-ferrous metals.

Alloy An alloy is a mixture of two or more metals. Brass is an alloy of copper and zinc, usually in the ratio 70% to 30% or 60% to 40%.

Corrosion The destruction of a metal by chemical action. Most corrosion takes place when a metal is in contact with moisture (see also mild steel and zinc).

Thermoplastic polymers These may be repeatedly warmed and cooled without appreciable changes occurring in the properties of the material. They are good insulators, but give off toxic fumes when burned. They have a flexible quality when operated up to a maximum temperature of 70°C but should not be flexed when the air temperature is near 0°C, otherwise they may crack. Polyvinylchloride (PVC) used for cable insulation is a thermoplastic polymer.

Thermosetting polymers Once heated and formed, products made from thermosetting polymers are fixed rigidly. Plug tops, socket outlets and switch plates are made from this material.

Rubber is a tough elastic substance made from the sap of tropical plants. It is a good insulator, but degrades and becomes brittle when exposed to sunlight.

Synthetic rubber is manufactured, as opposed to being produced naturally. Synthetic or artificial rubber is carefully manufactured to have all the good qualities of natural rubber – flexibility, good insulation and suitability for use over a wide range of temperatures.

Silicon rubber Introducing organic compounds into synthetic rubber produces a good insulating material which is flexible over a wide range of temperatures and which retains its insulating properties even when burned. These properties make it ideal for cables used in fire alarm installations such as FP200 cables.

Magnesium oxide The conductors of mineral insulated metal sheathed (MICC) cables are insulated with compressed magnesium oxide, a white chalk-like substance which is heat-resistant and a good insulator and lasts for many years. The magnesium oxide insulation, copper conductors and sheath, often additionally manufactured with various external sheaths to provide further protection from corrosion and weather, produce a cable designed for long-life and high-temperature installations. However, the magnesium oxide is very hygroscopic, which means that it attracts moisture and, therefore, the cable must be terminated with a special moisture-excluding seal, as shown in Fig. 2.55.

COPPER

Copper is extracted from an ore which is mined in South Africa, North America, Australia and Chile. For electrical purposes it is refined to about 98.8% pure copper, the impurities being extracted from the ore by smelting and electrolysis. It is a very good conductor, is non-magnetic and offers considerable resistance to atmospheric corrosion. Copper toughens with work, but may be annealed, or softened, by heating to dull red before quenching.

Copper forms the largest portion of the alloy brass, and is used in the manufacture of electrical cables, domestic heating systems, refrigerator tubes and vehicle radiators. An attractive soft reddish brown metal, copper is easily worked and is also used to manufacture decorative articles and jewellery.

ALUMINIUM

Aluminium is a grey-white metal obtained from the mineral bauxite which is found in the USA, Germany and the Russian Federation. It is a very good conductor, is non-magnetic, offers very good resistance to atmospheric corrosion and is notable for its extreme softness and lightness. It is used in the manufacture of power cables. The overhead cables of the National Grid are made of an aluminium conductor reinforced by a core of steel. Copper conductors would be too heavy to support themselves between the pylons. Lightness and resistance to corrosion make aluminium an ideal metal for the manufacture of cooking pots and food containers.

Aluminium alloys retain the corrosion resistance properties of pure aluminium with an increase in strength. The alloys are cast into cylinder heads and gearboxes for motorcars, and switch-boxes and luminaires for electrical installations. Special processes and fluxes have now been developed which allow aluminium to be welded and soldered.

BRASS

Brass is a non-ferrous alloy of copper and zinc which is easily cast. Because it is harder than copper or aluminium it is easily machined. It is a good conductor and is highly resistant to corrosion. For these reasons it is often used in the electrical and plumbing trades. Taps, valves, pipes, electrical terminals, plug top pins and terminal glands for steel wire armour (SWA) and MI cables are some of the many applications.

Brass is an attractive yellow metal which is also used for decorative household articles and jewellery. The combined properties of being an attractive metal which is highly resistant to corrosion make it a popular metal for ships' furnishings.

CAST STEEL

Cast steel is also called tool steel or high carbon steel. It is an alloy of iron and carbon which is melted in air-tight crucibles and then poured into moulds to form ingots. These ingots are then rolled or pressed into various shapes from which the finished products are made. Cast steel can be hardened and tempered and is therefore ideal for manufacturing tools (see also Chapter 5). Hammer heads, pliers, wire cutters, chisels, files and many machine parts are also made from cast steel.

MILD STEEL

Mild steel is also an alloy of iron and carbon but contains much less carbon than cast steel. It can be filed, drilled or sawn quite easily and may be bent when hot or cold, but repeated cold bending may cause it to fracture. In moist conditions corrosion takes place rapidly unless the metal is protected. Mild steel is the most widely used metal in the world, having considerable strength and rigidity without being brittle. Ships, bridges, girders, motorcar bodies, bicycles, nails, screws, conduit, trunking, tray and SWA are all made of mild steel.

ZINC

Zinc is a non-ferrous metal which is used mainly to protect steel against corrosion and in making the alloy brass. Mild steel coated with zinc is sometimes called *galvanized steel*, and this coating considerably improves steel's resistance to corrosion. Conduit, trunking, tray, steel wire armour, outside luminaires and electricity pylons are made of galvanized steel.

Construction of cables

Most cables can be considered to be constructed in three parts: the *conductor* which must be of a suitable cross-section to carry the load current; the *insulation*, which has a colour or number code for identification; and the *outer sheath* which may contain some means of providing protection from mechanical damage.

The conductors of a cable are made of either copper or aluminium and may be stranded or solid. Solid conductors are only used in fixed wiring installations and may be shaped in larger cables. Stranded conductors are more flexible and conductor sizes from $4.0\,\text{mm}^2$ to $25\,\text{mm}^2$ contain seven strands. A $10\,\text{mm}^2$ conductor, for example, has seven 1.35 mm diameter strands which collectively make up the $10\,\text{mm}^2$ cross-sectional area of the cable. Conductors above $25\,\text{mm}^2$ have more than seven strands, depending upon the size of the cable. Flexible cords have multiple strands of very fine wire, as fine as one strand of human hair. This gives the cable its very flexible quality.

New wiring colours

Twenty-eight years ago the United Kingdom agreed to adopt the European colour code for flexible cords,

that is, brown for live or phase conductor, blue for the neutral conductor and green combined with yellow for earth conductors. However, no similar harmonisation was proposed for non-flexible cables used for fixed wiring. These were to remain as red for live or phase conductor, black for the neutral conductor and green combined with yellow for earth conductors.

On the 31st of March 2004 the IEE published Amendment No. 2 to BS 7671:2001 which specified new cable core colours for all fixed wiring in United Kingdom electrical installations. These new core colours will 'harmonise' the United Kingdom with the practice in mainland Europe.

Existing fixed cable core colours

■ *Single Phase* red phase conductors, black neutral conductors, and green combined with yellow for earth conductors.
■ *Three Phase* red, yellow and blue phase conductors, black neutral conductors and green combined with yellow for earth conductors.

These core colours must *not* be used after 31st of March 2006.

New (harmonised) fixed cable core colours

■ *Single Phase* brown phase conductors, blue neutral conductors and green combined with yellow for earth conductors (just like the existing flexible cords)
■ *Three Phase* brown, black and grey phase conductors, blue neutral conductors and green combined with yellow for earth conductors.

These core colours *may* be used from 31st of March 2004.

Extensions or alterations to existing *single phase* installations do not require marking at the interface between the old and new fixed wiring colours. However, a warning notice must be fixed at the consumer unit or distribution fuse board which states:

Caution – this installation has wiring colours to two versions of BS 7671. Great care should be taken before undertaking extensions, alterations or repair that all conductors are correctly identified.

Alterations to *three phase* installations must be marked at the interface L1,L2,L3 for the phases and N for the neutral. Both new and old cables must be marked. These markings are preferred to coloured tape and a caution notice is again required at the distribution board.

PVC INSULATED AND SHEATHED CABLES

Domestic and commercial installations use this cable, which may be clipped direct to a surface, sunk in plaster or installed in conduit or trunking. It is the simplest and least expensive cable. Figure 2.53 shows a sketch of a twin and earth cable.

The conductors are covered with a colour-coded PVC insulation and then contained singly or with others in a PVC outer sheath.

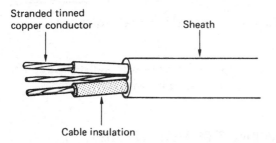

Fig. 2.53 A twin and earth PVC insulated and sheathed cable.

PVC/SWA CABLE

PVC insulated steel wire armour cables are used for wiring underground between buildings, for main supplies to dwellings, rising submains and industrial installations. They are used where some mechanical protection of the cable conductors is required.

The conductors are covered with colour-coded PVC insulation and then contained either singly or with others in a PVC sheath (see Fig. 2.54). Around this sheath is placed an armour protection of steel wires twisted along the length of the cable, and a final PVC sheath covering the steel wires protects them from corrosion. The armour sheath also provides the circuit protective conductor (CPC) and the cable is simply terminated using a compression gland.

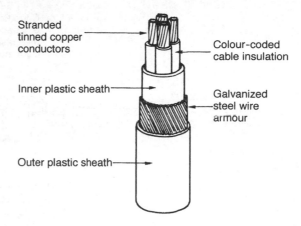

Fig. 2.54 A four-core PVC/SWA cable.

MI CABLE

A mineral insulated (MI) cable has a seamless copper sheath which makes it waterproof and fire- and corrosion-resistant. These characteristics often make it the only cable choice for hazardous or high-temperature installations such as oil refineries and chemical works, boiler-houses and furnaces, petrol pump and fire alarm installations.

The cable has a small overall diameter when compared to alternative cables and may be supplied as bare copper or with a PVC oversheath. It is colour-coded orange for general electrical wiring, white for emergency lighting or red for fire alarm wiring. The copper outer sheath provides the CPC, and the cable is terminated with a pot and sealed with compound and a compression gland (see Fig. 2.55).

The copper conductors are embedded in a white powder, magnesium oxide, which is non-ageing and non-combustible, but which is hygroscopic, which means that it readily absorbs moisture from the surrounding air, unless adequately terminated. The termination of an MI cable is a complicated process requiring the electrician to demonstrate a high level of practical skill and expertise for the termination to be successful.

FP 200 cable

FP 200 cable is similar in appearance to an MI cable in that it is a circular tube, or the shape of a pencil, and is available with a red or white sheath. However, it is much simpler to use and terminate than an MI cable.

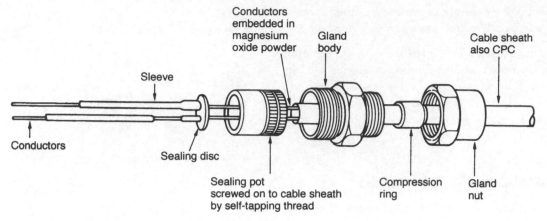

Fig. 2.55 MI cable with terminating seal and gland.

The cable is available with either solid or stranded conductors that are insulated with 'insudite' a fire resistant insulation material. The conductors are then screened, by wrapping an aluminium tape around the insulated conductors, that is, between the insulated conductors and the outer sheath. This aluminium tape screen is applied metal side down and in contact with the bare circuit protective conductor.

The sheath is circular and made of a robust thermoplastic low smoke, zero halogen material.

FP 200 is available in 2, 3, 4, 7, 12 and 19 cores with a conductor size range from 1.0 mm to 4.0 mm. The core colours are: two core, red and black, three core, red, yellow and blue and four core, black, red, yellow and blue.

The cable is as easy to use as a PVC insulated and sheathed cable. No special terminations are required, the cable may be terminated through a grommet into a knock out box or terminated through a simple compression gland.

The cable is a fire resistant cable, primarily intended for use in fire alarms and emergency lighting installations or it may be embedded in plaster.

HIGH-VOLTAGE POWER CABLES

The cables used for high-voltage power distribution require termination and installation expertise beyond the normal experience of a contracting electrician. The regulations covering high-voltage distribution are beyond the scope of the IEE regulations for electrical installations. Operating at voltages in excess of 33 kV and delivering thousands of kilowatts, these cables are

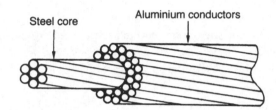

Fig. 2.56 132 kV overhead cable construction.

either suspended out of reach on pylons or buried in the ground in carefully constructed trenches.

HIGH-VOLTAGE OVERHEAD CABLES

Suspended from cable towers or pylons, overhead cables must be light, flexible and strong.

The cable is constructed of stranded aluminium conductors formed around a core of steel stranded conductors (see Fig. 2.56). The aluminium conductors carry the current and the steel core provides the tensile strength required to suspend the cable between pylons. The cable is not insulated since it is placed out of reach and insulation would only add to the weight of the cable.

HIGH-VOLTAGE UNDERGROUND CABLES

High-voltage cables are only buried underground in special circumstances when overhead cables would be unsuitable, for example, because they might spoil a view of natural beauty (see Chapter 1). Underground cables are very expensive because they are much more complicated to manufacture than overhead cables. In transporting vast quantities of power, heat is generated within the cable. This heat is removed by passing oil through the cable to expansion points, where the oil is cooled. The system is similar to the water cooling of an internal combustion engine. Figure 2.57 shows a typical high voltage cable construction.

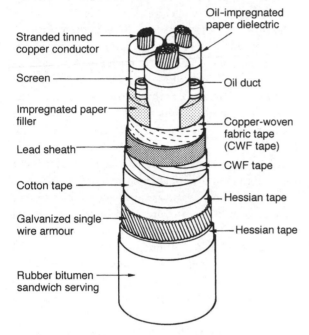

Fig. 2.57   132 kV underground cable construction.

The conductors may be aluminium or copper, solid or stranded. They are insulated with oil-impregnated brown paper wrapped in layers around the conductors. The oil ducts allow the oil to flow through the cable, removing excess heat. The whole cable within the lead sheath is saturated with oil, which is a good insulator. The lead sheath keeps the oil in and moisture out of the cable, and this is supported by the copper-woven fabric tape. The cable is protected by steel wire armouring, which has bitumen or PVC serving over it to protect the armour sheath from corrosion. The termination and installation of these cables is a very specialized job, undertaken by the supply authorities only.

Generation of electricity

The electricity-generating companies are required by Parliament to develop and maintain an efficient and economical supply of electricity for all parts of the United Kingdom and to take account of the effects which proposals might have upon the natural beauty of the countryside and buildings or objects of special interest. Efficiency does not always go hand in hand with the preservation of natural beauty, but the generating companies do take their responsibilities very seriously.

Modern power stations are very large, both physically and electrically (2000 MW), and cannot be hidden from view. But, by careful siting, careful selection of building materials, suitable ground modelling and tree planting, acceptable results can be obtained.

CONVENTIONAL POWER STATIONS

To keep the modern coal-fired power stations operating requires about 100 million tonnes of coal each year. Therefore, new coal-fired and oil-fired stations are built close to fuel sources in order to reduce transport costs and so that heavy coal-train traffic does not impose upon passenger railway routes.

NUCLEAR POWER STATIONS

One of the advantages of nuclear power stations is that they involve no difficult fuel transport problems. One modern power station fuel element has the energy equivalent of 1800 tonnes of coal. A reactor contains about 25 000 fuel elements, and these are all changed in small batches over a period of 4 or 5 years.

Nuclear power plants are built to the highest safety standards and during the last 30 years have gained a great deal of practical operating experience and as a result maintain a very high safety record. There are 14 operational nuclear power stations in England and Wales, but, public opinion has turned against nuclear power and so, as nuclear power stations reach the end of their designated lives, they are not being replaced. Back up power for windmills and other renewable sources will therefore have to be provided by expensive fossil fuels (gas) imported by pipeline from Russia, Algeria and the Middle East.

COOLING WATER

Excluding fuel, the other essential requirement of all power stations is an adequate supply of cooling water.

A 2000 MW station requires about 225 million litres of water per hour. Such quantities of water can only be obtained from the sea or major river estuaries. Inland power stations use water recirculated through cooling towers.

WATER DISCHARGE

Cooling towers emit plumes of visible water vapour through evaporation. Although unattractive, the plume does not pollute the atmosphere and quickly evaporates.

The cooling water discharged into the sea or river from a large power station may have been raised in temperature by 10°C. Extensive ecological research has confirmed that there are no adverse effects from warm water discharge and the evidence would suggest that fish can benefit from the warmed water.

CHIMNEY DISCHARGE

The burning of any kind of fuel produces pollutants, but modern power stations burn coal and oil in the least harmful way due to highly efficient boilers. Power stations are not sole polluters since industrial chimneys emit carbon dioxide (CO_2) and sulphur dioxide (SO_2) gases. The most controversial effect of the sulphur dioxide gas is the turning of rainfall over industrial areas into dilute sulphuric acid which attacks the stonework on buildings and inhibits the growth of plants. This has encouraged power stations to build tall chimneys to discharge the gases high into the atmosphere and so avoid harmful concentrations at ground level. However, unfortunately this does not provide a simple solution since the Norwegian and Swedish authorities, who are downwind of the UK, are now complaining that acid rain falling on their territory originates from UK power stations.

ASH DISPOSAL

The coal used in power station boilers is pulverized into very fine particles before burning. Consequently, the ash is a very fine powder which is removed from the flue gases by grit arrestors. The generating board's marketing officers have developed many uses for waste ash including the filling and formation of roads, motorways and airfield runways, the manufacture of building blocks and bricks and the reclamation of wasteland at reasonable costs.

RADIOACTIVE WASTE DISPOSAL

As the uranium fuel elements are 'burned' in the reactor they become less effective and are therefore replaced.

The spent fuel elements are transported in shielded containers to the reprocessing plant of the Windscale Works at Sellafield in Cumbria, where they are stripped of their cans and the radioactive products prepared for safe storage.

Radioactivity is the process by which an unstable atom emits high-energy particles until it becomes stable. The rate at which radioactive nuclei decay is characterized by their 'half-life', that is, the time taken for half the nuclei in a given sample to decay. For example, an unstable variety of xenon 138 has a half-life of 17 minutes, while plutonium 239 has a half-life of 24 000 years.

Short-lived radioactive waste is stored under water until it is safe for disposal like normal industrial waste. Some low-activity waste is encased in concrete and dumped deep in the ocean under strict international supervision. The long-term highly active waste products are stored at the Windscale Works as a concentrated liquid in stainless steel vessels surrounded by concrete shielding. A dozen such vessels, approximately equal to the volume of two average-sized houses, contain all the highly active waste produced by the UK since the 1950s, but storage as a non-reactive solid would be preferable in the long term. A process is being developed for converting the liquid waste into glass blocks which, when they have been cased in stainless steel containers, will be disposed of away from the human environment, probably by burying them deep in stable rock formations on land or under the deep ocean bed. It has been estimated that after about 1000 years the radioactivity of the glassified waste will have fallen to a safe level.

POWER GENERATION IN ENGLAND AND WALES

The generating companies National Power, PowerGen and Nuclear Electric supply more than 200 000 000 MWh of electricity, meeting a peak maximum demand of about 45 000 MW from a total installed capacity of approximately 50 000 MW. In 2004 about 72% of this energy is supplied from fossil fuel stations, about 24% from nuclear stations, less than 1% from hydroelectric installations and 3% from wind power.

National Power and PowerGen are heavily dependent upon fossil fuels (coal, oil and gas), but we know from respectable scientific sources that the fossil fuel era is drawing to a close. Popular estimates suggest that gas and oil will reach peak production in about the year 2060, while British coal reserves will last until 2200 at the present rate of consumption. The generating companies must give consideration to other ways of generating electricity so that coal, oil and gas might be conserved in the next century and hence the great interest in renewable energy supplies such as wind and wave power.

POWER GENERATION IN SCOTLAND

The generation of electricity in Scotland is the joint responsibility of the Scottish Power and Hydroelectric companies. They supply about 22 000 000 MWh of electricity, meeting a peak maximum demand of about 1600 MW. This electricity is generated by nuclear and fossil fuel power stations, 59 hydroelectric power stations and two pumped water storage schemes, while the Western Isles and Shetland are supplied by diesel power generators. The smaller islands are connected to the mainland by submarine cable and the whole network consists of 8800 km of 132 kV and 270 kV transmission cables.

The mountainous regions and relative isolation of many consumers in Scotland have given the electricity boards an ideal opportunity to pioneer many renewable energy schemes. In addition to the very successful hydroelectric schemes, Orkney has successfully operated an experimental 250 kW and 300 kW wind generator, and a 3 MW wind aerogenerator was connected to the board's system in 1987 (see Fig. 2.58).

POWER GENERATION IN NORTHERN IRELAND

The generation of electricity in Northern Ireland is the responsibility of the Northern Ireland Electricity Service. This supplies a little over 5 000 000 MWh of electricity, meeting a peak demand of about 1300 MW

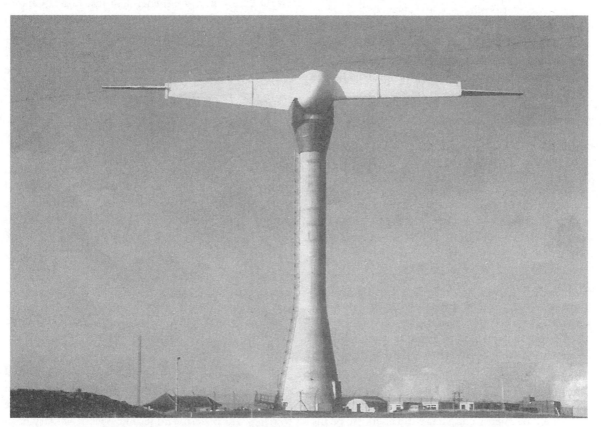

Fig. 2.58 A 3 MW wind generator, Burghar Hill, Orkney (by kind permission of the North of Scotland Hydroelectric Board).

from an installed capacity of 2050 MW. This electricity is generated by four fossil fuel stations and transmitted over 1250 km of 275 kV and 110 kV lines and cables to approximately 1.2 million customers.

The power stations operated by the Northern Ireland Electricity Service are heavily dependent upon oil and at a time when oil prices are uncertain there is a desire to have alternative fuel sources available for the long-term planning and stability of the industry. To this end the Kilroot oil-fired power station is being converted to burn either oil or coal, whichever is the cheaper fuel. This will reduce the dependence on oil by the Northern Ireland Electricity Service from 90% to 70% by the new millennium.

HYDROELECTRIC POWER

Water power makes a useful contribution to the energy needs of Scotland, but the possibilities of building similar stations in England are very limited since there are no high mountains or vast reserves of water that make these schemes attractive. However, hydroelectric power stations have many advantages over fossil fuel and nuclear stations. They can be brought up to full load very quickly and are non-polluting. Water is a so-called 'renewable' energy source and the low running costs may make them more attractive as the cost of other forms of energy increases.

WIND POWER

Modern wind machines will be very different from the traditional windmill of the last century which gave no more power than a car engine. Very large structures are needed to extract worthwhile amounts of energy from the wind. Modern wind generators are about as tall as electricity pylons, with a three-blade aeroplane-type propellor to catch the wind and turn the generator. They are usually sited together in groups of about 20 generators in what are known as 'wind energy farms'.

Each modern wind turbine generates about 600 kW of electricity; a wind energy farm of 20 generators will, therefore, generate 12 MW – a useful contribution to the National Grid, using a naturally occurring, renewable, non-polluting primary source of energy. The Department of Energy considers wind energy to be the most promising of the renewable energy sources.

The Countryside Commission, the government's adviser on land use, has calculated that to achieve a target of generating 10% of the total electricity supply by wind power by the year 2010 will require 40 000 generators of the present size. At the time of writing 2004 we have 89 operational wind farms mostly in Wales and Cornwall, with 16 more to be built by 2006 but this is a promising start to achieving this target and off-shore wind farms are now at an experimental stage.

Wind power is an endless renewable source of energy, is safer than nuclear power and provides none of the polluting emissions associated with fossil fuel. However, wind farms are, by necessity, sited in some of the wildest and most beautiful landscapes in the UK, such as the west coast of Scotland, the north Pennines, the Lake District, Wales and Cornwall. Siting wind energy farms in these areas of outstanding natural beauty has outraged conservationists. The Ramblers' Association, the Council for National Parks and the Council for the Protection of Rural England, together with its sister bodies in Scotland and Wales, have joined forces to call for tougher government controls on the location of wind energy farms. Prince Charles The Prince of Wales reluctantly joined the debate (Telegraph 08-08-2004) saying that he was in favour of renewable energy sources but believed that wind farms are 'a horrendous blot on the landscape'. He believes that if they are to be built at all, they should be constructed well out to sea.

The Department of Energy has calculated that 10 000 wind machines could provide the energy equivalent of 8 million tonnes of coal per year. While this is a worthwhile saving of fossil fuel, opponents point out the obvious disadvantages of wind machines, among them the need to maintain the energy supply during periods of calm, which means that wind machines can only ever supplement a more conventional electricity supply.

HARNESSING THE SEA

Great Britain is a small island surrounded by water. Surely we could harness some of the energy contained in the tides or waves?

In March 1985 the British government committed £220 000 to studies of a barrage across the Severn Estuary as a source of tidal power and a new traffic route to relieve the Severn Bridge. The Severn Estuary has a tidal range of up to 15 m – the largest in Europe – and a reasonable shape for building a barrage or dam across it. This would allow the basin to fill with water as the tide rose, and then allow the impounded water to flow out through electricity-generating turbines as the tide fell.

A scheme proposed by the team that designed and built the Thames barrier suggests that a 14 km barrage could supply as much energy as four or five nuclear power stations at two-thirds the cost per kilowatt. The impounded water would also provide a water sports area suitable for boating, water-skiing, swimming and angling.

Unfortunately such a scheme would produce power geared to the lunar cycle of the tides and not to the demands of industry and commerce. The tidal barrier might have disastrous ecological consequences upon numerous wildfowl and wading bird species by submerging the mud-flats which now provide winter shelter for these birds. Therefore the value of the power produced would need to be balanced against the possible consequences.

Marine Current Turbines Ltd., are currently carrying out research and development of submersed turbines which will rotate by exploiting the principle of flowing water in general and tidal streams in particular.

The general principle is that an 11 m diameter water turbine is lowered into the sea down a steal column drilled into the sea bed. The tidal movement of the water then rotates the turbine and generates electricity.

The prototype machines have been submersed in the sea off Lynmouth in Devon and they hope to have a reliable machine generating between 750 and 1200 kW by the year 2005.

Britain has no commercially produced wave-power electricity but a prototype wave machine built by Ocean Power Delivery of Edinburgh has successfully been installed off the Ovkney coast. It began generating 750 kW of electricity for the National Grid during August 2004. At Bergen in Norway a wave-energy power station came into operation in November 1985, and France has successfully operated a 240 MW tidal power station at Rance in Brittany for the past 25 years.

SOLAR ENERGY

On a bright sunny day the sun's radiation represents a power input of about 1 kW per square metre. A 2000 MW power station occupies about 1.5 square miles, that is 4 million square metres. Therefore about 4000 MW of solar radiation falls on to the land occupied by the power station. If we could cover the same area with solar cells having an efficiency of 50% we could generate the same quantity of electricity without burning any fuel. Unfortunately, mass-produced solar cells have a conversion efficiency of about 12% and therefore to actually produce 2000 MW would require about 7 square miles of solar cells, which is quite unacceptable on a small island like Great Britain.

While there seems to be a possibility of using solar power directly, the efficiency of solar cells must be greatly improved before a commercial solar station becomes a possibility. There is also the problem of maintaining the supply through the night and during dull days in the winter, when demand is higher than in summer.

Electricity today

Our future energy requirements and resources are difficult to predict but we do need independence from fossil fuels which, at the present rate of consumption, are predicted to run out by the first quarter of this millennium. It is therefore realistic to plan future systems which are flexible, have large safety margins, and give consideration to the renewable sources.

The generating companies will continue to meet our demand for electricity with fossil fuel and nuclear stations despite the worries caused by the Chernobyl power station disaster in April 1986. Research will continue into the renewable energy sources such as wind, wave, tidal and solar, but most of these alternative energy sources are in a very early stage of development and consequently the immediate future will undoubtedly see us remain dependent upon nuclear energy and fossil fuels.

Britain has large coal reserves, but these will be increasingly required for many uses other than electricity generation – to make fertilizers and chemicals, to produce oil substitutes and to manufacture town gas as natural gas becomes scarce.

Energy from fossil fuels made the industrial revolution possible and has enabled science and technology to reach a level of development at which the exploitation of other energy sources is within our capability. Electricity is now well established as a means of harnessing energy and bringing it to the service of man.

Electricity will continue into the twenty-first century because of its many advantages: it is easy to make, distribute, and control, and, when properly installed, is safe for anyone to use.

Using the laws discovered by Michael Faraday in 1831 (the laws of electromagnetic induction described earlier in this Chapter, electricity may be generated in commercial quantities and distributed to its end use simply by connecting supply and consumer together with suitable cables. The supply may be controlled manually with a switch or automatically with a thermostat or circuit breaker. The wheels of industry are driven by electric motors, and sophisticated artificial heating and lighting installations make it possible for industry and commerce to work long hours in safety and comfort. Modern homes use microwave cookers to prepare meals quickly, freezers to store food safely and conveniently, and electric cleaners to speed household cleaning chores. In our leisure time we watch television and listen to music on stereo hi-fi music centres. While home computers are used to run simple programs and play electronic games, business and commerce use large-capacity computers to store data, word-process and 'number-crunch'. All the high technology which we today take for granted is dependent upon a reliable and secure electricity supply.

Distribution of electricity

Electricity is generated in modern power stations at 25 kV and fed through transformers to the consumer over a complex network of cables known as the National Grid system. This is a network of cables, mostly at a very high voltage, suspended from transmission towers, linking together the 175 power stations and millions of consumers. There are approximately 5000 miles of high-voltage transmission lines in England and Wales, running mostly through the countryside.

Man-made structures erected in rural areas often give rise to concern, but every effort is made to route the overhead lines away from areas where they might spoil a fine view. There is full consultation with local authorities and interested parties as to the route which lines will take. Farmers are paid a small fee for having transmission line towers on their land. Over the years many different tower designs and colours have been tried, but for the conditions in the United Kingdom,

galvanized steel lattice towers are considered the least conspicuous and most efficient.

For those who consider transmission towers unsightly, the obvious suggestion might be to run all cables underground. In areas of exceptional beauty this is done, but underground cables are about 16 times more expensive than the equivalent overhead lines. The cost of running the largest lines underground is about £3 million per mile compared with about £200 000 overhead. On long transmission lines the losses can be high, but by raising the operating voltage and therefore reducing the current for a given power, the I^2R losses are reduced, the cable diameter is reduced and the overall efficiency of transmission is increased. In order to standardize equipment, standard voltages are used. These are:

- 400 kV and 275 kV for the super Grid;
- 132 kV for the original Grid;
- 66 kV and 33 kV for secondary transmission;
- 11 kV for high-voltage distribution;
- 400 V for commercial consumer supplies;
- 230 V for domestic consumer supplies.

A diagrammatic representation showing the distribution of power is given in Fig. 2.59

All local distribution in the UK is by underground cables from substations placed close to the load centre and supplied at 11 kV. Transformers in these local substations reduce the voltage to 400 V, three-phase and neutral distributor cables connect this supply to consumers. Connecting to one-phase and neutral of a three-phase 400 V supply gives a 230 V single-phase supply suitable for domestic consumers.

When single-phase loads are supplied from a three-phase supply, as shown in Fig. 2.60, the load should be 'balanced' across the phases. That is, the load should be equally distributed across the three phases so that each phase carries approximately the same current. This prevents any one phase being overloaded.

CONSUMER'S MAINS EQUIPMENT

The consumer's mains equipment is normally fixed close to the point at which the supply cable enters the building. To meet the requirements of the IEE Regulations it must provide:

- protection against electric shock (Chapter 41);
- protection against overcurrent (Chapter 43);
- isolation and switching (Chapter 46).

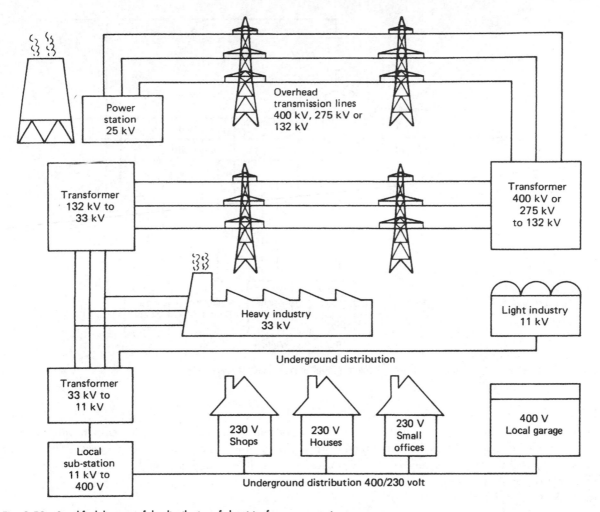

Fig. 2.59 Simplified diagram of the distribution of electricity from power station to consumer.

Protection against electric shock is provided by insulating and placing live parts out of reach in suitable enclosures, earthing and bonding metal work and providing fuses or circuit breakers so that the supply is automatically disconnected under fault conditions.

To provide overcurrent protection it is necessary to provide a device which will disconnect the supply automatically before the overload current can cause a rise in temperature which would damage the installation. A fuse or miniature circuit breaker (MCB) would meet this requirement.

An isolator is a mechanical device which is operated manually and is provided so that the whole of the installation, one circuit or one piece of equipment may be cut off from the live supply. In addition, a means of switching off for maintenance or emergency switching must be provided. A switch may provide the means of isolation, but an isolator differs from a switch in that it is intended to be opened when the circuit concerned is not carrying current. Its purpose is to ensure the safety of those working on the circuit by making dead those parts which are live in normal service. One device may provide both isolation and switching provided that the characteristics of the device meet the Regulations for both functions. The switching of electrically operated equipment in normal service is referred to as functional switching.

Circuits are controlled by switchgear which is assembled so that the circuit may be operated safely under normal conditions, isolated automatically under fault conditions, or isolated manually for safe maintenance. These requirements are met by good workmanship and the installation of proper materials such as switches, isolators, fuses or circuit breakers. The equipment

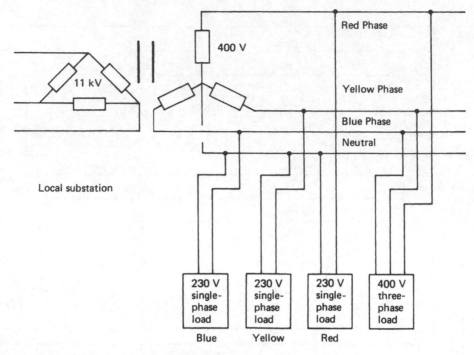

Fig. 2.60 Simplified diagram of the distribution from local substation to single-phase supply.

belonging to the supply authority is sealed to prevent unauthorized entry, because if connection were made to the supply before the meter, the energy used by the consumer would not be recorded on the meter. Figures 2.61 and 2.62 show connections and equipment for different installation situations.

GRADED PROTECTION DEVICES FOR EFFECTIVE DISCRIMINATION

The main fuse at the consumer's service position protects the incoming cable from short circuits at the mains position and against overload and short circuits in the final circuits. However, in a properly designed installation there will be other protective devices between the main fuse and the consumer's equipment because of the subdivision of the installation, and it is these which should operate before the main fuse. The current rating of the fuses or circuit breakers should be so graded that when a fault occurs only the device nearest to the fault will operate. The other devices should not operate, so that the supply to healthy circuits is not impaired.

If a fault occurs in the appliance shown in Fig. 2.63, only the plug top fuse should blow if the circuit fuses

have been correctly graded. Similarly, if a fault occurs on the cable feeding the socket circuit, only that fuse in the fuse board should blow leaving the lights, cooker and bell operating normally.

ELECTRICAL BONDING TO EARTH

The purpose of the bonding regulations is to keep all the exposed metalwork of an installation at the same earth potential as the metalwork of the electrical installation, so that no currents can flow and cause an electric shock. For a current to flow there must be a difference of potential between two points, but if the points are joined together there can be no potential difference. This bonding or linking together of the exposed metal parts of an installation is known as 'equipotential bonding'.

MAIN EQUIPOTENTIAL BONDING

Where earthed electrical equipment may come into contact with the metalwork of other services, they too must be effectively connected to the main earthing terminal of the installation (IEE Regulation 130–04).

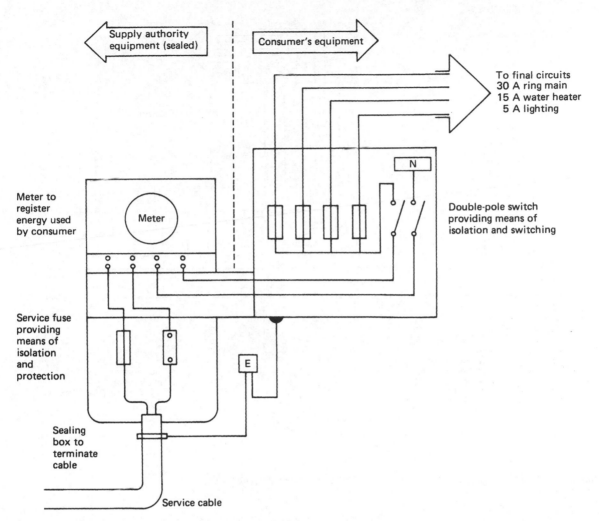

Fig. 2.61 Simplified diagram of connections and equipment at a domestic service position.

Other services are described by IEE Regulation 413–02–02 as:

- main water pipes;
- main gas pipes;
- other service pipes and ducting;
- risers of central heating and air conditioning systems;
- exposed metal parts of the building structure;
- lightning protective conductors.

Main equipotential bonding should be made to gas and water services at their point of entry into the building, as shown in Fig. 2.64, using insulated bonding conductors of not less than half the cross-section of the incoming main earthing conductor. The minimum permitted size is 6 mm² but the cross-section need not exceed 25 mm² (IEE Regulation 547–02–01). The bonding clamp must be fitted on the consumer's side of the gas meter between the outlet union, before any branch pipework but within 600 mm of the meter. A permanent label must also be fixed at or near the point of connection of the bonding conductor with the words 'Safety Electrical Connection – Do Not Remove' (IEE Regulation 514–13–01). Supplementary bonding is described in Chapter 4 of this volume.

ELECTRICAL SHOCK AND OVERLOAD PROTECTION

Electric shock is normally caused either by touching a conductive part that is normally live, called *direct contact*, or by touching an exposed conductive part made

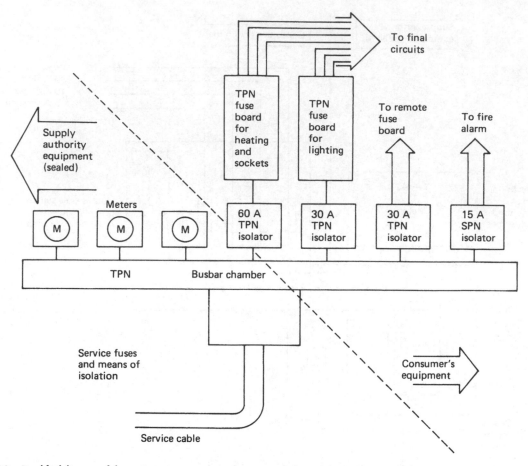

Fig. 2.62 Simplified diagram of the equipment at an industrial or commercial service position.

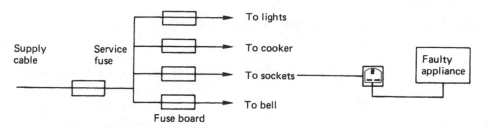

Fig. 2.63 Subdivision of circuits in a domestic position.

live by a fault, called *indirect contact*. The touch voltage curve in Fig. 2.65 shows that a person in contact with 230 V must be released from this danger in 40 ms if harmful effects are to be avoided. Similarly, a person in contact with 400 V must be released in 15 ms to avoid being harmed.

In general, protection against direct contact is achieved by insulating live parts. Protection against indirect contact is achieved by earthed equipotential

bonding and automatic disconnection of the supply in the event of a fault occurring. Separated extra-low-voltage supplies (SELV) provide protection against both direct and indirect contact.

Part 4 of the IEE Regulations deals with the application of protective measures for safety and Chapter 53 with the regulations for switching devices or switchgear required for protection, isolation and switching of a consumer's installation.

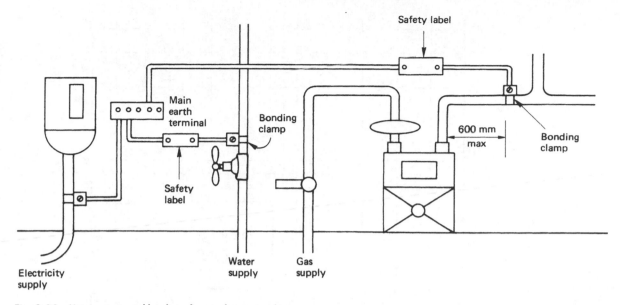

Fig. 2.64 Main equipotential bonding of gas and water supplies.

The consumer's main switchgear must be readily accessible to the consumer and be able to:

- isolate the complete installation from the supply,
- protect against overcurrent and
- cut off the current in the event of a serious fault occurring.

Protection against overcurrent

Excessive current may flow in a circuit as a result of an overload or a short circuit. An overload or overcurrent is defined as a current which exceeds the rated value in an otherwise healthy circuit. A short circuit is an overcurrent resulting from a fault of negligible impedance between live conductors having a difference in potential under normal operating conditions. Overload currents usually occur in a circuit because it is abused by the consumer or because it has been badly designed or modified by the installer. Short circuits usually occur as a result of an accident which could not have been predicted before the event.

An overload may result in currents of two or three times the rated current flowing in the circuit, while short-circuit currents may be hundreds of times greater than the rated current. In both cases the basic requirement for protection is that the circuit should be interrupted before the fault causes a temperature

rise which might damage the insulation, terminations, joints or the surroundings of the conductors. If the device used for overload protection is also capable of breaking a prospective short-circuit current safely, then one device may be used to give protection from both faults (Regulation 432–02–01). Devices which offer protection from overcurrent are:

- semi-enclosed fuses manufactured to BS 3036,
- cartridge fuses manufactured to BS 1361 and BS 1362,
- high breaking capacity fuses (HBC fuses) manufactured to BS 88,
- miniature circuit breakers (MCBs) manufactured to BS 3871.

SEMI-ENCLOSED FUSES (BS 3036)

The semi-enclosed fuse consists of a fuse wire, called the fuse element, secured between two screw terminals in a fuse carrier. The fuse element is connected in series with the load and the thickness of the element is sufficient to carry the normal rated circuit current. When a fault occurs an overcurrent flows and the fuse element becomes hot and melts or 'blows'.

The designs of the fuse carrier and base are also important. They must not allow the heat generated from an overcurrent to dissipate too quickly from the element, otherwise a larger current would be required

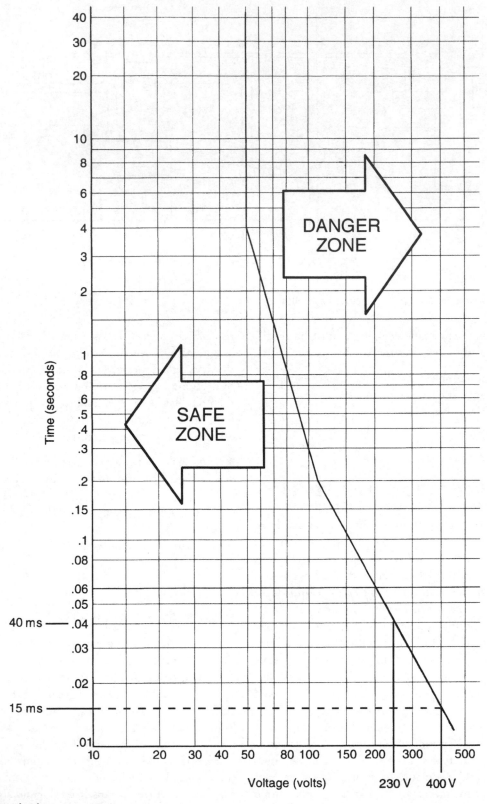

Fig. 2.65 Touch voltage curve.

to 'blow' the fuse. Also if over-enclosed, heat will not escape and the fuse will 'blow' at a lower current. This type of fuse is illustrated in Fig. 2.66. The fuse element should consist of a single strand of plain or tinned copper wire having a diameter appropriate to the current rating as given in Table 2.5.

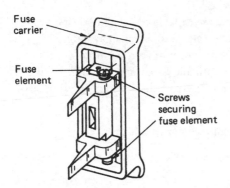

Fig. 2.66 A semi-enclosed fuse.

Table 2.5 Size of fuse element

Current rating (A)	Wire diameter (mm)
5	0.20
10	0.35
15	0.50
20	0.60
30	0.85

Advantages of semi-enclosed fuses

■ They are very cheap compared with other protective devices both to install and to replace.
■ There are no mechanical moving parts.
■ It is easy to identify a 'blown fuse'.

Disadvantages of semi-enclosed fuses

■ The fuse element may be replaced with wire of the wrong size either deliberately or by accident.
■ The fuse element weakens with age due to oxidization, which may result in a failure under normal operating conditions.
■ The circuit cannot be restored quickly since the fuse element requires screw fixing.
■ They have low breaking capacity since, in the event of a severe fault, the fault current may vaporize the fuse element and continue to flow in the form of an arc across the fuse terminals.

■ There is a danger from scattering hot metal if the fuse carrier is inserted into the base when the circuit is faulty.

CARTRIDGE FUSES (BS 1361)

The cartridge fuse breaks a faulty circuit in the same way as a semi-enclosed fuse, but its construction eliminates some of the disadvantages experienced with an open-fuse element.

The fuse element is encased in a glass or ceramic tube and secured to end-caps which are firmly attached to the body of the fuse so that they do not blow off when the fuse operates. Cartridge fuse construction is illustrated in Fig. 2.67. With larger-size cartridge fuses, lugs or tags are sometimes brazed on to the end-caps to fix the fuse cartridge mechanically to the carrier. They may also be filled with quartz sand to absorb and extinguish the energy of the arc when the cartridge is brought into operation.

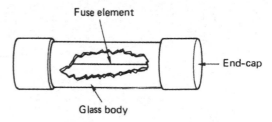

Fig. 2.67 A cartridge fuse.

Advantages of cartridge fuses

■ They have no mechanical moving parts.
■ The declared rating is accurate.
■ The element does not weaken with age.
■ They have small physical size and no external arcing which permits their use in plug tops and small fuse carriers.
■ Their operation is more rapid than semi-enclosed fuses. Operating time is inversely proportional to the fault current.

Disadvantages of cartridge fuses

■ They are more expensive to replace than rewirable fuse elements.
■ They can be replaced with an incorrect cartridge.
■ The cartridge may be shorted out by wire or silver foil in extreme cases of bad practice.

■ They are not suitable where extremely high fault currents may develop.

HIGH BREAKING CAPACITY FUSES (BS 88)

As the name might imply, these cartridge fuses are for protecting circuits where extremely high fault currents may develop such as on industrial installations or distribution systems.

The fuse element consists of several parallel strips of pure silver encased in a substantial ceramic cylinder, the ends of which are sealed with tinned brass end-caps incorporating fixing lugs. The cartridge is filled with silica sand to ensure quick are extraction. Incorporated on the body is an indicating device to show when the fuse has blown. HBC fuse construction is shown in Fig. 2.68.

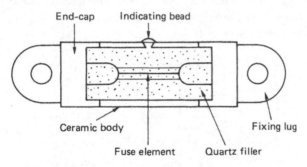

Fig. 2.68 HBC fuse.

Advantages of HBC fuses

■ They have no mechanical moving parts.
■ The declared rating is accurate.
■ The element does not weaken with age.
■ Their operation is very rapid under fault conditions.
■ They are capable of breaking very heavy fault currents safely.
■ They are capable of discriminating between a persistent fault and a transient fault such as the large starting current taken by motors.
■ It is difficult to confuse cartridges since different ratings are made to different physical sizes.

Disadvantages of HBC fuses

■ They are very expensive compared to semi-enclosed fuses.

MINIATURE CIRCUIT BREAKERS (BS 3871)

The disadvantage of all fuses is that when they have operated they must be replaced. An MCB overcomes this problem since it is an automatic switch which opens in the event of an excessive current flowing in the circuit and can be closed when the circuit returns to normal.

An MCB of the type shown in Fig. 2.69 incorporates a thermal and magnetic tripping device. The load current flows through the thermal and the electromagnetic

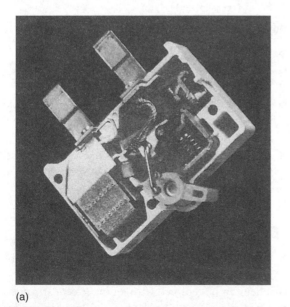

(a)

(b)

Fig. 2.69 (a) Interior view of Wylex 'plug-in' MCB; (b) 'plug-in' MCB fits any standard Wylex consumer's unit.

mechanisms. In normal operation the current is insufficient to operate either device, but when an overload occurs, the bimetal strip heats up, bends and trips the mechanism. The time taken for this action to occur provides an MCB with the ability to discriminate between an overload which persists for a very short time, for example the starting current of a motor, and an overload due to a fault. The device only trips when a fault current occurs. This slow operating time is ideal for overloads but when a short circuit occurs it is important to break the faulty circuit very quickly. This is achieved by the coil electromagnetic device.

When a large fault current (above about eight times the rated current) flows through the coil a strong magnetic flux is set up which trips the mechanisms almost instantly. The circuit can be restored when the fault is removed by pressing the ON toggle. This latches the various mechanisms within the MCB and 'makes' the switch contact. The toggle switch can also be used to disconnect the circuit for maintenance or isolation or to test the MCB for satisfactory operation. The simplified diagram in Fig. 2.70 shows the various parts within an MCB.

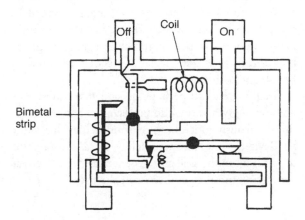

Fig. 2.70 A simplified diagram of an MCB.

Advantages of MCBs

- Tripping characteristics and therefore circuit protection are set by installer.
- The circuit protection is difficult to interfere with.
- The circuit is provided with discrimination.
- A faulty circuit may be easily and quickly restored.
- The supply may be safely restored by an unskilled operator.

Disadvantages of MCBs

- They are very expensive compared to rewirable fuses.
- They contain mechanical moving parts and therefore require regular testing to ensure satisfactory operation under fault conditions.

FUSING FACTOR

The speed with which a protective device will operate under fault conditions gives an indication of the level of protection being offered by that device. This level of protection or fusing performance is given by the fusing factor of the device:

$$\text{Fusing factor} = \frac{\text{Minimum fusing current}}{\text{Current rating}}$$

The minimum fusing current of a device is the current which will cause the fuse or MCB to blow or trip in a given time (BS 88 gives this operating time as 4 hours). The current rating of a device is the current which it will carry continuously without deteriorating.

Thus, a 10 A fuse which operates when 15 A flows will have a fusing factor of $15 \div 10 = 1.5$.

Since the protective device must carry the rated current it follows that the fusing factor must always be greater than one. The closer the fusing factor is to one, the better is the protection offered by that device. The fusing factors of the protective devices previously considered are:

- semi-enclosed fuses: between 1.5 and 2,
- cartridge fuses: between 1.25 and 1.75,
- HBC fuses: less than 1.25,
- MCBs: less than 1.5.

In order to give protection to the conductors of an installation:

- the current rating of the protective device must be equal to or less than the current carrying capacity of the conductor;
- the current causing the protective device to operate must not be greater than 1.45 times the current carrying capacity of the conductor to be protected.

The current carrying capacities of cables given in the tables of Appendix 4 of the IEE Regulations assume that the circuit will comply with these requirements and that the circuit protective device will have a fusing factor of 1.45 or less. Cartridge fuses, HBC fuses and MCBs do have a fusing factor less than 1.45 and therefore when this type of protection is afforded the current carrying capacities of cables may be read directly from the tables.

However, semi-enclosed fuses can have a fusing factor of 2. The wiring regulations require that the rated current of a rewirable fuse must not exceed 0.725 times the current carrying capacity of the conductor it is to protect. This factor is derived as follows:

The maximum fusing factor of a rewirable fuse is 2.

Now, if I_n = current rating of the protective device

I_z = current carrying capacity of conductor

I_2 = current causing the protective device to operate.

Then $I_2 = 2 I_n \leqslant 1.45 I_z$

therefore $I_n \leqslant \dfrac{1.45 \, I_z}{2}$

or $I_n \leqslant 0.725 \, I_z$

When rewirable fuses are used, the current carrying capacity of the cables given in the tables is reduced by a factor of 0.725, as detailed in Appendix 4 item 5 of the Regulations.

Earth leakage protection

When it is required to provide the very best protection from electric shock and fire risk, earth fault protection devices are incorporated in the installation. The object of the Regulations concerning these devices, 413–02–15, 471–08–01 and 471–16–02, is to remove an earth fault current in less than 5 seconds, and limit the voltage which might appear on any exposed metal parts under fault conditions to 50 V. They will continue to provide adequate protection throughout the life of the installation even if the earthing conditions deteriorate. This is in direct contrast to the protection provided by overcurrent devices, which require a low resistance earth loop impedance path.

THE RESIDUAL CURRENT DEVICE (RCD)

The basic circuit for a single-phase RCD is shown in Fig. 2.71. The load current is fed through two equal and opposing coils wound on to a common transformer

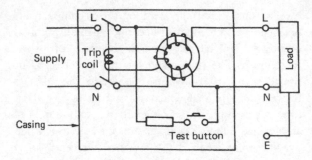

Fig. 2.71   Construction of a residual current device.

core. The phase and neutral currents in a healthy circuit produce equal and opposing fluxes in the transformer core, which induces no voltage in the tripping coil. However, if more current flows in the phase conductor than in the neutral conductor as a result of a fault between live and earth, an out-of-balance flux will result in an emf being induced in the trip coil which will open the double pole switch and isolate the load. Modern RCDs have tripping sensitivities between 10 and 30 mA, and therefore a faulty circuit can be isolated before the lower lethal limit to human beings (about 50 mA) is reached.

Consumer units can now be supplied which incorporate an RCD, so that any equipment supplied by the consumer unit outside the zone created by the main equipotential bonding, such as a garage or greenhouse, can have the special protection required by Regulation 471–08–01.

Finally, it should perhaps be said that a foolproof method of giving protection to people or animals who simultaneously touch both live and neutral has yet to be devised. The ultimate safety of an installation depends upon the skill and experience of the electrical contractor and the good sense of the user.

Hand and power tools

A craftsman earns his living by hiring out his skills or selling products made using his skills and expertise. He shapes his environment, mostly for the better, improving the living standards of himself and others.

Tools extend the limited physical responses of the human body and therefore good-quality, sharp tools are important to a craftsman. An electrician is no less a craftsman than a wood carver. Both must work with a high degree of skill and expertise and both must have sympathy and respect for the materials which they use. Modern electrical installations using new materials are lasting longer than 50 years. Therefore they must be properly installed. Good design, good workmanship and the use of proper materials are essential if the installation is to comply with the relevant regulations, and reliably and safely meet the requirements of the customer for over half a century.

An electrician must develop a number of basic craft skills particular to his own trade, but he also requires some of the skills used in many other trades. An electrician's tool-kit will reflect both the specific and general nature of the work.

The basic tools required by an electrician are those used in the stripping and connecting of conductors.

These are pliers, side cutters, knife and an assortment of screwdrivers, as shown in Fig. 2.72.

The tools required in addition to these basic implements will depend upon the type of installation work being undertaken. When wiring new houses or rewiring old ones, the additional tools required are those usually associated with a bricklayer and joiner. Examples are shown in Fig. 2.73.

When working on industrial installations, installing conduit and trunking, the additional tools required by an electrician would more normally be those associated with a fitter or sheet-metal fabricator, and examples are shown in Fig. 2.74.

Where special tools are required, for example those required to terminate mineral insulated (MI) cables or

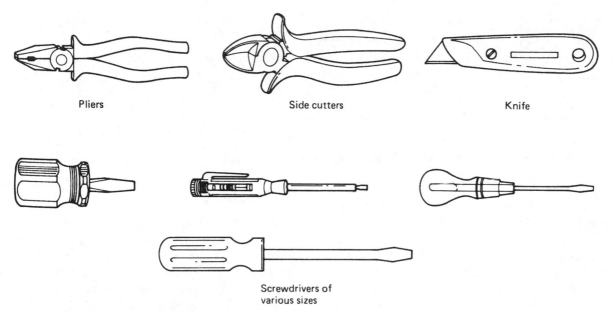

Pliers   Side cutters   Knife

Screwdrivers of various sizes

Fig. 2.72 The tools used for making electrical connections.

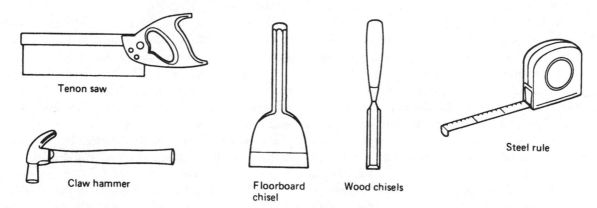

Tenon saw

Claw hammer

Floorboard chisel

Wood chisels

Steel rule

Fig. 2.73 Some additional tools required by an electrician engaged in house wiring.

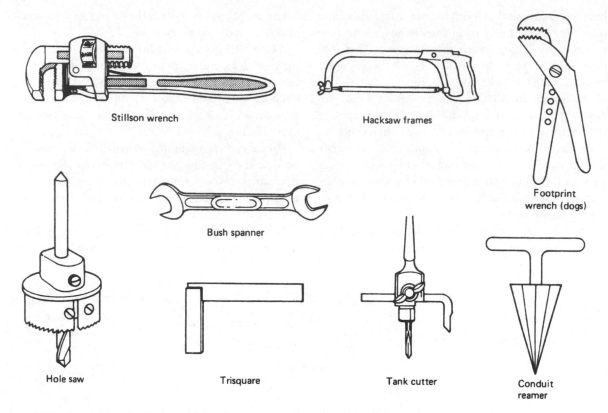

Stillson wrench

Hacksaw frames

Footprint
wrench (dogs)

Bush spanner

Hole saw

Trisquare

Tank cutter

Conduit
reamer

Fig. 2.74 Some additional tools required by an electrician engaged in industrial installations.

the bending and cutting tools for conduit and cable trays as shown in Fig. 2.75, they will often be provided by an employer but most hand-tools are provided by the electrician himself.

In general, good-quality tools last longer and stay sharper than those of inferior quality, but tools are very expensive to buy. A good set of tools can be assembled over the training period if the basic tools are bought first and the extended tool-kit acquired one tool at a time.

Another name for an installation electrician is a 'journeyman' electrician and, as the name implies, an electrician must be mobile and prepared to carry his tools from one job to another. Therefore, a good tool-box is an essential early investment, so that the right tools for the job can be easily transported.

Tools should be cared for and maintained in good condition if they are to be used efficiently and remain serviceable. Screwdrivers should have a flat squared off end and wood chisels should be very sharp. Access to a grindstone will help an electrician to maintain his tools in first-class condition. Additionally, wood chisels will require sharpening on an oilstone to give them a very sharp edge.

ELECTRICAL TOOLS

Portable electrical tools can reduce much of the hard work for any tradesman and increase his productivity. Electrical tools should be maintained in a good condition and be appropriate for the purpose for which they are used. The use of reduced voltage double insulation or an RCD can further increase safety without any loss of productivity. Some useful electrical tools are shown in Fig. 2.76.

Electric drills are probably used most frequently of all electrical tools. They may be used to drill metal or wood. Wire brushes are made which fit into the drill chuck for cleaning the metal. Variable-speed electric drills, which incorporate a vibrator, will also drill brick and concrete as easily as wood when fitted with a masonry drill bit.

Hammer drills give between two and three thousand impacts per minute and are used for drilling concrete walls and floors.

Cordless electric drills are also available which incorporate a rechargeable battery, usually in the handle. They offer the convenience of electric drilling when an

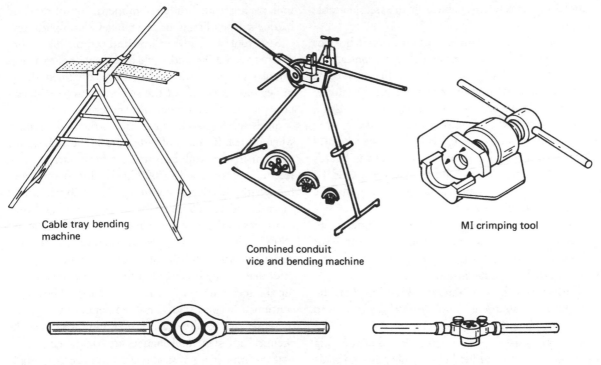

Cable tray bending machine

Combined conduit vice and bending machine

MI crimping tool

Conduit stocks and dies: two views

Fig. 2.75 Some special tools required by an electrician engaged in industrial installations.

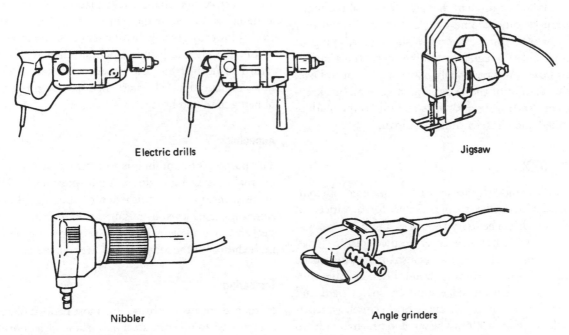

Electric drills

Jigsaw

Nibbler

Angle grinders

Fig. 2.76 Electrical hand-tools.

electrical supply is not available or if an extension cable is impractical.

Angle grinders are useful for cutting chases in brick or concrete. The discs are interchangeable. Silicon carbide discs are suitable for cutting slate, marble, tiles, brick and concrete, and aluminium oxide discs for cutting iron and steel such as conduit and trucking.

Jigsaws can be fitted with wood or metal cutting blades. With a wood cutting blade fitted they are useful for cutting across floorboards and skirting boards or any other application where a padsaw would be used. With a metal cutting blade fitted they may be used to cut trunking.

When a lot of trunking work is to be undertaken, an electric nibbler is a worthwhile investment. This nibbles out the sheet metal, is easily controllable and is one alternative to the jigsaw.

All tools must be used safely and sensibly. Cutting tools should be sharpened and screwdrivers ground to a sharp square end on a grindstone.

It is particularly important to check that the plug top and cables of hand held electrically powered tools and extension leads are in good condition. Damaged plug tops and cables must be repaired before you use them. All electrical power tools of 110 V and 230 V must be tested with a portable appliance tester (PAT) in accordance with the company's procedures, but probably at least once each year.

Tools and equipment that are left lying about in the workplace can become damaged or stolen and may also be the cause of people slipping, tripping or falling. Tidy up regularly and put power tools back in their boxes. You personally may have no control over the condition of the workplace in general, but keeping your own work area clean and tidy is the mark of a skilled and conscientious craftsman.

STEEL TOOLS

To understand why tools need to undergo different processes to make them suitable for a variety of purposes, it is necessary to look at the processes and the properties that they impart to the tools.

The hand-tools used by a craftsman are often made from steel. To cut through wood, brick or cable conductors, the cutting edge must be sharp, hard and tough, but the head of, say, a brick chisel, which is struck by a hammer, must be soft and tough in order to prevent cracking and splintering. Consequently, the

tools we use require different properties of softness and hardness depending upon the tool and its application.

Steel tools are alloys of iron and carbon. They contain between 0.7% and 1.7% of carbon and, to a large extent, the proportion of carbon contained determines the properties of the steel; the higher the carbon content, the stronger and harder the steel.

Steels undergo chemical and structural changes when heated. If a piece of carbon steel is heated steadily its temperature will rise at a uniform rate until it reaches approximately 700°C. At this point, even though the heating is continued, the temperature of the steel remains constant for a short period and then continues to rise at a slower rate until it reaches about 780°C. This pause in the temperature rise and the slowing down of the rate of temperature increase indicates that energy is being absorbed by the steel, bringing about chemical and structural changes. If heating is continued beyond 780°C the temperature will rise at the same rate as when first heated, indicating that the chemical and structural changes are completed.

If the same piece of steel is allowed to cool naturally after heating, the action is reversed and the metal returns to its normal composition.

Hardening

When the temperature of a steel tool is raised chemical and structural changes take place within the steel as described above. If the temperature of the heated steel is lowered quickly, by quenching in clean water or oil, the chemical and structural changes cannot return to normal but are 'frozen' or locked in their new configuration. The steel then has the properties of being hard and brittle.

Annealing

The purpose of annealing is to soften steel or relieve internal stresses and strains set up by previous working or use, making the finished tool more malleable. Annealing is achieved by raising the temperature of the steel and then allowing it to cool very slowly, probably in the hearth of the forge used to heat the metal.

Tempering

Hardened steel is too brittle for most hand-tools and is tempered to give the steel back some of its normal toughness and ductility. To temper small articles, one surface of the hardened steel is polished and then heated slowly. The polished surface will change colour as heat is absorbed because thin films of oxide form at different temperatures. Table 2.6 shows the connection between temperature and colour as the steel is quenched from its tempering temperature.

Table 2.6   Colour indication of temperature

Colour	Approximate temperature °C	Type of article to be tempered to this temperature
Pale straw	220–230	Metal turning tools
Dark straw	240–245	Taps, dies and drills
Yellow brown	250–255	Wood turning tools
Brown, just turning purple	260–265	Chisels and axes
Purple	270–280	Cold chisels, punches and knives
Blue	290–300	Screwdrivers

Some tools, cold chisels and punches for example, require a fairly soft tough head but a hardened and tempered cutting edge or point. In these cases the hardening and tempering can be carried out at the same time.

The working or cutting end and half the tool length are heated to the hardening temperature, which is indicated when the metal glows cherry red. The tool is then removed from the heat and the working end quenched up to half the distance heated. After quenching, the end of the tool is quickly polished and the tempering colours observed as the heat from the unquenched portion travels through the metal by conduction. When the required colour reaches the end, the whole tool is quenched.

Whether oil or water is used for quenching depends on the type of tool being tempered. Water quenching produces a very hard steel but it is liable to cause cracks and distortion. Oil quenching is less liable to cause these defects but produces a slightly softer steel. A faster and more even rate of cooling can be obtained if the steel is moved about in the cooling liquid. If only part of the steel is to be hardened, that part should be moved up and down in the liquid to avoid a sharp boundary between the soft and hard portions.

Safe working practice

Every year thousands of people have accidents at their place of work despite the legal requirements laid down by the Health and Safety Executive. Many people

recover quickly but an accident at work can result in permanent harm or even death.

At the very least, injuries hurt individuals. They may prevent you from doing the things you enjoy in your space time and they cost a lot of money, to you in loss of earnings and to your employer in loss of production and possibly damage to equipment. Your place of work may look harmless but it can be dangerous.

If there are five or more people employed by your company then the company must have its own safety policy as described earlier in this chapter. This must spell out the organisation and arrangements which have been put in place to ensure that you and your workmates are working in a safe place.

Your employer must also have carried out an assessment on the risks to your health and safety in the place where you are working. You should be told about the safety policy and risk assessment, for example you may have been given a relevant leaflet when you started work. We will discuss Risk Assessment in some detail in Chapter 3 of this book.

You have a responsibility under the Health and Safety at Work Act to:

■ learn how to work safely and to follow company procedures of work;
■ obey all safety rules, notices and signs;
■ not interfere with or misuse anything provided for safety;
■ report anything that seems damaged, faulty or dangerous;
■ behave sensibly, not play practical jokes and not distract other people at work;
■ walk sensibly and not run around the workplace;
■ use the prescribed walkways;
■ drive only those vehicles for which you have been properly trained and passed the necessary test;
■ not wear jewellery which could become caught in moving parts if you are using machinery at work;
■ always wear appropriate clothing and PPE if necessary.

Tidiness

Slips, trips and falls are still the major cause of accidents at work. To help prevent them:

■ keep work areas clean and tidy;
■ keep walkways clear;

- do not leave objects lying around blocking up walkways;
- clean up spills or wet patches on the floor straight away;

Personal hygiene

In the work environment, dirt and contact with chemicals and cleaning fluids may make you feel ill or cause unpleasant skin complaints. Therefore, you should always:

- wear appropriate personal protective equipment;
- wash your hands after using the toilet, after work and before you eat a meal, using soap and water or appropriate cleaners;
- dry your hands with the towel or dryer provided; do not use rags or your clothes;
- use barrier creams when they are provided to protect your skin;
- obtain medical advice about any skin complaint such as rashes, blisters or ulcers and tell your supervisor of the problems being experienced.

Ergonomics

Ergonomics is the scientific study of the efficiency of workers at work and the conditions required for them to achieve their maximum efficiency. To be efficient at work, workers must follow the workplace rules as described above. However, they must equally be *provided* with the tools and equipment which will enable them to be efficient in an environment which is safe and clean. For example:

- Hygiene and welfare facilities must be in place and appropriate for the workforce, such as toilets and washbasins, soap and towels, lockers, drinking water and adequate facilities for taking food and refreshment.
- Premises, furniture and fittings must be kept clean and spillage must be cleaned up.
- Floors and gangways must be clear and safe to use.
- Equipment being used by workers must be designed for good health, such as seats and machine controls designed for best control and posture.
- Don't lift a load manually if it is more appropriate to use a mechanical aid.

- The workplace must be a safe place to work or, where particular hazards exist, they must be clearly identified.

We will look at the hazards associated with working above ground and good manual handling techniques in Chapter 3 of this book.

Exercises

1 Solar energy heats the earth by:
   (a) conduction only
   (b) convection only
   (c) radiation only
   (d) conduction and convection.
2 The heat which is transferred to a room from an oil-filled radiator is by:
   (a) conduction only
   (b) convection only
   (c) radiation only
   (d) radiation and convection.
3 A temperature of 25° on the Celsius scale is equal to a temperature on the Kelvin scale of:
   (a) $-248\,\mathrm{K}$
   (b) $187\,\mathrm{K}$
   (c) $248\,\mathrm{K}$
   (d) $298\,\mathrm{K}$.
4 The absolute zero, $0\,\mathrm{K}$, is equal to a temperature reading on the Celsius scale of:
   (a) $-273°\mathrm{C}$
   (b) $-100°\mathrm{C}$
   (c) $32°\mathrm{C}$
   (d) $212°\mathrm{C}$.
5 The pre-set temperature of an immersion heater is maintained by a:
   (a) thermometer
   (b) rheostat
   (c) thermostat
   (d) simmerstat.
6 The temperature of a boiling ring is controlled by a:
   (a) thermometer
   (b) rheostat
   (c) thermostat
   (d) simmerstat.
7 The SI units of length, resistance, and power are:
   (a) millimetre, ohm, kilowatt
   (b) centimetre, ohm, watt
   (c) metre, ohm, watt
   (d) kilometre, ohm, kilowatt.

8 The current taken by a $10\,\Omega$ resistor when connected to a 230 V supply is:
 (a) 41 mA
 (b) 2.3 A
 (c) 23 A
 (d) 230 A.

9 The resistance of an element which takes 12 A from a 230 V supply is:
 (a) $2.88\,\Omega$
 (b) $5\,\Omega$
 (c) $12.24\,\Omega$
 (d) $19.16\,\Omega$.

10 A $12\,\Omega$ lamp was found to be taking a current of 2 A at full brilliance. The voltage across the lamp under these conditions was:
 (a) 6 V
 (b) 12 V
 (c) 24 V
 (d) 240 V.

11 The resistance of 100 m of $1\,mm^2$ cross-section copper cable of resistivity $7.5 \times 10^{-9}\,\Omega m$ will be:
 (a) $1.75\,m\Omega$
 (b) $1.75\,\Omega$
 (c) $17.5\,\Omega$
 (d) $17.5\,k\Omega$.

12 The resistance of a motor field winding at 0°C was found to be $120\,\Omega$. Find its new resistance at 20°C if the temperature coefficient of the winding is $0.004\,\Omega/\Omega$°C.
 (a) $116.08\,\Omega$
 (b) $120.004\,\Omega$
 (c) $121.08\,\Omega$
 (d) $140.004\,\Omega$.

13 The resistance of a motor field winding was found to be $120\,\Omega$ at an ambient temperature of 20°C. If the temperature coefficient of resistance is $0.004\,\Omega/\Omega$°C the resistance of the winding at 60°C will be approximately:
 (a) $102\,\Omega$
 (b) $120\,\Omega$
 (c) $130\,\Omega$
 (d) $138\,\Omega$.

14 A capacitor is charged by a steady current of 5 mA for 10 s. The total charge stored on the capacitor will be:
 (a) 5 mC
 (b) 50 mC
 (c) 5 C
 (d) 50 C.

15 When 100 V was connected to a $20\,\mu F$ capacitor the charge stored was:
 (a) 2 mC
 (b) 5 mC
 (c) 20 mC
 (d) 100 mC.

16 An air dielectric capacitor is often used:
 (a) for power-factor correction of fluorescents
 (b) for tuning circuits
 (c) when correct polarity connections are essential
 (d) when only a very small physical size can be accommodated by the circuit enclosure.

17 An electrolytic capacitor:
 (a) is used for power-factor correction in fluorescents
 (b) is used for tuning circuits
 (c) must only be connected to the correct polarity
 (d) has a small capacitance for a large physical size.

18 A paper dielectric capacitor is often used:
 (a) for power-factor correction in fluorescents
 (b) for tuning circuits
 (c) when correct polarity connections are essential
 (d) when only a small physical size can be accommodated in the circuit enclosure.

19 A current flowing through a solenoid sets up a magnetic flux. If an iron core is added to the solenoid while the current is maintained at a constant value the magnetic flux will:
 (a) remain constant
 (b) totally collapse
 (c) decrease in strength
 (d) increase in strength.

20 Resistors of $6\,\Omega$ and $3\,\Omega$ are connected in series. The combined resistance value will be:
 (a) $2\,\Omega$
 (b) $3.6\,\Omega$
 (c) $6.3\,\Omega$
 (d) $9\,\Omega$.

21 Resistors of $3\,\Omega$ and $6\,\Omega$ are connected in parallel. The equivalent resistance will be:
 (a) $2\,\Omega$
 (b) $3.6\,\Omega$
 (c) $6.3\,\Omega$
 (d) $9\,\Omega$.

22 Three resistors of 24, 40 and $60\,\Omega$ are connected in series. The total resistance will be:
 (a) $12\,\Omega$
 (b) $26.4\,\Omega$
 (c) $44\,\Omega$
 (d) $124\,\Omega$.

23 Resistors of 24, 40 and 60 Ω are connected together in parallel. The effective resistance of this combination will be:
(a) 12 Ω
(b) 26.4 Ω
(c) 44 Ω
(d) 124 Ω.

24 Two identical resistors are connected in series across a 12 V battery. The voltage drop across each resistor will be:
(a) 2 V
(b) 3 V
(c) 6 V
(d) 12 V.

25 Two identical resistors are connected in parallel across a 24 V battery. The voltage drop across each resistor will be:
(a) 6 V
(b) 12 V
(c) 24 V
(d) 48 V.

26 A 6 Ω resistor is connected in series with a 12 Ω resistor across a 36 V supply. The current flowing through the 6 Ω resistor will be:
(a) 2 A
(b) 3 A
(c) 6 A
(d) 9 A.

27 A 6 Ω resistor is connected in parallel with a 12 Ω resistor across a 36 V supply. The current flowing through the 12 Ω resistor will be:
(a) 2 A
(b) 3 A
(c) 6 A
(d) 9 A.

28 The total power dissipated by a 6 Ω and 12 Ω resistor connected in parallel across a 36 V supply will be:
(a) 72 W
(b) 324 W
(c) 576 W
(d) 648 W.

29 Three resistors are connected in series and a current of 10 A flows when they are connected to a 100 V supply. If another resistor of 10 Ω is connected in series with the three series resistors the current carried by this resistor will be:
(a) 4 A
(b) 5 A

(c) 10 A
(d) 100 A.

30 The rms value of a sinusoidal waveform whose maximum value is 100 V will be:
(a) 63.7 V
(b) 7.071 V
(c) 100 V
(d) 100.67 V.

31 The average value of a sinusoidal alternating current whose maximum value is 10 A will be:
(a) 6.37 A
(b) 7.071 A
(c) 10 A
(d) 10.67 A.

32 Capacitors of 24, 40 and 60 μF are connected in series. The equivalent capacitance will be:
(a) 12 μF
(b) 44 μF
(c) 76 μF
(d) 124 μF

33 Capacitors of 24, 40 and 60 μF are connected in parallel. The total capacitance will be:
(a) 12 μF
(b) 44 μF
(c) 76 μF
(d) 124 μF.

34 If we assume the acceleration due to gravity to be 10 m/s^2 a 50 kg bag of cement falling to the ground will exert a force of:
(a) 5 N
(b) 50 N
(c) 100 N
(d) 500 N.

35 The work done by a man carrying a 50 kg bag of cement up a 10 m ladder, assuming the acceleration due to gravity to be 10 m/s^2, will be:
(a) 50 J
(b) 500 J
(c) 5000 J
(d) 10 000 J.

36 A building hoist is to be used to raise sixty 50 kg bags of cement to the top of a 100 m high building in 1 minute. Assuming the acceleration due to gravity to be 10 m/s^2, the size of the hoist motor would be:
(a) 10 kW
(b) 50 kW
(c) 60 kW
(d) 100 kW.

37 A passenger lift has the capacity to raise 500 kg at the rate of 2 m/s. Assuming the acceleration due to gravity to be 10 m/s^2, the rating of the lift motor will be:
(a) 5 kW
(b) 10 kW
(c) 50 kW
(d) 100 kW.

38 A conduit bending machine operates on the principle of a lever of the
(a) first class
(b) second class
(c) third class
(d) fourth class.

39 A sack truck is a machine which operates on the principle of a lever of the
(a) first class
(b) second class
(c) third class
(d) fourth class.

40 A vehicle with a high centre of gravity will exhibit
(a) high stability
(b) low stability
(c) neutral equilibrium
(d) stable equilibrium.

41 A block and tackle pulley system contains three pulleys in the upper sheath and two in the lower sheath to which the load is attached. The velocity ratio of the system is:
(a) 2
(b) 3
(c) 5
(d) 6.

42 The MA and VR of a particular machine were found to be 3 and 4, respectively. The efficiency of the machine is:
(a) 12%
(b) 13.3%
(c) 70%
(d) 75%.

43 A machine of 60% efficiency has an MA of 3 and therefore the VR is:
(a) 5
(b) 6
(c) 7
(d) 8.

44 The efficiency of an 8 pulley wheel system is found to be 75% and therefore the MA is:
(a) 3

(b) 4
(c) 5
(d) 6.

45 Which particular machine would be most suitable for raising a 200 kg lift motor 9 m up a 12 m lift shaft:
(a) a screw jack
(b) a lever
(c) a pulley system
(d) a wedge.

46 Which, if any, of the following machines use the principle of an inclined plane in their operation:
(a) the lever
(b) the screw jack
(c) the pulley system
(d) none of the above.

47 Any motor converts:
(a) electrical energy to power
(b) electrical energy to mechanical energy
(c) mechanical energy to power
(d) mechanical energy to electrical energy.

48 Any generator converts:
(a) electrical energy to power
(b) electrical energy to mechanical energy
(c) mechanical energy to power
(d) mechanical energy to electrical energy.

49 An electric motor operates on the principle of:
(a) Ohm's law
(b) the three effects of an electric current
(c) the forces acting upon a conductor in a magnetic field
(d) a cage rotor induction motor.

50 A centre zero voltmeter is connected to the commutator of a simple single-loop generator. If the loop remains stationary the meter will:
(a) read zero
(b) give a positive value
(c) give a negative value
(d) oscillate about the zero point.

51 A centre zero voltmeter is connected to the slip rings of a simple single-loop generator. If the loop is rotated through one complete revolution the meter will:
(a) continue to read zero
(b) move from zero to positive and back to zero twice
(c) move from zero to positive, then to negative and finally return to zero

(d) move from zero to positive and maintain the positive value.

52 An oscilloscope is connected to the slip rings of a simple generator. If the coil is continuously rotated the oscilloscope will show:
(a) a zero voltage
(b) a positive d.c. voltage
(c) a unidirectional waveform
(d) a sinusoidal waveform.

53 A series d.c. motor has the characteristic of:
(a) constant speed about 5% below synchronous speed
(b) start winding 90° out of phase with the run winding
(c) low starting torque but almost constant speed
(d) high starting torque and a speed which varies with load.

54 A shunt motor has the characteristic of:
(a) constant speed about 5% below synchronous speed
(b) start winding 90° out of phase with the run winding
(c) low starting torque but almost constant speed
(d) high starting torque and a speed which varies with load.

55 One advantage of all d.c. machines is:
(a) that they are almost indestructable
(b) that starters are never required
(c) that they may be operated on a.c. or d.c. supplies
(d) the ease with which speed may be controlled.

56 One advantage of a series d.c. motor is that:
(a) it is almost indestructable
(b) starters are never required
(c) it may be operated on a.c. or d.c. supplies
(d) speed is constant at all loads.

57 A d.c. shunt motor would normally be used for a:
(a) domestic oven fan motor
(b) portable electric drill motor
(c) constant speed lathe motor
(d) record turntable drive motor.

58 A d.c. series motor would normally be used for a:
(a) domestic oven fan motor
(b) portable electric drill motor
(c) constant speed lathe motor
(d) record turntable drive motor.

59 The core of a transformer is laminated to:
(a) reduce cost
(b) reduce copper losses

(c) reduce hysteresis loss
(d) reduce eddy current loss.

60 The transformation ratio of a step-down transformer is 20:1. If the primary voltage is 230 V the secondary voltage will be:
(a) 2.3 V
(b) 11.5 V
(c) 20 V
(d) 23 V.

61 The cables which can best withstand high temperatures are:
(a) MI cables
(b) PVC cables with asbestos oversleeves
(c) PVC/SWA cables
(d) PVC cables in galvanized conduit.

62 All modern power stations require two of the following for their efficient operation:
(a) an adequate supply of fuel
(b) access to a main rail link
(c) a large, highly skilled labour force
(d) lots of cooling water.

63 Which of the following energy sources may be classified as 'renewable energy'?
(a) coal
(b) oil
(c) nuclear
(d) hydro.

64 Electricity is generated in a modern power station at:
(a) 230 V
(b) 400 V
(c) 25 V
(d) 132 kV.

65 Electricity is distributed on the National Grid at:
(a) 230 V
(b) 400 V
(c) 25 kV
(d) 132 kV.

66 The highest transmission line voltage in Britain is:
(a) 240 kV
(b) 415 kV
(c) 400 kV
(d) 1000 kV.

67 The voltage used for transmission on the Grid is transformed to a very high voltage because:
(a) this increases the line current
(b) the p.f. of the line is improved at high values
(c) the line resistance is increased
(d) the line efficiency is increased.

68 The national transmission network in the UK is known as:
   (a) National Power
   (b) the National Grid System
   (c) the National Coal Board
   (d) the British Transmission System.

69 The generation of electricity in England and Wales is the responsibility of:
   (a) the generating companies
   (b) regional electricity companies
   (c) the National Coal Board
   (d) Parliament.

70 The distribution of electricity to individual consumers is the responsibility of:
   (a) the generating companies
   (b) regional electricity companies
   (c) the National Coal Board
   (d) Parliament.

71 A 230 V single-phase supply may be obtained from a 400 V three-phase supply by connecting the load between:
   (a) any line and neutral
   (b) any line and live
   (c) any two phases
   (d) neutral and earth.

72 The phase to neutral voltage of a 400 V, three-phase, four-wire supply is:
   (a) 230 V
   (b) 400 V
   (c) 11 kV
   (d) 132 kV.

73 The distribution transformer is the main source of supply to a domestic dwelling. This transformer:
   (a) generates the required electrical power
   (b) is star-connected to provide a single-phase supply
   (c) provides a d.c. supply
   (d) steps up the voltage to 230 V.

74 Discriminative operation of the excess current protection device of an installation means that when a fault occurs:
   (a) the supply is cut off from the installation
   (b) the main protection device opens
   (c) the faulty circuit only is disconnected from the supply
   (d) the distribution fuse board only is disconnected from the supply.

75 Overload or overcurrent protection is offered by a:
   (a) transistor
   (b) transformer
   (c) functional switch
   (d) circuit breaker.

76 The current rating of a protective device is the current which:
   (a) it will carry continuously without deterioration
   (b) will cause the device to operate
   (c) will cause the device to operate within 4 hours
   (d) is equal to the fusing factor.

77 Describe briefly how heat is transferred by conduction, convection and radiation.

78 Describe the operation of a thermostat and a simmerstat, and give one practical application for each.

79 Describe and give one practical example of the three effects of an electric current.

80 Describe with sketches the meaning of the terms *frequency* and *period* as applied to an a.c. wave-form.

81 A lift motor is to be used to raise a constant load of 2000 kg at a speed of 0.3 m/s. The motor is supplied at 400 V and works at a power-factor of 0.8 lagging. Find the current taken by the motor, assuming the acceleration due to gravity to be 9.81 m/s^2.

82 With the aid of sketches, describe the construction of:
   (a) a double-wound transformer
   (b) an auto-transformer.
   State the losses which occur in a transformer.

83 Describe what is meant by acid rainfall, and how it causes problems. Who is claimed to be responsible for creating acid rainfall, and what steps are being taken to eliminate it?

84 Describe how waste products from nuclear power stations are transported for reprocessing. How are long-term highly active waste products stored safely?

85 Describe the main advantages of generating electricity by nuclear power stations.

86 Describe how electricity may be generated by one renewable energy source. State the advantages and disadvantages over present methods of generating commercial quantities of electricity.

87 State five advantages of using electricity as a means of harnessing energy.

88 Describe how electricity is distributed from the power station to a domestic consumer's terminals. State the voltages used at each stage and describe the advantages and disadvantages of overhead and underground cables.

89 With the aid of a sketch, describe the construction of a semi-enclosed fuse and a cartridge fuse. State the advantages and disadvantages of each type of fuse and identify one typical application for each device.

90 With the aid of a sketch, describe the construction of a cartridge fuse and an HBC fuse. State the advantages of each type of fuse and identify one typical application for each device.

91 Describe the operation of a MCB:
   (a) when carrying the rated current
   (b) during overcurrent conditions
   (c) during short-circuit conditions.
   State the advantages of an MCB compared with a fuse.

92 Describe the operation of an RCD and state one application for this device.

HEALTH AND SAFETY AND ELECTRICAL PRINCIPLES

—

HEALTH AND SAFETY APPLICATIONS
Hazard risk assessment

In Chapter 1 of this book we looked at some of the health and safety rules and regulations. In particular we now know that the Health and Safety at Work Act is the most important piece of recent legislation because it places responsibilities for safety at work on both employers and employees. This responsibility is enforceable by law. We know what the regulations say about the control of substances, which might be hazardous to our health at work, because we briefly looked at the COSHH Regulations 2002 in Chapter 1. We also know that if there is a risk to health and safety at work our employer must provide personal protective equipment (PPE) free of charge, for us to use so that we are safe at work. The law is in place, we all apply the principles of health and safety at work and we always wear the appropriate PPE, so what are the risks? Well, getting injured at work is not a pleasant subject to think about but each year about 300 people in Great Britain lose their lives at work. In addition, there are about 158 000 non-fatal injuries reported to the Health and Safety Executive each year and an estimated 2.2 million people suffer ill health caused by, or made worse by, work. It is a mistake to believe that these things only happen in dangerous occupations such as deep sea diving, mining and quarrying, fishing industry, tunnelling and fire-fighting or that it only happens in exceptional circumstances such as would never happen in your workplace. This is not the case. Some basic

thinking and acting beforehand, could have prevented most of these accident statistics, from happening.

The most common causes of accidents are:

- slips, trips and falls;
- manual handling, that is moving objects by hand;
- using equipment, machinery or tools;
- storage of goods and materials which then become unstable;
- fire;
- electricity;
- mechanical handling.

To control the risk of an accident we usually:

- eliminate the cause;
- substitute a procedure or product with less risk;
- enclose the dangerous situation;
- put guards around the hazard;
- use safe systems of work;
- supervise, train and give information to staff;
- if the hazard cannot be removed or minimised then provide PPE.

Let us now look at the application of some of the procedures that make the workplace a safer place to work but first of all I want to explain what I mean when I use the words hazard and risk.

HAZARD AND RISK

A hazard is something with the 'potential' to cause harm, for example, chemicals, electricity or working above ground.

A risk is the 'likelihood' of harm actually being done.

Competent persons are often referred to in the Health and Safety at Work Regulations, but who is 'competent'? For the purposes of the Act, a competent person is anyone who has the necessary technical skills, training and expertise to safely carry out the particular activity. Therefore, a competent person dealing with a hazardous situation reduces the risk.

Think about your workplace and at each stage of what you do, think about what might go wrong. Some simple activities may be hazardous. Here are some typical activities where accidents might happen.

Typical activity	Potential hazard
Receiving materials	Lifting and carrying
Stacking and storing	Falling materials
Movement of people	Slips, trips and falls
Building maintenance	Working at heights or in confined spaces
Movement of vehicles	Collisions

How high are the risks? Think about what might be the worst result, is it a broken finger or someone suffering permanent lung damage or being killed? How likely is it to happen? How often is that type of work carried out and how close do people get to the hazard? How likely is it that something will go wrong?

How many people might be injured if things go wrong. Might this also include people who do not work for your company?

Employers of more than five people must document the risks at work and the process is known as Hazard Risk Assessment.

HAZARD RISK ASSESSMENT – THE PROCESS

The Management of Health and Safety at Work Regulations 1999 tells us that employers must systematically examine the workplace, the work activity and the management of safety in the establishment through a process of risk assessments. A record of all significant risk assessment findings must be kept in a safe place and be made available to an HSE Inspector if required. Information based on the risk assessment findings must be communicated to relevant staff and if changes in work behaviour patterns are recommended in the interests of safety, then they must be put in place.

So risk assessment must form a part of any employer's robust policy of health and safety. However, an employer only needs to 'formally' assess the significant risks. He is not expected to assess the trivial and minor types of household risks. Staff are expected to read and to act upon these formal risk assessments and they are unlikely to do so enthusiastically if the file is full of trivia. An assessment of risk is nothing more than a careful examination of what, in your work, could cause harm to people. It is a record that shows whether sufficient precautions have been taken to prevent harm.

The HSE recommends five steps to any risk assessment.

Step 1
Look at what might reasonably be expected to cause harm. Ignore the trivial and concentrate only on significant hazards that could result in serious harm or injury. Manufacturers data sheets or instructions can also help you spot hazards and put risks in their true perspective.

Step 2
Decide who might be harmed and how. Think about people who might not be in the workplace all the time – cleaners, visitors, contractors or maintenance personnel. Include members of the public or people who share the workplace. Is there a chance that they could be injured by activities taking place in the workplace.

Step 3
Evaluate what is the risk arising from an identified hazard. Is it adequately controlled or should more be done? Even after precautions have been put in place, some risk may remain. What you have to decide, for each significant hazard, is whether this remaining risk is low, medium or high. First of all, ask yourself if you have done all the things that the law says you have got to do. For example, there are legal requirements on the prevention of access to dangerous machinery. Then ask yourself whether generally accepted industry standards are in place, but do not stop there – think for yourself, because the law also says that you must do what is reasonably practicable to keep the workplace safe. Your real aim is to make all risks small by adding precautions, if necessary.

If you find that something needs to be done, ask yourself:

(a) Can I get rid of this hazard altogether?
(b) If not, how can I control the risk so that harm is unlikely?

Only use personal protective equipment (PPE) when there is nothing else that you can reasonably do.

If the work that you do varies a lot, or if there is movement between one site and another, select those hazards which you can reasonably foresee, the ones that apply to most jobs and assess the risks for them. After that, if you spot any unusual hazards when you get on site, take what action seems necessary.

Step 4

Record your findings and say what you are going to do about risks that are not adequately controlled. If there are fewer than five employees you do not need to write anything down but if there are five or more employees, the significant findings of the risk assessment must be recorded. This means writing down the more significant hazards and assessing if they are adequately controlled and recording your most important conclusions. Most employers have a standard risk assessment form which they use such as that shown in Fig. 3.1 but any format is suitable. The important thing is to make a record.

There is no need to show how the assessment was made, providing you can show that:

1 a proper check was made,
2 you asked those who might be affected,
3 you dealt with all obvious and significant hazards,
4 the precautions are reasonable and the remaining risk is low,
5 you informed your employees about your findings.

Risk assessments need to be *suitable* and *sufficient*, not perfect. The two main points are:

1 Are the precautions reasonable?
2 Is there a record to show that a proper check was made?

File away the written Assessment in a dedicated file for future reference or use. It can help if an HSE Inspector questions the company's precautions or if the company becomes involved in any legal action. It shows that the company has done what the law requires.

Step 5

Review the assessments from time to time and revise them if necessary.

COMPLETING A RISK ASSESSMENT

When completing a risk assessment such as that shown in Fig. 3.1, do not be over complicated. In most firms in the commercial, service and light industrial sector, the hazards are few and simple. Checking them is commonsense but necessary.

Step 1

List only hazards which you could reasonably expect to result in significant harm under the conditions prevailing in your workplace. Use the following examples as a guide:

- Slipping or tripping hazards, e.g. from poorly maintained or partly installed floors and stairs.
- Fire, e.g. from flammable materials you might be using, such as solvents.
- Chemicals, e.g. from battery acid.
- Moving parts of machinery, e.g. blades.
- Rotating parts of handtools, e.g. drills.
- Accidental discharge of cartridge operated tools.
- High pressure air from airlines, e.g. air powered tools.
- Pressure systems, e.g. steam boilers.
- Vehicles, e.g. fork lift trucks.
- Electricity, e.g. faulty tools and equipment.
- Dust, e.g. from grinding operations or thermal insulation.
- Fumes, e.g. from welding.
- Manual handling, e.g. lifting, moving or supporting loads.
- Noise levels too high, e.g. machinery.
- Poor lighting levels, e.g. working in temporary or enclosed spaces.
- Low temperatures, e.g. working outdoors or in refrigeration plant.
- High temperatures, e.g. working in boiler rooms or furnaces.

Step 2

Decide who might be harmed, do not list individuals by name. Just think about groups of people doing similar work or who might be affected by your work:

- Office staff.
- Electricians.
- Maintenance personnel.
- Other contractors on site.
- Operators of equipment.
- Cleaners.
- Members of the public.

Pay particular attention to those who may be more vulnerable, such as

- staff with disabilities,
- visitors,
- young or inexperienced staff,
- people working in isolation or enclosed spaces.

HAZARD RISK ASSESSMENT	FLASH-BANG ELECTRICAL CO.
For Company name or site: _____ Address: _____ _____	Assessment undertaken by:_____ Signed: _____ Date: _____
STEP 5 Assessment review date: _____	
STEP 1 List the hazards here	STEP 2 Decide who might be harmed
STEP 3 Evaluate (what is) the risk – is it adequately controlled? State risk level as low, medium or high	STEP 4 Further action – what else is required to control any risk identified as medium or high?

Fig. 3.1 Hazard risk assessment standard form.

Step 3

Calculate what is the risk – is it adequately controlled? Have you already taken precautions to protect against the hazards which you have listed in Step 1. For example,

■ have you provided adequate information to staff,
■ have you provided training or instruction.

Do the precautions already taken

■ meet the legal standards required,
■ comply with recognised industrial practice,
■ represent good practice,
■ reduce the risk as far as is reasonably practicable.

If you can answer 'yes' to the above points then the risks are adequately controlled, but you need to state the precautions you have put in place. You can refer to company procedures, company rules, company practices etc., in giving this information. For example, if we consider there might be a risk of electric shock from using electrical power tools, then the risk of a shock will be *less* if the company policy is to PAT test all power tools each year and to fit a label to the tool showing that it has been tested for electrical safety. If the stated company procedure is to use battery drills whenever possible, or 110 V drills when this is not possible, and to *never* use 230 V drills, then this again will reduce the risk. If a policy such as this is written down in the company Safety Policy Statement, then you can simply refer to the appropriate section of the Safety Policy Statement and the level of risk will be low. (Note: PAT testing is described in Chapter 2 of Advanced Electrical Installation Work.)

Step 4

Further action – what more could be done to reduce those risks which were found to be inadequately controlled?

You will need to give priority to those risks that affect large numbers of people or which could result in serious harm. Senior managers should apply the principles below when taking action, if possible in the following order:

1 Remove the risk completely.
2 Try a less risky option.
3 Prevent access to the hazard (e.g. by guarding).
4 Organise work differently in order to reduce exposure to the hazard.

5 Issue PPE.
6 Provide welfare facilities (e.g. washing facilities for removal of contamination and first aid).

Any hazard identified by a risk assessment as *high risk* must be brought to the attention of the person responsible for Health and Safety within the company. Ideally, in Step 4 of the Risk Assessment you should be writing 'No further action is required. The risks are under control and identified as low risk'.

The assessor may use as many standard Hazard Risk Assessment forms, such as that shown in Fig. 3.1, as the assessment requires. Upon completion they should be stapled together or placed in a plastic wallet and stored in the dedicated file.

You might like to carry out a risk assessment on a situation you are familiar with at work, using the standard form of Fig. 3.1, or your employer's standard forms. Alternatively you might like to complete the VDU workstation risk assessment checklist given in the next section.

We all use computers, and you might find it interesting to carry out a risk assessment of the computer workstation you use most, either at home, work or college, just for fun and to get an idea of how to carry out a risk assessment.

VDU operation hazards

Those who work at Supermarket checkouts, assemble equipment or components, or work for long periods at a visual display unit (VDU) and keyboard can be at risk because of the repetitive nature of the work. The hazard associated with these activities is a medical condition called 'upper limb disorders'. The term covers a number of related medical conditions.

HEALTH AND SAFETY (DISPLAY SCREEN EQUIPMENT) REGULATIONS 1992

To encourage employers to protect the health of their workers and reduce the risks associated with VDU work, the Health and Safety Executive (HSE) have introduced the Health and Safety (Display Screen Equipment) Regulations 1992. The regulations came into force on 1st January 1993, and employers who use standard office VDUs must show that they have taken steps to comply with the regulations.

So who is affected by the regulations? The regulations identify employees who use VDU equipment as 'users' if they:

■ use a VDU more or less continuously on most days;
■ use a VDU more or less continuously for periods of an hour or more each day;
■ need to transfer information quickly to or from the screen;
■ need to apply high levels of attention or concentration to information displayed on a screen;
■ are very dependent upon VDUs or have little choice about using them.

All VDU users must be trained to use the equipment safely and protect themselves from upper limb disorders, temporary eyestrain, headaches, fatigue and stress.

To comply with the regulations an employer must:

■ train users of VDU equipment and those who will carry out a risk assessment;
■ carry out a workstation risk assessment;
■ plan changes of activities or breaks for users;
■ provide eye and eyesight testing for users;
■ make sure new workstations comply with the regulations in the future;
■ give users information on the above.

User training

Good user training will normally cover the following topics:

■ the operating hazards and risks as describe above;
■ the importance of good posture and changing position as shown in Fig. 3.2;
■ how to adjust furniture to avoid risks;
■ how to organise the workstation to avoid awkward or repeated stretching movements;

■ how to avoid reflections and glare on the monitor screen;
■ how to organise working routines so that there is a change of activity or a break;
■ how to adjust and clean the monitor screen;
■ how a user might contribute to a workstation risk assessment;
■ who to contact if problems arise.

When carrying out user training, the trainer might want to consider using a video, a computer based training programme, discussions or seminars or the HSE employee leaflet *Working with VDUs* which can be obtained from the address given in Appendix B.

Workstation risk assessment

A simple way to carry out a workstation risk assessment is to use a checklist such as that shown later in this section. Users can work through the checklist themselves. They know what the problems at their workstation are and whether they are comfortable or not. A trainer/assessor should then check the completed checklist and resolve the problems which the user cannot solve. For example, users may not know how the adjustment mechanism actually operates on their chair – a shorter user may benefit from a footrest as shown in Fig. 3.2, or document holder may be more convenient for word processing users as shown in Fig. 3.3.

Breaks

Breaking up long spells of display screen work helps to prevent fatigue and upper limb problems. Where possible encourage VDU users to carry out other tasks such as taking telephone calls, filing and photocopying.

Fig. 3.2 Examples of good posture when using VDU equipment.

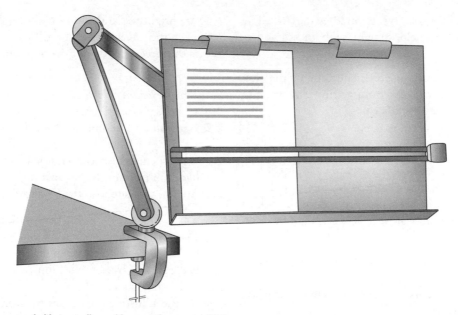

Fig. 3.3 A document holder typically used by a word processing VDU operator.

Otherwise, plan for users to take breaks away from the VDU screen if possible. The length of break required is not fixed by the law; the time will vary depending upon the work being done. Breaks should be taken before users become tired and short frequent breaks are better than longer infrequent ones.

Eye and eyesight testing

VDU users and those who are to become users of VDU equipment can request an eye and eyesight test that is free of charge to them. If the test shows that they need glasses specifically to carry out their VDU work, then their employer must pay for a basic pair of frames and lenses. Users are also entitled to further tests at regular intervals but if the user's normal glasses are suitable for VDU work, then the employer is not required to pay for them.

Workstations

Make sure that new workstations comply with the regulations when:

- major changes to the workstation display screen equipment, furniture or software are made;
- new users start work or change workstations;
- workstations are re-sited;
- the nature of the work changes considerably.

Users, trainers and assessors should focus on those aspects which have changed. For example:

- if the location of the workstation has changed, is the lighting adequate, is lighting or sunlight now reflecting off the display unit?
- different users have different needs – replacing a tall user with a short user may mean that a footrest is required;
- users working from a number of source documents will need more desk space than users who are word processing.

A risk assessment should always be carried out on a new workstation or when a new operator takes over a workstation. Some questions cannot be answered until a user has had an opportunity to try the workstation. For example, does the user find the layout comfortable to operate, are there reflections on the screen at different times of the day as the sun moves around the building?

To be comfortable the operator should adjust the chair and equipment so that:

- Arms are horizontal and eyes are roughly at the height of the top of the VDU casing.
- Hands can rest on the work surface in front of the keyboard with fingers outstretched over the keys.
- Feet are placed flat on the floor – too much pressure on the backs of legs and knees may mean that a footrest is needed.

- The small of the back is supported by the chair. The back should be held straight with the shoulders relaxed.

The arms on the chair or obstructions under the desk must not prevent the user from getting close enough to the keyboard comfortably.

Information

Good employers, who comply with the Display Screen Equipment Regulations, should let their employees know what care has been taken to reduce the risk to their health and safety at work. Users should be given information on:

- the health and safety relating to their particular workstations;
- the risk assessments carried out and the steps taken to reduce risks;
- the recommended break times and changes in activity to reduce risks;
- the company procedures for obtaining eye and eyesight tests.

This information might be communicated to workers by:

- telling staff, for example, as part of an induction programme;
- circulating a booklet or leaflet to relevant staff;
- putting the information on a noticeboard;
- using a computer based information system, providing staff are trained in their use.

VDU WORKSTATION RISK ASSESSMENT CHECKLIST

Using a checklist such as that shown below or the more extensive checklist shown in the HSE book 'VDUs, An Easy Guide to the Regulations' is one way to assess workstation risks. You don't have to, but many employers find it a convenient method.

Risk factors are grouped under five headings and to each question the user should initially give a simple yes/no response. A 'yes' response means that no further action is necessary but a 'no' response will indicate that further follow-up action is required to reduce or eliminate risks to a user.

1. Is the display screen image clear?
1.1 Are the characters readable?   Y/N
1.2 Is the image free of flicker and movement? Y/N

1.3 Are brightness and contrast adjustable?   Y/N
1.4 Does the screen swivel and tilt?   Y/N
1.5 Is the screen free from glare and reflections?
                                                                   Y/N

2. Is the keyboard comfortable?
2.1 Is the keyboard tiltable?   Y/N
2.2 Can you find a comfortable keyboard position?
                                                                   Y/N
2.3 Is there enough space to rest your   Y/N
      hands in front of the keyboard?
2.4 Are the characters on the keys easily readable?
                                                                   Y/N

3. Does the furniture fit the work and user?
3.1 Is the work surface large enough?   Y/N
3.2 Is the surface free of reflections?   Y/N
3.3 Is the chair stable?   Y/N
3.4 Do the adjustment mechanisms work?   Y/N
3.5 Are you comfortable?   Y/N

4. Is the surrounding environment risk free?
4.1 Is there enough room to change position and
      vary movement?   Y/N
4.2 Are levels of light, heat and noise comfortable?
                                                                   Y/N
4.3 Does the air feel comfortable in terms of
      temperature and humidity?   Y/N

5. Is the software user friendly?
5.1 Can you comfortably use the software?   Y/N
5.2 Is the software suitable for the work task? Y/N
5.3 Have you had enough training?   Y/N

A copy of all risk assessments carried out should be placed in a dedicated file which can then be held by the trainer/assessor or other responsible person.

A copy of the full checklist can be found in the publication 'VDUs, an Easy Guide to the Regulations'. Other relevant publications include 'Display Screen Equipment Work and Guidance on Regulations L26' and 'Industry Advisory (General) leaflet IND(G) 36(L) 1993 Working with VDUs'. These and other Health and Safety Publications are available from the HSE; the address is given in Appendix B.

Manual handling

Manual handling is lifting, transporting or supporting loads by hand or by bodily force. The load might be

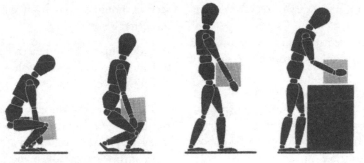

Fig. 3.4 Correct manual lifting and carrying procedure.

any heavy object, a printer, a visual display unit, a box of tools or a stepladder. Whatever the heavy object is, it must be moved thoughtfully and carefully, using appropriate lifting techniques if personal pain and injury are to be avoided. Many people hurt their back, arms and feet, and over one third of all three day reported injuries submitted to the HSE each year are the result of manual handling.

When lifting heavy loads, correct lifting procedures must be adopted to avoid back injuries. Figure 3.4 demonstrates the technique. Do not lift objects from the floor with the back bent and the legs straight as this causes excessive stress on the spine. Always lift with the back straight and the legs bent so that the powerful leg muscles do the lifting work. Bend at the hips and knees to get down to the level of the object being lifted, positioning the body as close to the object as possible. Grasp the object firmly and, keeping the back straight and the head erect, use the leg muscles to raise in a smooth movement. Carry the load close to the body. When putting the object down, keep the back straight and bend at the hips and knees, reversing the lifting procedure. A bad lifting technique will result in sprains, strains and pains. There have been too many injuries over the years resulting from bad manual handling techniques. The problem has become so serious that the Health and Safety Executive has introduced new legislation under the Health and Safety at Work Act 1974, the Manual Handling Operations Regulations 1992. Publications such as *Getting to Grips with Manual Handling* can be obtained from HSE Books; the address and Infoline are given in Appendix B.

Where a job involves considerable manual handling, employers must now train employees in the correct lifting procedures and provide the appropriate equipment necessary to promote the safe manual handling of loads.

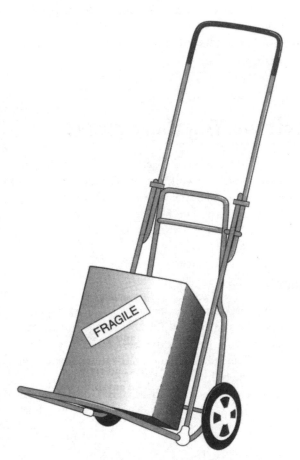

Fig. 3.5 Always use a mechanical aid to transport a load when available.

Consider some 'good practice' when lifting loads.

■ Do not lift the load manually if it is more appropriate to use a mechanical aid. Only lift or carry what you can easily manage.
■ Always use a trolley, wheelbarrow or truck such as those shown in Fig. 3.5 when these are available.

- Plan ahead to avoid unnecessary or repeated movement of loads.
- Take account of the centre of gravity of the load when lifting – the weight acts through the centre of gravity.
- Never leave a suspended load unsupervised.
- Always lift and lower loads gently.
- Clear obstacles out of the lifting area.
- Use the manual lifting techniques described above and avoid sudden or jerky movements.
- Use gloves when manual handling to avoid injury from rough or sharp edges.
- Take special care when moving loads wrapped in grease or bubble-wrap.
- Never move a load over other people or walk under a suspended load.

Safe working above ground

Working above ground level creates added dangers and slows down the work rate of the electrician. Every precaution should be taken to ensure that the working platform is appropriate for the purpose and in good condition.

LADDERS

It is advisable to inspect the ladder before climbing it. It should be straight and firm. All rungs and tie rods should be in position and there should be no cracks in the stiles. The ladder should not be painted since the paint may be hiding defects.

Extension ladders should be erected in the closed position and extended one section at a time. Each section should overlap by at least the number of rungs indicated below:

- Ladder up to 4.8 m length – 2 rungs overlap.
- Ladder up to 6.0 m length – 3 rungs overlap.
- Ladder over 6.0 m length – 4 rungs overlap.

The angle of the ladder to the building should be in the proportion 4 up to 1 out or 75° as shown in Fig. 3.6. The ladder should be lashed at the top and bottom to prevent unwanted movement and placed on firm and level ground. When ladders provide access to a roof or working platform the ladder must extend at least 1 m or 5 rungs above the landing place.

Short ladders may be carried by one person resting the ladder on the shoulder, but longer ladders should be carried by two people, one at each end, to avoid accidents when turning corners.

Long ladders or extension ladders should be erected by two people as shown in Fig. 3.7. One person stands

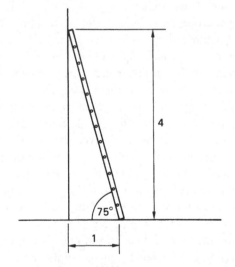

Fig. 3.6 A correctly erected ladder.

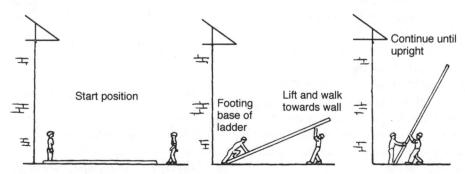

Fig. 3.7 Correct procedure for erecting long or extension ladder.

on or 'foots' the ladder, while the other person lifts and walks under the ladder towards the walls. When the ladder is upright it can be positioned in the correct place, at the correct angle and secured before being climbed.

TRESTLE SCAFFOLD

Figure 3.8 shows a trestle scaffold. Two pairs of trestles spanned by scaffolding boards provide a simple working platform. The platform must be at least two boards or 450 mm wide. At least one-third of the trestle must be above the working platform. If the platform is more than 2 m above the ground, toe-boards and guardrails must be fitted, and a separate ladder provided for access. The boards which form the working platform should be of equal length and not overhang the trestles by more than four times their own thickness. The maximum span of boards between trestles is:

- 1.3 m for boards 40 mm thick
- 2.5 m for boards 50 mm thick

Trestles which are higher than 3.6 m must be tied to the building to give them stability. Where anyone can fall more than 4.5 m from the working platform, trestles may not be used.

Fig. 3.8 A trestle scaffold.

MOBILE SCAFFOLD TOWERS

Mobile scaffold towers may be constructed of basic scaffold components or made from light alloy tube. The tower is built up by slotting the sections together until the required height is reached. A scaffold tower is shown in Fig. 3.9.

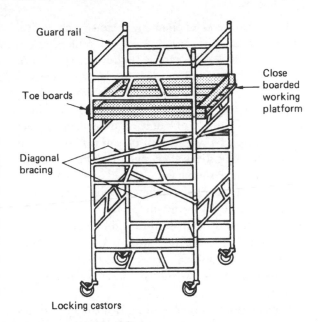

Fig. 3.9 A mobile scaffold tower.

If the working platform is above 2 m from the ground it must be closed-boarded and fitted with guardrails and toeboards. When the platform is being used, all four wheels must be locked. The platform must not be moved unless it is clear of tools, equipment and workers and should be pushed at the base of the tower and not at the top.

The stability of the tower depends upon the ratio of the base width to tower height. A ratio of base to height of 1 : 3 gives good stability. Outriggers can be used to increase stability by effectively increasing the base width. If outriggers are used then they must be fitted diagonally across all four corners of the tower and not on one side only. The tower must not be built more than 12 m high unless it has been specially designed for that purpose. Any tower higher than 9 m should be secured to the structure of the building to increase stability.

Access to the working platform of a scaffold tower should be by a ladder securely fastened vertically to the tower. Ladders must never be leaned against a tower since this might push the tower over.

Secure electrical isolation

Electric shock occurs when a person becomes part of the electrical circuit. The level or intensity of the shock will depend upon many factors, such as age, fitness and the circumstances in which the shock is received. The lethal level is approximately 50 mA, above which muscles contract, the heart flutters and breathing stops. A shock above the 50 mA level is therefore fatal unless the person is quickly separated from the supply. Below 50 mA only an unpleasant tingling sensation may be experienced or you may be thrown across a room or shocked enough to fall from a roof or ladder, but the resulting fall may lead to serious injury.

To prevent people receiving an electric shock accidentally, all circuits contain protective devices. All exposed metal is earthed, fuses and miniature circuit breakers (MCBs) are designed to trip under fault conditions and residual current devices (RCDs) are designed to trip below the fatal level as described in Chapter 2.

Construction workers and particularly electricians do receive electric shocks, usually as a result of carelessness or unforeseen circumstances. As an electrician working on electrical equipment you must always make sure that the equipment is switched off or electrically isolated before commencing work. Every circuit must be provided with a means of isolation (IEE Regulation 130–06–01). When working on portable equipment or desk top units it is often simply a matter of unplugging the equipment from the adjacent supply. Larger pieces of equipment, and electrical machines may require isolating at the local isolator switch before work commences. To deter anyone from re-connecting the supply while work is being carried out on equipment, a sign 'Danger – Electrician at Work' should be displayed on the isolator and the isolation 'secured' with a small padlock or the fuses removed so that no-one can re-connect whilst work is being carried out on that piece of equipment. The Electricity at Work Regulations 1989 are very specific at Regulation 12(1) that we must ensure the disconnection and separation of electrical equipment from every source of supply and that this disconnection and separation is secure. Where a test instrument or voltage indicator is used to prove the supply dead, Regulation 4(3) of the Electricity at Work Regulations 1989 recommends that the following procedure is adopted.

1 First connect the test device such as that shown in Fig. 3.10 to the supply which is to be isolated. The test device should indicate mains voltage.

2 Next, isolate the supply and observe that the test device now reads zero volts.

3 Then connect the same test device to a known live supply or proving unit such as that shown in Fig. 3.11 to 'prove' that the tester is still working correctly.

4 Finally secure the isolation and place warning signs; only then should work commence.

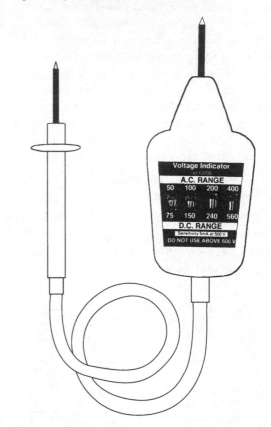

Fig. 3.10 Typical voltage indicator.

The test device being used by the electrician must incorporate safe test leads which comply with the Health and Safety Executive Guidance Note 38 on electrical test equipment. These leads should incorporate barriers to prevent the user touching live terminals when testing and incorporating a protective fuse and be well insulated and robust, such as those shown in Fig. 3.12.

To isolate a piece of equipment or individual circuit successfully, competently, safely and in accordance with all the relevant regulations, we must follow a procedure such as that given by the flow diagram of Fig. 3.13. Start at the top and work down the flow diagram.

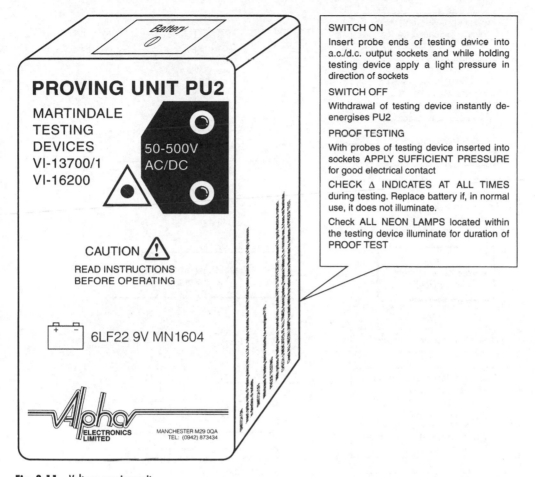

Fig. 3.11 Voltage proving unit.

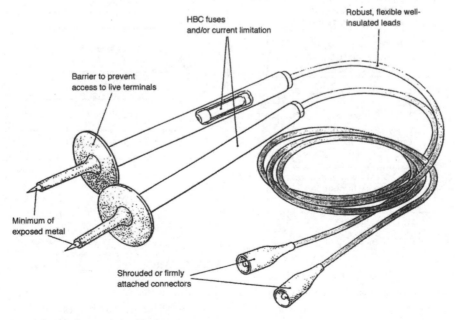

Fig. 3.12 Recommended type of test probe and leads.

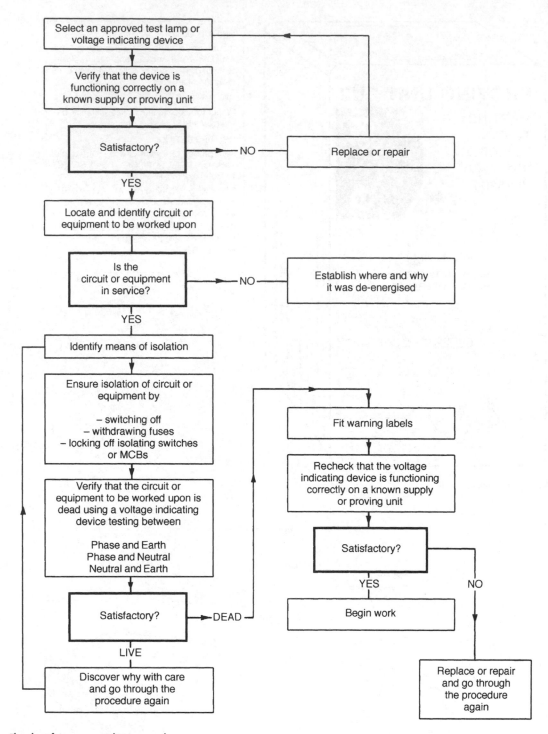

Fig. 3.13 Flowchart for a secure isolation procedure.

When the heavy outlined boxes are reached, pause and ask yourself whether everything is satisfactory up to this point. If the answer is 'yes', move on. If the answer is 'no', go back as indicated by the diagram.

LIVE TESTING

The Electricity at Work Regulations 1989 at Regulation 4(3) tell us that it is preferable that supplies be made dead before work commences. However, it does acknowledge that some work, such as fault finding and testing, may require the electrical equipment to remain energised. Therefore, if the fault finding and testing can only be successfully carried out live then the person carrying out the fault diagnosis must:

■ be trained so that they understand the equipment and the potential hazards of working live and can, therefore, be deemed 'competent' to carry out that activity;
■ only use approved test equipment;
■ set up appropriate warning notices and barriers so that the work activity does not create a situation dangerous to others.

While live testing may be required by workers in the electrotechnical industries in order to find the fault, live repair work must not be carried out. The individual circuit or piece of equipment must first be isolated before work commences in order to comply with the Electricity at Work Regulations 1989.

Disposing of waste

We have said many times in this book so far, that having a good attitude to health and safety, working conscientiously and neatly, keeping passageways clear and regularly tidying up the workplace is the sign of a good and competent craftsman. But what do you do with the rubbish that the working environment produces? Well, all the packaging material for electrical fittings and accessories usually goes into either your employer's skip or the skip on site designated for that purpose. All the off-cuts of conduit, trunking and tray also go into the skip. In fact, most of the general site debris will probably go into the skip and the waste disposal company will take the skip contents to a designated local council land fill area for safe disposal.

The part coils of cable and any other re-useable leftover lengths of conduit, trunking or tray will be taken back to your employer's stores area. Here it will be stored for future use and the returned quantities deducted from the costs allocated to that job.

What goes into the skip for normal disposal into a land fill site is usually a matter of common sense. However, some substances require special consideration and disposal. We will now look at asbestos and large quantities of used fluorescent tubes.

Asbestos is a mineral found in many rock formations. When separated it becomes a fluffy, fibrous material with many uses. It was used extensively in the construction industry during the 1960's and 70's for roofing material, ceiling and floor tiles, fire resistant board for doors and partitions, for thermal insulation and commercial and industrial pipe lagging.

In the buildings where it was installed some 40 years ago, when left alone, it does not represent a health hazard, but those buildings are increasingly becoming in need of renovation and modernisation. It is in the dismantling and breaking up of these asbestos materials that the health hazard increases. Asbestos is a serious health hazard if the dust is inhaled. The tiny asbestos particles find their way into delicate lung tissue and remain embedded for life, causing constant irritation and eventually, serious lung disease.

Working with asbestos materials is not a job for anyone in the electrotechnical industry. If asbestos is present in situations or buildings where you are expected to work, it should be removed by a specialist contractor before your work commences. Specialist contractors, who will wear fully protective suits and use breathing apparatus, are the only people who can safely and responsibly carry out the removal of asbestos. They will wrap the asbestos in thick plastic bags and store them temporarily in a covered and locked skip. This material is then disposed of in a special land fill site with other toxic industrial waste materials and the site monitored by the local authority for the foreseeable future.

There is a lot of work for electrical contractors in my part of the country, updating and improving the lighting in government buildings and schools. This work often involves removing the old fluorescent fittings, hanging on chains or fixed to beams and installing a suspended ceiling and an appropriate number of recessed modular fluorescent fittings. So what do we do with the old fittings? Well, the fittings are made of sheet steel, a couple of plastic lampholders, a little cable, a starter and ballast. All of these materials can go into the ordinary skip. However, the fluorescent tubes contain a little mercury and fluorescent powder with toxic elements, which cannot be disposed of in the normal land fill sites. The responsible way to dispose of fluorescent tubes is by grinding them up into small pieces using a 'lamp crusher', which looks very much like a garden waste shredder. The crushed lamp contents falls into a heavy duty plastic bag, which is sealed and disposed of along with the asbestos, material and other industrial waste in special land fill sites.

The COSHH Regulations have encouraged specialist companies to set up businesses dealing with the responsible disposal of toxic waste material. Specialist companies have systems and procedures, which meet the relevant regulation, and they would usually give an electrical company a certificate to say that they had disposed of particular waste material responsibly. The cost of this service is then passed on to the customer. These days, large employers and local authorities insist that waste is disposed of properly.

The Environmental Health Officer at your local council offices will always give advice and point you in the direction of specialist companies dealing with toxic waste disposal.

Electrical Principles
Alternating current theory

In Chapter 2 of this book at Fig. 2.13 and Fig. 2.40 we looked at the generation of an a.c. waveform and the calculation of average and rms values. In this section we will first of all consider the theoretical circuits of pure resistance, inductance and capacitance acting alone in an a.c. circuit before going on to consider the practical circuits of resistance, inductance and capacitance acting together. Let us first define some of our terms of reference.

RESISTANCE

In any circuit, *resistance* is defined as opposition to current flow. From Ohm's law

$$R = \frac{V_R}{I_R} \, (\Omega)$$

However, in an a.c. circuit, resistance is only part of the opposition to current flow. The inductance and capacitance of an a.c. circuit also cause an opposition to current flow, which we call *reactance*.

Inductive reactance (X_L) is the opposition to an a.c. current in an inductive circuit. It causes the current in the circuit to lag behind the applied voltage, as shown in Fig. 3.14. It is given by the formula

$$X_L = 2\pi f L \, (\Omega)$$

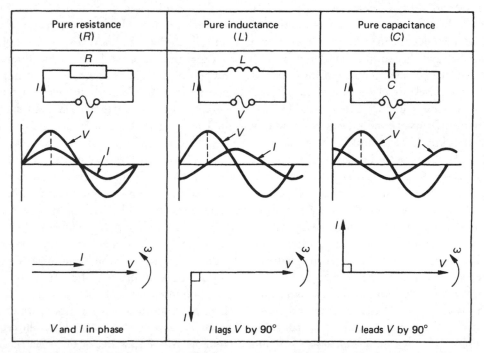

Fig. 3.14 Voltage and current relationships in resistive, capacitive and inductive circuits.

where

$\pi = 3.142$ a constant
$f =$ the frequency of the supply
$L =$ the inductance of the circuit

or by

$$X_L = \frac{V_L}{I_L}$$

Capacitive reactance (X_C) is the opposition to an a.c. current in a capacitive circuit. It causes the current in the circuit to lead ahead of the voltage, as shown in Fig. 3.14. It is given by the formula

$$X_C = \frac{1}{2\pi f C} \ (\Omega)$$

where π and f are defined as before and C is the capacitance of the circuit. It can also be expressed as

$$X_C = \frac{V_C}{I_C}$$

EXAMPLE

Calculate the reactance of a 150 μF capacitor and a 0.05 H inductor if they were separately connected to the 50 Hz mains supply.
For capacitive reactance

$$X_C = \frac{1}{2\pi f C}$$

where $f = 50$ Hz and $C = 150$ μF $= 150 \times 10^{-6}$ F.

$$\therefore X_C = \frac{1}{2 \times 3.142 \times 50 \text{ Hz} \times 150 \times 10^{-6}} = 21.2 \ \Omega$$

For inductive reactance,

$$X_L = 2\pi f L$$

where $f = 50$ Hz and $L = 0.05$ H.

$$\therefore X_L = 2 \times 3.142 \times 50 \text{ Hz} \times 0.05 \text{ H} = 15.7 \ \Omega$$

IMPEDANCE

The total opposition to current flow in an a.c. circuit is called impedance and given the symbol Z. Thus impedance is the combined opposition to current flow of the resistance, inductive reactance and capacitive reactance of the circuit and can be calculated from the formula

$$Z = \sqrt{R^2 + X^2} \ (\Omega)$$

or

$$Z = \frac{V_T}{I_T}$$

EXAMPLE 1

Calculate the impedance when a 5 Ω resistor is connected in series with a 12 Ω inductive reactance.

$$Z = \sqrt{R^2 + X_L^2} \ (\Omega)$$
$$\therefore Z = \sqrt{5^2 + 12^2}$$
$$Z = \sqrt{25 + 144}$$
$$Z = \sqrt{169}$$
$$Z = 13 \ \Omega$$

EXAMPLE 2

Calculate the impedance when a 48 Ω resistor is connected in series with a 55 Ω capacitive reactance.

$$Z = \sqrt{R^2 + X_C^2} \ (\Omega)$$
$$\therefore Z = \sqrt{48^2 + 55^2}$$
$$Z = \sqrt{2304 + 3025}$$
$$Z = \sqrt{5329}$$
$$Z = 73 \ \Omega$$

RESISTANCE, INDUCTANCE AND CAPACITANCE IN AN a.c. CIRCUIT

When a resistor only is connected to an a.c. circuit the current and voltage waveforms remain together, starting and finishing at the same time. We say that the waveforms are *in phase*.

When a pure inductor is connected to an a.c. circuit the current lags behind the voltage waveform by an angle of 90°. We say that the current *lags* the voltage by 90°. When a pure capacitor is connected to an a.c. circuit the current *leads* the voltage by an angle of 90°. These various effects can be observed on an oscilloscope, but the circuit diagram, waveform diagram and phasor diagram for each circuit are shown in Fig. 3.14.

Phasor diagrams

Phasor diagrams and a.c. circuits are an inseparable combination. Phasor diagrams allow us to produce a model or picture of the circuit under consideration which helps us to understand the circuit. A phasor is a straight line, having definite length and direction, which represents to scale the magnitude and direction of a quantity such as a current, voltage or impedance.

To find the combined effect of two quantities we combine their phasors by adding the beginning of the second phasor to the end of the first. The combined effect of the two quantities is shown by the resultant phasor, which is measured from the original zero position to the end of the last phasor.

EXAMPLE

Find by phasor addition the combined effect of currents A and B acting in a circuit. Current A has a value of 4 A, and current B a value of 3 A, leading A by 90°. We usually assume phasors to rotate anticlockwise and so the complete diagram will be as shown in Fig. 3.15. Choose a scale of, for example, 1 A = 1 cm and draw the phasors to scale, i.e. A = 4 cm and B = 3 cm, leading A by 90°.

The magnitude of the resultant phasor can be measured from the phasor diagram and is found to be 5 A acting at a phase angle ϕ of about 37° leading A. We therefore say that the combined effect of currents A and B is a current of 5 A at an angle of 37° leading A.

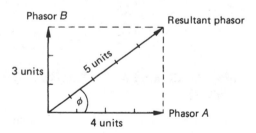

Fig. 3.15 The phasor addition of currents A and B.

Phase angle ϕ

In an a.c. circuit containing resistance only, such as a heating circuit, the voltage and current are in phase, which means that they reach their peak and zero values together, as shown in Fig. 3.16(a).

In an a.c. circuit containing inductance, such as a motor or discharge lighting circuit, the current often reaches its maximum value after the voltage, which means that the current and voltage are out of phase with each other, as shown in Fig. 3.16(b). The phase difference, measured in degrees between the current and voltage, is called the phase angle of the circuit, and is denoted by the symbol ϕ, the lower-case Greek letter phi.

When circuits contain two or more separate elements, such as RL, RC or RLC, the phase angle between the total voltage and total current will be neither 0° nor 90° but will be determined by the relative values of resistance and reactance in the circuit. In Fig. 3.17 the phase angle between applied voltage and current is some angle ϕ.

ALTERNATING CURRENT SERIES CIRCUIT

In a circuit containing a resistor and inductor connected in series as shown in Fig. 3.17, the current I will flow through the resistor and the inductor causing the voltage V_R to be dropped across the resistor and V_L to be dropped across the inductor. The sum of these voltages will be equal to the total voltage V_T but because this is an a.c. circuit the voltages must be added by phasor addition. The result is shown in Fig. 3.17, where V_R is drawn to scale and in phase with the current and V_L is drawn to scale and leading the current by 90°. The phasor addition of these two voltages gives us the magnitude and direction of V_T, which leads the current by some angle ϕ.

In a circuit containing a resistor and capacitor connected in series as shown in Fig. 3.18, the current I will flow through the resistor and capacitor causing voltage drops V_R and V_C. The voltage V_R will be in phase with the current and V_C will lag the current by 90°. The phasor addition of these voltages is equal to the total voltage V_T which, as can be seen in Fig. 3.18, is lagging the current by some angle ϕ.

THE IMPEDANCE TRIANGLE

We have now established the general shape of the phasor diagram for a series a.c. circuit. Figures 3.17 and 3.18 show the voltage phasors, but we know that $V_R = IR$, $V_L = IX_L$, $V_C = IX_C$ and $V_T = IZ$, and therefore the phasor diagrams (a) and (b) of Fig. 3.19 must be equal. From Fig. 3.19(b), by the theorem of Pythagoras, we have

$$(IZ)^2 = (IR)^2 + (IX)^2$$

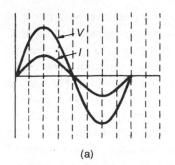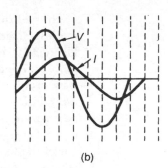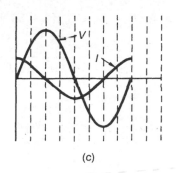

Fig. 3.16 Phase relationship of a.c. waveform: (a) V and in I phase, phase angle $\phi = 0°$ and power factor $= \cos\phi = 1$; (b) V and I displaced by $45°$, $\phi = 45°$ and p.f. $= 0.707$; (c) V and I displaced by $90°$, $\phi = 90°$ and p.f. $= 0$.

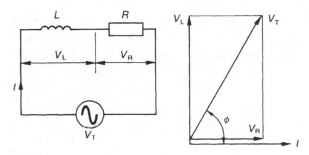

Fig. 3.17 A series RL circuit and phasor diagram.

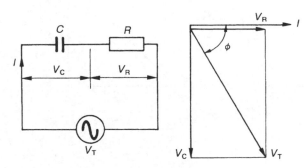

Fig. 3.18 A series RC circuit and phasor diagram.

$$I^2 Z^2 = I^2 R^2 + I^2 X^2$$

If we now divide throughout by I^2 we have

$$Z^2 = R^2 + X^2$$
$$\text{or} \quad Z = \sqrt{R^2 + X^2} \ \Omega$$

The phasor diagram can be simplified to the impedance triangle given in Fig. 3.19(c).

EXAMPLE 1

A coil of $0.15\,\text{H}$ is connected in series with a $50\,\Omega$ resistor across a $100\,\text{V}$ $50\,\text{Hz}$ supply. Calculate (a) the reactance of the coil, (b) the impedance of the circuit, and (c) the current.

For (a),

$$X_L = 2\pi fL \ (\Omega)$$
$$\therefore X_L = 2 \times 3.142 \times 50\,\text{Hz} \times 0.15\,\text{H} = 47.1\,\Omega$$

For (b),

$$Z = \sqrt{R^2 + X^2} \ (\Omega)$$
$$\therefore Z = \sqrt{(50\,\Omega)^2 + (47.1\,\Omega)^2} = 68.69\,\Omega$$

For (c),

$$I = V/Z (\text{A})$$
$$\therefore I = \frac{100\,V}{68.69\,\Omega} = 1.46\,\text{A}$$

EXAMPLE 2

A $60\,\mu\text{F}$ capacitor is connected in series with a $100\,\Omega$ resistor across a $230\,\text{V}$ $50\,\text{Hz}$ supply. Calculate (a) the reactance of the capacitor, (b) the impedance of the circuit, and (c) the current.

For (a),

$$X_C = \frac{1}{2\pi fC} \ (\Omega)$$

$$\therefore X_C = \frac{1}{2\pi \times 50\,\text{Hz} \times 60 \times 10^{-6}\,\text{F}} = 53.05\,\Omega$$

For (b),

$$Z = \sqrt{R^2 + X^2} \ (\Omega)$$
$$\therefore Z = \sqrt{(100 \ \Omega)^2 + (53.05 \ \Omega)^2} = 113.2 \ \Omega$$

For (c),

$$I = V/Z \ (A)$$
$$\therefore I = \frac{230 \ V}{113.2 \ \Omega} = 2.03 \ A$$

POWER AND POWER FACTOR

Power factor (p.f.) is defined as the cosine of the phase angle between the current and voltage:

$$p.f. = \cos \phi$$

If the current lags the voltage as shown in Fig. 3.17, we say that the p.f. is lagging, and if the current leads the voltage as shown in Fig. 3.18, the p.f. is said to be leading. From the trigonometry of the impedance triangle shown in Fig. 3.19, p.f. is also equal to

$$p.f. = \cos \phi = \frac{R}{Z} = \frac{V_R}{V_T}$$

The electrical power in a circuit is the product of the instantaneous values of the voltage and current. Figure 3.20 shows the voltage and current waveform for a pure inductor and pure capacitor. The power waveform is obtained from the product of V and I at every instant in the cycle. It can be seen that the power waveform reverses every quarter cycle, indicating that energy is alternately being fed into and taken out of the inductor and capacitor. When considered over one complete cycle, the positive and negative portions are equal, showing that the average power consumed by a pure inductor or capacitor is zero. This shows that inductors and capacitors store energy during one part of the voltage cycle and feed it back into the supply later in the cycle. Inductors store energy as a magnetic field and capacitors as an electric field.

In an electric circuit more power is taken from the supply than is fed back into it, since some power is dissipated by the resistance of the circuit, and therefore

$$P = I^2 R \ (W)$$

In any d.c. circuit the power consumed is given by the product of the voltage and current, because in a

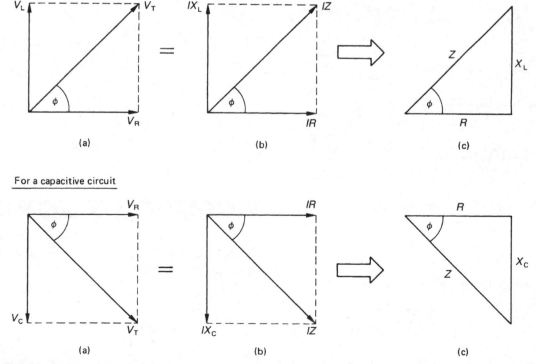

For an inductive circuit

For a capacitive circuit

Fig. 3.19 Phasor diagram and impedance triangle.

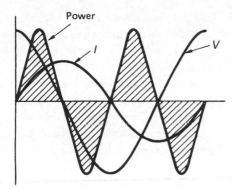

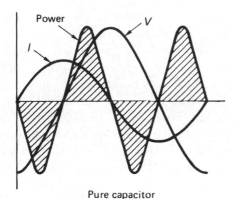

Fig. 3.20 Waveform for the a.c. power in purely inductive and purely capacitive circuits.

d.c. circuit voltage and current are in phase. In an a.c. circuit the power consumed is given by the product of the current and that part of the voltage which is in phase with the current. The in-phase component of the voltage is given by $V \cos \phi$, and so power can also be given by the equation

$$P = VI \cos \phi \text{ (W)}$$

EXAMPLE 1

A coil has a resistance of 30 Ω and a reactance of 40 Ω when connected to a 250 V supply. Calculate (a) the impedance, (b) the current, (c) the p.f., and (d) the power.

For (a),

$$Z = \sqrt{R^2 + X^2} \text{ (Ω)}$$
$$\therefore Z = \sqrt{(30\,Ω)^2 + (40\,Ω)^2} = 50\,Ω$$

For (b),

$$I = V / Z \text{(A)}$$
$$\therefore I = \frac{250\,V}{50\,Ω} = 5\,A$$

For (c),

$$\text{p.f.} = \cos \phi = \frac{R}{Z}$$
$$\therefore \text{p.f.} = \frac{30\,Ω}{50\,Ω} = 0.6 \text{ lagging}$$

For (d),

$$P = VI \cos \phi \text{ (W)}$$
$$\therefore P = 250\,V \times 5\,A \times 0.6 = 750\,W$$

EXAMPLE 2

A capacitor of reactance 12 Ω is connected in series with a 9 Ω resistor across a 150 V supply. Calculate (a) the impedance of the circuit, (b) the current, (c) the p.f., and (d) the power.

For (a),

$$Z = \sqrt{R^2 + X^2} \text{ (Ω)}$$
$$\therefore Z = \sqrt{(9\,Ω)^2 + (12\,Ω)^2} = 15\,Ω$$

For (b),

$$I = V / Z \text{(A)}$$
$$\therefore I = \frac{150\,V}{15\,Ω} = 10\,A$$

For (c),

$$\text{p.f.} = \cos \phi = \frac{R}{Z}$$
$$\therefore \text{p.f.} = \frac{9\,Ω}{15\,Ω} = 0.6 \text{ leading}$$

For (d),

$$P = VI \cos \phi \text{ (W)}$$
$$\therefore P = 150\,V \times 10\,A \times 0.6 = 900\,W$$

The power factor of most industrial loads is lagging because the machines and discharge lighting used in industry are mostly inductive. This causes an additional magnetizing current to be drawn from the supply, which does not produce power, but does need to be supplied, making supply cables larger.

EXAMPLE 3

A 230 V supply feeds three 1.84 kW loads with power factors of 1, 0.8 and 0.4. Calculate the current at each power factor.

The current is given by

$$I = \frac{P}{V \cos \phi}$$

where $P = 1.84\,kW = 1840\,W$ and $V = 230\,V$. If the p.f. is 1, then

$$I = \frac{1840 \text{ W}}{230 \text{ V} \times 1} = 8 \text{ A}$$

For a p.f. of 0.8,

$$I = \frac{1840 \text{ W}}{230 \text{ V} \times 0.8} = 10 \text{ A}$$

For a p.f. of 0.4,

$$I = \frac{1840 \text{ W}}{230 \text{ V} \times 0.4} = 20 \text{ A}$$

It can be seen from these calculations that a 1.84 kW load supplied at a power factor of 0.4 would require a 20 A cable, while the same load at unity power factor could be supplied with an 8 A cable. There may also be the problem of higher voltage drops in the supply cables. As a result, the supply companies encourage installation engineers to improve their power factor to a value close to 1 and sometimes charge penalties if the power factor falls below 0.8.

Power-factor improvement

Most installations have a low power factor because of the inductive nature of the load. A capacitor has the opposite effect of an inductor, and so it seems reasonable to add a capacitor to a load which is known to have a lower power factor.

Figure 3.21(a) shows an industrial load with a low power factor. If a capacitor is connected in parallel with the load, the capacitor current I_C leads the applied voltage by 90°. When this current is added to the load current the resultant current has a much improved power factor, as can be seen in Fig. 3.21(b).

Capacitors may be connected across the main busbars of industrial loads in order to provide power-factor improvement, but smaller capacitors may also be connected across an individual piece of equipment, as is the case for fluorescent light fittings.

PARALLEL CIRCUITS

In practice, most electrical installations consist of a number of circuits connected in parallel to form a network. The branches of the parallel network may consist of one component or two or more components connected in series. You should now have an appreciation of series circuits and we will now consider two branch parallel circuits. In a parallel circuit the supply voltage is applied to each of the network branches. Voltage is used as the reference when drawing phasor diagrams and the currents are added by phasor addition.

In a parallel circuit containing a pure resistor and inductor as shown in Fig. 3.22, the current flowing through the resistive branch will be in phase with the voltage and the current flowing in the inductive branch will be 90° lagging the voltage. The phasor addition of these currents will give the total current drawn

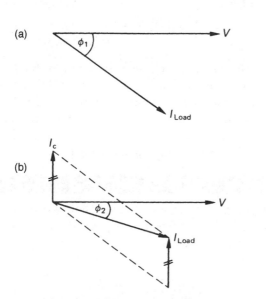

Fig. 3.21 Power-factor improvement using capacitors.

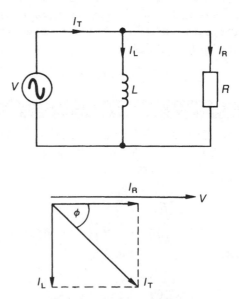

Fig. 3.22 A parallel RLC circuit and phasor diagram.

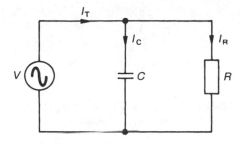

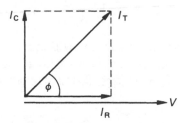

Fig. 3.23 A parallel RC circuit and phasor diagram.

from the supply and its phase angle as shown in the phasor diagram of Fig. 3.22.

In a parallel circuit containing a pure resistor and capacitor connected in parallel, as shown in Fig. 3.23, the current flowing through the resistive branch will be in phase with the voltage and the current in the capacitive branch will lead the voltage by 90°. The phasor addition of these currents will give the total current and its phase angle, as shown in Fig. 3.23.

EXAMPLE 1

A pure inductor of 100 mH is connected in parallel with a 30 Ω resistor to a 230 V 50 Hz supply. Calculate the branch currents and the supply current.

$$I_R = \frac{V}{R} \ (A)$$

$$\therefore I_R = \frac{230 \, V}{30 \, \Omega} = 7.67 \, A$$

$$X_L = 2\pi f L \ (\Omega)$$

$$\therefore X_L = 2 \times 3.142 \times 50 \, Hz \times 100 \times 10^{-3} \, H$$

$$X_L = 31.42 \, \Omega$$

$$I_L = \frac{V}{X_L} \ (A)$$

$$\therefore I_L = \frac{230 \, V}{31.42 \, \Omega} = 7.32 \, A$$

From the trigonometry of the phasor diagram in Fig. 3.22, the total current is given by

$$I_T = \sqrt{I_R^2 + I_L^2} \ (A)$$

$$\therefore I_T = \sqrt{(7.67 \, A)^2 + (7.32 \, A)^2}$$

$$I_T = 10.60 \, A.$$

EXAMPLE 2

A pure capacitor of 60 μF is connected in parallel with a 40 Ω resistor across a 230 V 50 Hz supply. Calculate the branch currents and the supply currents.

$$I_R = \frac{V}{R} \ (A)$$

$$\therefore I_R = \frac{230 \, V}{40 \, \Omega} = 5.75 \, A$$

$$X_C = \frac{1}{2\pi f C} \ (\Omega)$$

$$\therefore X_C = \frac{1}{2 \times 3.142 \times 50 \, Hz \times 60 \times 10^{-6} \, F}$$

$$X_C = 53.05 \, \Omega$$

$$I_C = \frac{V}{X_C} \ (A)$$

$$\therefore I_C = \frac{230 \, V}{53.05 \, \Omega} = 4.34 \, A$$

From the trigonometry of the phasor diagram in Fig. 3.23, the total current is given by

$$I_T = \sqrt{I_C^2 + I_R^2} \ (A)$$

$$I_T = \sqrt{(5.75 \, A)^2 + (4.34 \, A)^2}$$

$$I_T = 7.2 \, A.$$

In considering these two examples we have assumed the capacitor and inductor to be pure. In practice the inductor will contain some resistance and the network may, therefore, be considered as a series RL branch connected in parallel with a capacitor as shown in Fig. 3.24.

EXAMPLE 3

A coil having a resistance of 50 Ω and inductance 318 mH is connected in parallel with 20 μF capacitor across a 230 V 50 Hz supply. Calculate

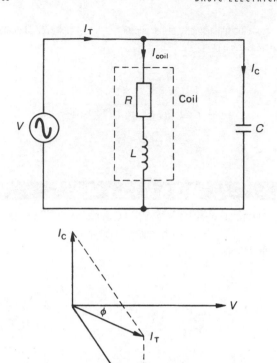

Fig. 3.24 A parallel circuit and phasor diagram.

the branch currents and the supply network. The circuit diagram for this network is shown in Fig. 3.24.

$$X_C = \frac{1}{2\pi fC} \ (\Omega)$$

$$\therefore X_C = \frac{1}{2 \times 3.142 \times 50 \, Hz \times 20 \times 10^{-6} \, F}$$

$$X_C = 159.2 \, \Omega$$

$$I_C = \frac{V}{X_C} \ (A)$$

$$\therefore I_C = \frac{230 \, V}{159.2 \, \Omega} = 1.45 \, A$$

$$X_L = 2\pi fL \ (\Omega)$$

$$\therefore X_L = 2 \times 3.142 \times 50 \, Hz \times 318 \times 10^{-3} \, H$$

$$X_L = 100 \, \Omega$$

$$Z_{coil} = \sqrt{R^2 + X_L^2} \ (\Omega)$$

$$\therefore Z_{coil} = \sqrt{(50 \, \Omega)^2 + (100 \, \Omega)^2}$$

$$Z_{coil} = 111.8 \, \Omega$$

$$I_{coil} = \frac{V}{Z} \ (A)$$

$$\therefore I_{coil} = \frac{230 \, V}{111.8 \, \Omega} = 2.06 \, A$$

The capacitor current will lead the supply voltage by 90°. The coil current will lag the voltage by an angle given by:

$$\phi_{coil} = \cos^{-1} \frac{R}{Z}$$

$$\therefore \phi_{coil} = \cos^{-1} \frac{50 \, \Omega}{111.8 \, \Omega} = 63.4°$$

The coil and capacitor currents can now be drawn to scale as shown in Fig. 3.25. The total current is the phasor addition of these currents and is found to be 1.0 A at 20° lagging from the phasor diagram.

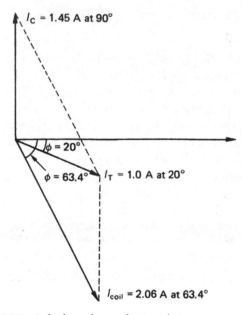

Fig. 3.25 Scale phasor diagram for Example 3.

THREE-PHASE a.c.

A three-phase voltage is generated in exactly the same way as a single-phase a.c. voltage. For a three-phase voltage three separate windings, each separated by 120°, are rotated in a magnetic field. The generated voltage will be three identical sinusoidal waveforms each separated by 120°, as shown in Fig. 3.26.

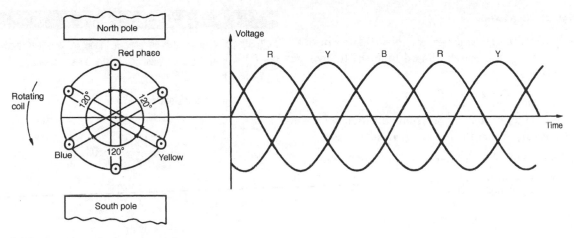

Fig. 3.26 Generation of a three-phase voltage.

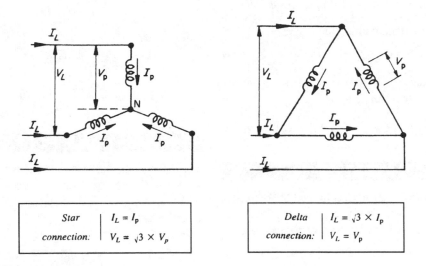

Star	$I_L = I_p$
connection:	$V_L = \sqrt{3} \times V_p$

Delta	$I_L = \sqrt{3} \times I_p$
connection:	$V_L = V_p$

Fig. 3.27 Star and delta connections.

Star and delta connections

The three phase windings may be star connected or delta connected as shown in Fig. 3.27. The important relationship between phase and line currents and voltages is also shown. The square root of 3 ($\sqrt{3}$) is simply a constant for three-phase circuits, and has a value of 1.732. The delta connection is used for electrical power transmission because only three conductors are required. Delta connection is also used to connect the windings of most three-phase motors because the phase windings are perfectly balanced and, therefore, do not require a neutral connection.

Making a star connection has the advantage that two voltages become available – a line voltage between any two phases, and a phase voltage between line and neutral which is connected to the star point.

In any star-connected system currents flow along the lines (I_L), through the load and return by the neutral conductor connected to the star point. In a *balanced* three-phase system all currents have the same value and when they are added up by phasor addition, we find the resultant current is zero. Therefore, no current flows in the neutral and the star point is at zero volts. The star point of the distribution transformer is earthed because earth is also at zero potential. A star-connected system is also called a three-phase four-wire system and allows us to connect single-phase loads to a three-phase system.

Three-phase power

We know from our single-phase theory earlier in this chapter that power can be found from the following formula:

$$\text{Power} = VI \cos \phi \ (\text{W})$$

In any balanced three-phase system, the total power is equal to three times the power in any one phase.

$$\therefore \ \text{Total three-phase power} = 3V_P I_P \cos \phi \ (\text{W}) \ (1)$$

Now for a star connection,

$$V_P = V_L / \sqrt{3} \quad \text{and} \quad I_L = I_P \qquad (2)$$

Substituting equation (2) into equation (1), we have

$$\text{Total three-phase power} = \sqrt{3} \ V_L L_L \cos \phi \ (\text{W})$$

Now consider a delta connection:

$$V_P = V_L \quad \text{and} \quad I_P = I_L / \sqrt{3} \qquad (3)$$

Substituting equation (3) into equation (1) we have, for any balanced three-phase load,

$$\text{Total three-phase power} = \sqrt{3} \ V_L L_L \cos \phi \ (\text{W})$$

EXAMPLE 1

A balanced star-connected three-phase load of $10\,\Omega$ per phase is supplied from a $400\,\text{V}$ $50\,\text{Hz}$ mains supply at unity power factor. Calculate (a) the phase voltage, (b) the line current and (c) the total power consumed.

For a star connection,

$$V_L = \sqrt{3} \ V_P \quad \text{and} \quad I_L = I_P$$

For (a),

$$V_P = V_L / \sqrt{3} \ (\text{V})$$

$$V_P = \frac{400\,\text{V}}{1.732} = 230.9\,\text{V}$$

For (b),

$$I_L = I_P = V_P / R_P (\text{A})$$

$$I_L = I_P = \frac{230.9\,\text{V}}{10\,\Omega} = 23.09\,\text{A}$$

For (c),

$$\text{Power} = \sqrt{3} \ V_L \ I_L \cos \phi \ (\text{W})$$

$$\text{Power} = 1.732 \times 400\,\text{V} \times 23.09\,\text{A} \times 1 = 16\,\text{kW}$$

EXAMPLE 2

A $20\,\text{kW}$ $400\,\text{V}$ balanced delta-connected load has a power factor of 0.8. Calculate (a) the line current and (b) the phase current.

We have that

$$\text{Three-phase power} = \sqrt{3} \ V_L \ I_L \cos \phi \ (\text{W})$$

For (a),

$$I_L = \frac{\text{Power}}{\sqrt{3} \ V_L \cos \phi} \ (\text{A})$$

$$\therefore I_L = \frac{20\,000\,\text{W}}{1.732 \times 400\,\text{V} \times 0.8}$$

$$I_L = 36.08\,(\text{A})$$

For delta connection,

$$I_L = \sqrt{3} \ I_P \ (\text{A})$$

Thus, for (b),

$$I_P = I_L \ \sqrt{3} \ (\text{A})$$

$$\therefore I_P = \frac{36.08\,\text{A}}{1.732} = 20.83\,\text{A}$$

EXAMPLE 3

Three identical loads each having a resistance of $30\,\Omega$ and inductive reactance of $40\,\Omega$ are connected first in star and then in delta to a $400\,\text{V}$ three-phase supply. Calculate the phase currents and line currents for each connection.

For each load,

$$Z = \sqrt{R^2 + X_L^2} \ (\Omega)$$

$$\therefore Z = \sqrt{30^2 + 40^2}$$

$$Z = \sqrt{2500}$$

For star connection,

$$V_L = \sqrt{3} \ V_P \quad \text{and} \quad I_L = I_P$$

$$V_P = V_L / \sqrt{3} \ (\text{V})$$

$$\therefore V_P = \frac{400\,\text{V}}{1.732} = 230.9\,\text{V}$$

$$I_P = V_P / Z_P \ (\text{A})$$

$$\therefore I_P = \frac{230.9\,\text{V}}{50\,\Omega} = 4.62\,\text{A}$$

$$I_P = I_L$$

therefore phase and line currents are both equal to $4.62\,\text{A}$.

For delta connection,

$$V_L = V_P \text{ and } I_L = \sqrt{3}\, I_P$$

$$V_L = V_P = 400\,V$$

$$I_P = V_P / Z_P \text{ (A)}$$

$$\therefore I_P = \frac{400\,V}{50\,\Omega} = 8\,A$$

$$I_L = \sqrt{3}\, I_P \text{ (A)}$$

$$\therefore I_L = 1.732 \times 8\,A = 13.86\,A$$

Magnetism

The basic rules of magnetism were laid down in Chapter 2 of this book. Here we will look at some of the laws of magnetism as they apply to electrical applications, such as generators, motors and transformers.

A current carrying conductor maintains a magnetic field around the conductor which is proportional to the current flowing. When this magnetic field interacts with another magnetic field, forces are exerted which describe the basic principles of electric motors. This is further discussed in Chapter 2 of this book and demonstrated by the diagrams shown in Fig. 2.1.

Michael Faraday demonstrated on 29 August 1831 that electricity could be produced by magnetism. He stated that 'when a conductor cuts or is cut by a magnetic field an emf is induced in that conductor. The amount of induced emf is proportional to the rate or speed at which the magnetic field cuts the conductor'. This basic principle laid down the laws of present-day electricity generation where a strong magnetic field is rotated inside a coil of wire to generate electricity.

This law can be translated into a formula as follows:

$$\text{Induced emf} = Blv \text{ (V)}$$

where B is the magnetic flux density, measured in tesla, to commemorate Nikola Tesla (1856–1943) a famous Yugoslav who invented the two-phase and three-phase alternator and motor; l is the length of conductor in the magnetic field, measured in metres; and v is the velocity or speed at which the conductor cuts the magnetic flux (measured in metres per second).

EXAMPLE

A 15 cm length of conductor is moved at 20 m/s through a magnetic field of flux density 2 T. Calculate the induced emf.

$$\text{emf} = Blv \text{ (V)}$$

$$\therefore \text{emf} = 2\,T \times 0.15 \times 20\,m/s$$

$$\text{emf} = 6\,V$$

INDUCTANCE

If a coil of wire is wound on to an iron core as shown in Fig. 3.28, a magnetic field will become established in the core when a current flows in the coil due to the switch being closed.

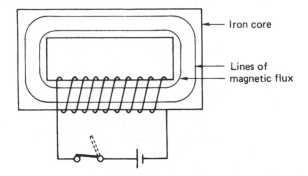

Fig. 3.28 An inductive coil or choke.

When the switch is opened the current stops flowing and, therefore, the magnetic flux collapses. The collapsing magnetic flux induces an emf into the coil and this voltage appears across the switch contacts. The effect is known as *self-inductance*, or just *inductance*, and is one property of any coil. The unit of inductance is the henry (symbol H), to commemorate the work of the American physicist Joseph Henry (1797–1878), and a circuit is said to possess an inductance of 1 henry when an emf of 1 volt is induced in the circuit by a current changing at the rate of 1 ampere per second.

Fluorescent light fittings contain a choke or inductive coil in series with the tube and starter lamp. The starter lamp switches on and off very quickly, causing rapid current changes which induce a large voltage across the tube electrodes sufficient to strike an arc in the tube.

When two separate coils are placed close together, as they are in a transformer, a current in one coil produces a magnetic flux which links with the second coil. This induces a voltage in the second coil and is the basic principle of the transformer action which is described later in this chapter. The two coils in this case are said to possess *mutual inductance*, as shown by Fig. 3.29. A mutual inductance of 1 henry exists between two coils when a uniformly varying current of 1 ampere per second in one coil produces an emf of 1 volt in the other coil.

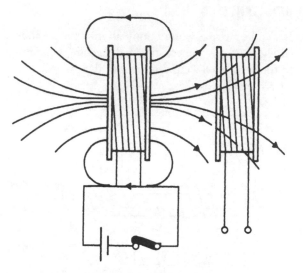

Fig. 3.29 Mutual induction between two coils.

The emf induced in a coil such as that shown in Fig. 3.28 is dependent upon the rate of change of magnetic flux and the number of turns on the coil. The average induced emf is, therefore, given by:–

$$\text{emf} = \frac{-(\Phi_2 - \Phi_1)}{t} \times N \text{ (V)}$$

where Φ is the magnetic flux measured in webers, to commemorate the work of the German physicist, Wilhelm Weber (1804–91), t is the time in seconds and N the number of turns. The minus sign indicates that the emf is a back emf opposing the rate of change of current as described later by Lenz's law.

EXAMPLE

The magnetic flux linking 2000 turns of electromagnetic relay changes from 0.6 mWb to 0.4 mWb in 50 ms. Calculate the average value of the induced emf.

$$\text{emf} = -\frac{(\Phi_2 - \Phi_1)}{t} \times N \text{ (V)}$$

$$\therefore \text{emf} = -\frac{(0.6 - 0.4) \times 10^{-3}}{50 \times 10^{-3}} \times 2000$$

$$\text{emf} = -8 \text{ V}$$

ENERGY STORED IN A MAGNETIC FIELD

When we open the switch of an inductive circuit such as that shown in Fig. 3.28 the magnetic flux collapses and produces an arc across the switch contacts. The arc is produced by the stored magnetic energy being discharged across the switch contacts. The stored magnetic energy (symbol W) is expressed in joules and given by the following formulae:

$$\text{Energy} = W = \tfrac{1}{2} LI^2 \text{ (J)}$$

where L is the inductance of the coil in henrys and I is the current flowing in amperes.

EXAMPLE

The field windings of a motor have an inductance of 3 H and carry a current of 2 A. Calculate the magnetic energy stored in the coils.

$$W = \tfrac{1}{2} LI^2 \text{ (J)}$$

$$W = \tfrac{1}{2} \times 3\,H \times (2\,A)^2$$

$$W = 6\,J$$

MAGNETIC HYSTERESIS

There are many different types of magnetic material and they all respond differently to being magnetized. Some materials magnetize easily, and some are difficult to magnetize. Some materials retain their magnetism, while others lose it. The properties of a magnetic sample may be examined in detail if we measure the flux density (B) of the material with increasing and then decreasing values of magnetic field strength (H). The result will look like the graphs shown in Fig. 3.30 and are called *hysteresis loops*.

The hysteresis effect causes the magnetic material to retain some of its magnetism after the magnetic field strength has been removed. The flux density remaining when H is zero is called the *residual flux density*. This residual flux density can be reduced to zero by applying a negative magnetic field strength ($-H$). The value of this demagnetizing force is called the *coercive force*.

When the material has been worked to magnetic saturation – that is, B_{max} and H_{max}, the residual flux density is called *remanence* and the coercive force is called *coercivity*. The value of coercivity varies enormously for different materials, being about 40 000 A/m for Alnico (an alloy of iron, aluminium and nickel,

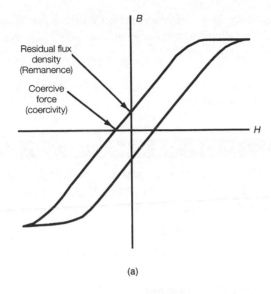

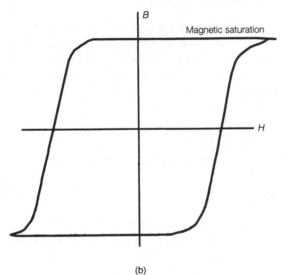

Fig. 3.30 Magnetic hysteresis loops: (a) electromagnetic material; (b) permanent magnetic material.

used for permanent magnets) and about 3 A/m for Mumetal (an alloy of nickel and iron).

Materials from which permanent magnets are made should have a high value of residual flux density and coercive force and, therefore, display a wide hysteresis loop, as shown by loop (b) in Fig. 3.30.

The core of an electromagnet is required to magnetize easily, and to lose its magnetism equally easily when switched off. Suitable materials will, therefore, have a low value of residual flux density and coercive force and, therefore, display a narrow hysteresis loop, as shown by loop (a) in Fig. 3.30.

The hysteresis effect causes an energy loss whenever the magnetic flux changes. This energy loss is converted to heat in the iron. The energy lost during a complete cycle of flux change is proportional to the area enclosed by the hysteresis loop.

When an iron core is subjected to alternating magnetization, as in a transformer, the energy loss occurs at every cycle and so constitutes a continuous power loss, and, therefore, for applications such as transformers, a material with a narrow hysteresis loop is required.

FLEMING'S RIGHT-HAND RULE

We have seen that electricity and magnetism are connected together by Faraday's laws of electromagnetic induction, which say that forces are exerted on current carrying conductors placed within a magnetic field and that conductors moving in a magnetic field have an emf induced in them. These laws have applications to electric motors and generators. When carrying out his experiments many years later, it occurred to Fleming that the thumb and first two fingers of the right hand could be used to predict the direction of the induced emf and so he formulated the following rule. Extend the thumb, first finger and second finger of the right hand so that they are all at right angles to each other. If the first finger is pointed in the direction of the magnetic field (north to south) and the thumb in the direction of the motion, then the second finger will point in the direction of the induced emf and current flow. This is shown by Fig. 3.31.

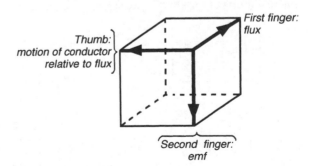

Fig. 3.31 Fleming's right-hand rule.

LENZ'S LAW

After the publication of Faraday's work on the production of electricity from magnetism in 1831, scientists in other countries repeated the experiments and built upon the basic principles adding to the total knowledge.

In Russia, Heinrich Lenz was able to publish another law of electromagnetic induction in 1834. This states that the direction of the induced emf always sets up a current opposing the motion which induced the emf.

This leads us to the concept of the back emf in a motor and the reason for the negative sign in the previous calculation on induced emf in the coil shown in Fig. 3.28.

Direct current motors

If a current carrying conductor is placed into the field of a permanent magnet as shown in Fig. 3.32 (c) a force F will be exerted on the conductor to push it out of the magnetic field.

To understand the force, let us consider each magnetic field acting alone. Figure 3.32 (a) shows the magnetic field due to the current carrying conductor only. Figure 3.32 (b) shows the magnetic field due to the permanent magnet in which is placed the conductor carrying no current. Figure 3.32 (c) shows the effect of the combined magnetic fields which are distorted and, because lines of magnetic flux never cross, but behave like stretched elastic bands, always trying to find the shorter distance between a north and south pole, the force F is exerted on the conductor, pushing it out of the permanent magnetic field.

This is the basic motor principle, and the force F is dependent upon the strength of the magnetic field B, the magnitude of the current flowing in the conductor

I and the length of conductor within the magnetic field l. The following equation expresses this relationship:

$$F = BlI \text{ (N)}$$

where B is in tesla, l is in metres, I is in amperes and F is in newtons.

EXAMPLE

A coil which is made up of a conductor some 15 m in length, lies at right angles to a magnetic field of strength 5 T. Calculate the force on the conductor when 15 A flows in the coil.

$$F = BlI \text{ (N)}$$
$$F = 5 \text{T} \times 15 \text{m} \times 15 \text{A} = 1125 \text{N}$$

PRACTICAL d.c. MOTORS

Practical motors are constructed as shown in Fig. 3.33. All d.c. motors contain a field winding wound on pole pieces attached to a steel yoke. The armature winding rotates between the poles and is connected to the commutator. Contact with the external circuit is made through carbon brushes rubbing on the commutator segments. Direct current motors are classified by the way in which the field and armature windings are connected, which may be in series or in parallel.

Series motor

The field and armature windings are connected in series and consequently share the same current. The series

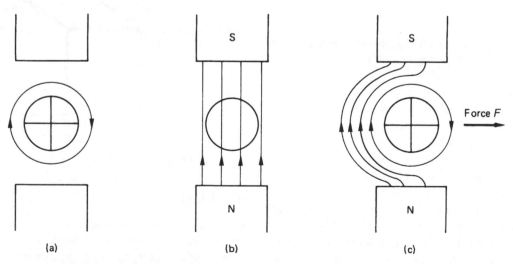

Fig. 3.32 Force on a conductor in a magnetic field.

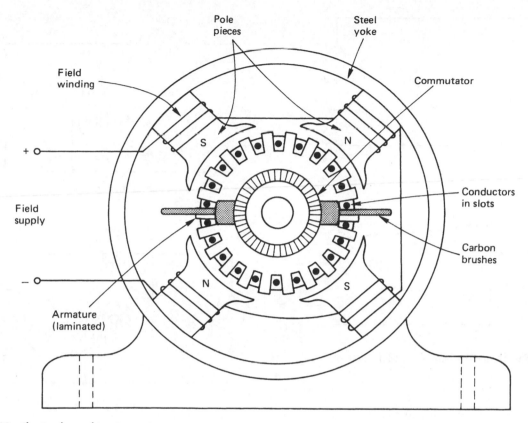

Fig. 3.33 Showing d.c. machine construction.

motor has the characteristics of a high starting torque but a speed which varies with load. Theoretically the motor would speed up to self-destruction, limited only by the windage of the rotating armature and friction, if the load were completely removed. Figure 3.34 shows series motor connections and characteristics. For this reason the motor is only suitable for direct coupling to a load, except in very small motors, such as vacuum cleaners and hand drills, and is ideally suited for applications where the machine must start on load, such as electric trains, cranes and hoists.

Reversal of rotation may be achieved by reversing the connections of either the field or armature windings but not both. This characteristic means that the machine will run on both a.c. or d.c. and is, therefore, sometimes referred to as a 'universal' motor.

Shunt motor

The field and armature windings are connected in parallel (see Fig. 3.35). Since the field winding is across the supply, the flux and motor speed are considered

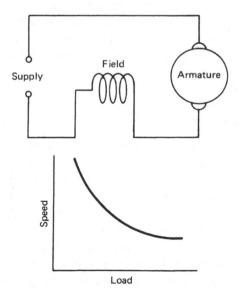

Fig. 3.34 Series motor connections and characteristics.

constant under normal conditions. In practice, however, as the load increases the field flux distorts and there is a small drop in speed of about 5% at full load, as shown in Fig. 3.35. The machine has a low starting

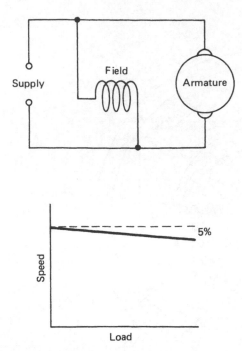

Fig. 3.35 Shunt motor connections and characteristics.

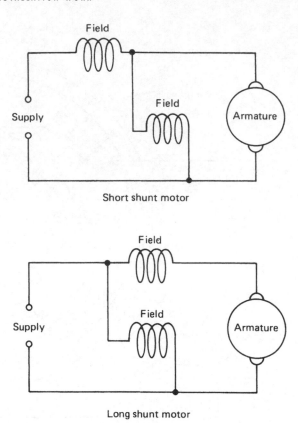

Fig. 3.36 Compound motor connections.

torque and it is advisable to start with the load disconnected. The shunt motor is a very desirable d.c. motor because of its constant speed characteristics. It is used for driving power tools, such as lathes and drills. Reversal of rotation may be achieved by reversing the connections to either the field or armature winding but not both.

Compound motor

The compound motor has two field windings – one in series with the armature and the other in parallel. If the field windings are connected so that the field flux acts in opposition, the machine is known as a *short shunt* and has the characteristics of a series motor. If the fields are connected so that the field flux is strengthened, the machine is known as a *long shunt* and has constant speed characteristics similar to a shunt motor. The arrangement of compound motor connections is given in Fig. 3.36. The compound motor may be designed to possess the best characteristics of both series and shunt motors, that is, good starting torque together with almost constant speed. Typical applications are for electric motors in steel rolling mills, where a constant speed is required under varying load conditions.

SPEED CONTROL OF d.c. MACHINES

One of the advantages of a d.c. machine is the ease with which the speed may be controlled. The speed of a d.c. motor is inversely proportional to the strength of the magnetic flux in the field winding. The magnetic flux in the field winding can be controlled by the field current and, as a result, controlling the field current will control the motor speed.

A variable resistor connected into the field circuit, as shown in Fig. 3.37, provides one method of controlling the field current and the motor speed. This method has the disadvantage that much of the input energy is dissipated in the variable resistor and an alternative, when an a.c. supply is available, is to use thyristor control.

BACK emf AND MOTOR STARTING

When the armature conductors cut the magnetic flux of the main field, an emf is induced in the armature, as described earlier on page 72. This induced emf is known as the back emf, since it acts in opposition to the supply voltage. During normal running, the back emf is always a little smaller than the supply voltage,

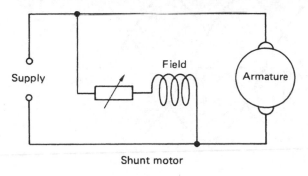

Field

Armature

Supply

Shunt motor

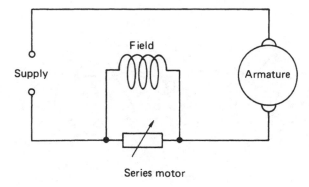

Field

Armature

Supply

Series motor

Fig. 3.37 Speed control of a d.c. motor.

and acts as a limit to the motor current. However, when the motor is first switched on, the back emf does not exist because the conductors are stationary and so a motor starter is required to limit the starting current to a safe value. This applies to all but the very smallest of motors and is achieved by connecting a resistor in series with the armature during starting, so that the resistance can be gradually reduced as the speed builds up.

The control switch of Fig. 3.38 is moved to the start position, which connects the variable resistors in series with the motor, thereby limiting the starting current. The control switch is moved progressively over the variable resistor contacts to the run position as the motor speed builds up. A practical motor starter is designed so that the control switch returns automatically to the 'off' position whenever the motor stops, so that the starting resistors are connected when the machine is once again switched on.

Three-phase a.c. motors

If a three-phase supply is connected to three separate windings equally distributed around the stationary

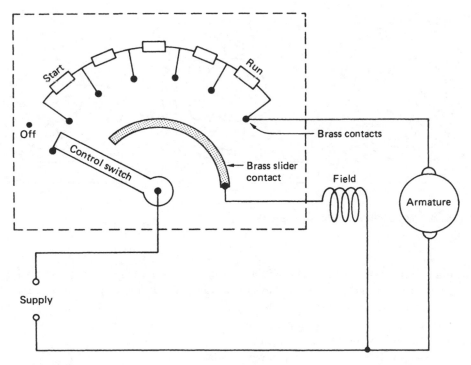

Fig. 3.38 A d.c. motor starting.

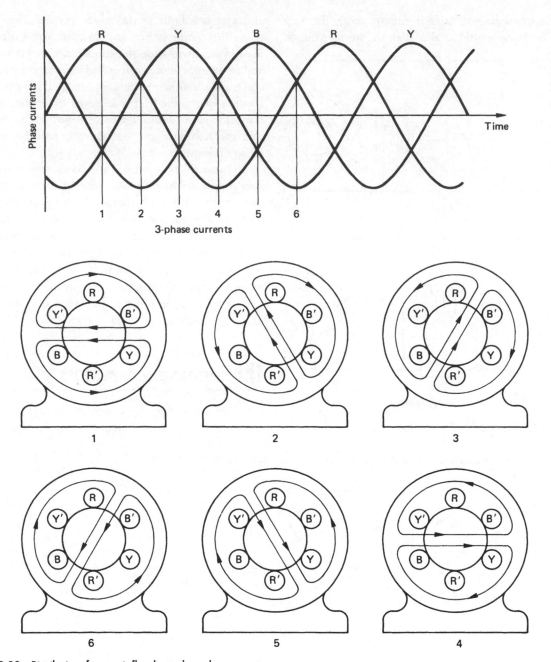

Fig. 3.39 Distribution of magnetic flux due to three-phase currents.

part or stator of an electrical machine, an alternating current circulates in the coils and establishes a magnetic flux. The magnetic field established by the three-phase currents travels in a clockwise direction around the stator, as can be seen by considering the various intervals of time 1 to 6 shown in Fig. 3.39. The three-phase supply establishes a rotating magnetic flux which rotates at the same speed as the supply frequency. This is called synchronous speed, denoted n_S:

$$n_S = \frac{f}{P} \quad \text{or} \quad N_S = \frac{60f}{P}$$

where

n_S is measured in revolutions per second

N_S is measured in revolutions per minute
f is the supply frequency measured in hertz
P is the number of pole pairs.

EXAMPLE

Calculate the synchronous speed of a four-pole machine connected to a 50 Hz mains supply.
  We have

$$n_S = \frac{f}{P} \text{ (rps)}$$

A four-pole machine has two pairs of poles:

$$\therefore n_S = \frac{50\ \text{Hz}}{2} = 25\ \text{rps}$$

$$\text{or}\quad N_S = \frac{60 \times 50\ \text{Hz}}{2} = 1500\ \text{rpm}$$

This rotating magnetic field is used to practical effect in the induction motor.

THREE-PHASE INDUCTION MOTOR

When a three-phase supply is connected to insulated coils set into slots in the inner surface of the stator or stationary part of an induction motor as shown in Fig. 3.40 (a), a rotating magnetic flux is produced. The rotating magnetic flux cuts the conductors of the rotor and induces an emf in the rotor conductors by Faraday's law, which states that when a conductor cuts or is cut by a magnetic field, an emf is induced in that conductor, the magnitude of which is proportional to the *rate* at which the conductor cuts or is cut by the magnetic flux. This induced emf causes rotor currents to flow and establish a magnetic flux which reacts with the stator flux and causes a force to be exerted on the rotor conductors, turning the rotor as shown in Fig. 3.40 (b).

The turning force or torque experienced by the rotor is produced by inducing an emf into the rotor conductors due to the *relative* motion between the conductors and the rotating field. The torque produces rotation in the same direction as the rotating magnetic field. At switch-on, the rotor speed increases until it approaches the speed of the rotating magnetic flux, that is, the synchronous speed. The faster the rotor revolves the less will be the difference in speed between the rotor and the rotating magnetic field. By Faraday's laws, this will result in less induced emf, less rotor current and less torque on the rotor. The rotor can never run at synchronous speed because, if it did so, there would be no induced emf, no current and no torque. The induction motor is called an asynchronous motor. In practice, the rotor runs at between 2% and 5% below the synchronous speed so that a torque can be maintained on the rotor which overcomes the rotor losses and the applied load.

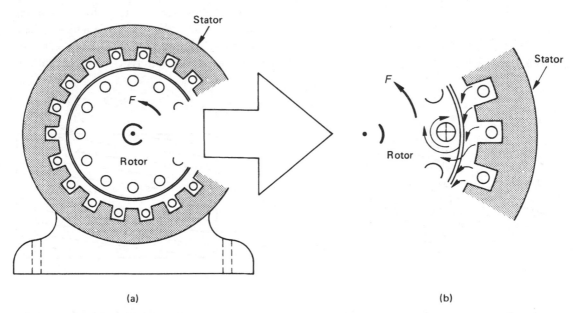

(a)   (b)

Fig. 3.40   Segment taken out of an induction motor to show turning force: (a) construction of an induction motor; (b) production of torque by magnetic fields.

The difference between the rotor speed and synchronous speed is called slip; the per-unit slip, denoted s, is given by

$$s = \frac{n_S - n}{n_S} = \frac{N_S - N}{N_S}$$

where
n_S = synchronous speed in revolutions per second
N_S = synchronous speed in revolutions per minute
n = rotor speed in revolutions per second
N = rotor speed in revolutions per minute.

The percentage slip is just the per-unit slip multiplied by 100.

EXAMPLE

A two-pole induction motor runs at 2880 rpm when connected to the 50 Hz mains supply. Calculate the percentage slip.

The synchronous speed is given by

$$N_S = \frac{60 \times f}{p} \text{ (rpm)}$$

$$\therefore N_S = \frac{60 \times 50 \text{ Hz}}{1} = 3000 \text{ rpm}$$

Thus the per-unit slip is

$$s = \frac{N_s - N}{N_s}$$

$$\therefore s = \frac{3000 \text{ rpm} - 2880 \text{ rpm}}{3000 \text{ rpm}}$$

$$s = 0.04.$$

So the percentage slip is $0.04 \times 100 = 4\%$.

ROTOR CONSTRUCTION

There are two types of induction motor rotor – the wound rotor and the cage rotor. The cage rotor consists of a laminated cylinder of silicon steel with copper or aluminium bars slotted in holes around the circumference and short-circuited at each end of the cylinder as shown in Fig. 3.41. In small motors the rotor is cast in aluminium. Better starting and quieter running are achieved if the bars are slightly skewed. This type of rotor is extremely robust and since there are no external connections there is no need for slip-rings or brushes. A machine fitted with a cage rotor does suffer from a low starting torque and the machine must be chosen which has a higher starting torque than the load, as shown by curve (b) in Fig. 3.42. A machine with the characteristic shown by curve (a) in Fig. 3.42 would not start since the load torque is greater than the machine starting torque.

Alternatively the load may be connected after the motor has been run up to full speed, or extra resistance can be added to a wound rotor through slip-rings and brushes since this improves the starting torque, as shown by curve (c) in Fig. 3.42. The wound rotor consists of a laminated cylinder of silicon steel with copper coils embedded in slots around the circumference. The windings may be connected in star or delta and the end connections brought out to slip rings mounted on the shaft. Connection by carbon brushes can then be made to an external resistance to improve starting, but once normal running speed is achieved the external resistance is short circuited. Therefore, the principle of operation for both types of rotor is the same.

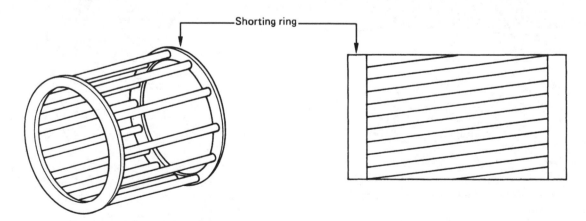

Shorting ring

Arrangement of conductor bars in a cage rotor

Skewed rotor conductors

Fig. 3.41   Construction of a cage rotor.

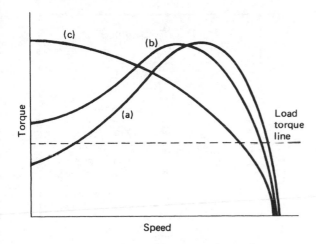

Fig. 3.42 Various speed–torque characteristics for an induction motor.

The cage induction motor has a small starting torque and should be used with light loads or started with the load disconnected. The speed is almost constant at about 5% less than synchronous speed. Its applications are for constant speed machines such as fans and pumps. Reversal of rotation is achieved by reversing any two of the stator winding connections.

THREE-PHASE SYNCHRONOUS MOTOR

If the rotor of a three-phase induction motor is removed and replaced with a simple magnetic compass, the compass needle will rotate in the same direction as the rotating magnetic field set up by the stator winding. That is, the compass needle will rotate at synchronous speed. This is the basic principle of operation of the synchronous motor.

In a practical machine, the rotor is supplied through slip-rings with a d.c. supply which sets up an electromagnet having north and south poles.

When the supply is initially switched on, the rotor will experience a force, first in one direction and then in the other direction every cycle as the stator flux rotates around the rotor at synchronous speed. Therefore, the synchronous motor is not self-starting. However, if the rotor is rotated at or near synchronous speed, then the stator and rotor poles of opposite polarity will 'lock together' producing a turning force or torque which will cause the rotor to rotate at synchronous speed.

If the rotor is slowed down and it comes out of synchronism, then the rotor will stop because the torque will be zero. The synchronous motor can, therefore, only be run at synchronous speed, which for a 50 Hz supply will be 3000, 1500, 1000 or 750 rpm depending upon the number of poles, as discussed earlier in this chapter.

A practical synchronous machine can be brought up to synchronous speed by either running it initially as an induction motor or by driving it up to synchronous speed by another motor called a 'pony motor'. Once the rotor achieves synchronous speed the pony motor is disconnected and the load applied to the synchronous motor.

With such a complicated method of starting the synchronous motor, it is clearly not likely to find applications which require frequent stopping and starting. However, the advantage of a synchronous motor is that it runs at a constant speed and operates at a leading power factor. It can, therefore, be used to improve a bad power factor while driving constant speed machines such as ventilation fans and pumping compressors.

Single-phase a.c. motors

A single-phase a.c. supply produces a pulsating magnetic field, not the rotating magnetic field produced by a three-phase supply. All a.c. motors require a rotating field to start. Therefore, single-phase a.c. motors have two windings which are electrically separated by about 90°. The two windings are known as the start and run windings. The magnetic fields produced by currents flowing through these out-of-phase windings create the rotating field and turning force required to start the motor. Once rotation is established, the pulsating field in the run winding is sufficient to maintain rotation and the start winding is disconnected by a centrifugal switch which operates when the motor has reached about 80% of the full load speed.

A cage rotor is used on single-phase a.c. motors, the turning force being produced in the way described previously for three-phase induction motors and shown in Fig. 3.40. Because both windings carry currents which are out of phase with each other, the motor is known as a 'split-phase' motor. The phase displacement between the currents in the windings is achieved in one of two ways:

■ by connecting a capacitor in series with the start winding, as shown in Fig. 3.43(a), which gives a

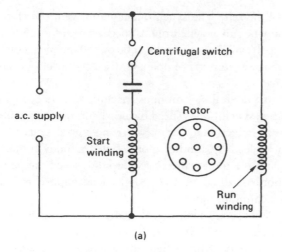

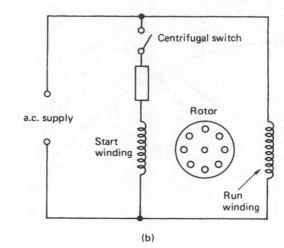

Fig. 3.43   Circuit diagram of: (a) capacitor split-phase motors; (b) resistance split-phase motors.

90° phase difference between the currents in the start and run windings;

■ by designing the start winding to have a high resistance and the run winding a high inductance, once again creating a 90° phase shift between the currents in each winding, as shown in Fig. 3.43(b).

When the motor is first switched on, the centrifugal switch is closed and the magnetic fields from the two coils produce the turning force required to run the rotor up to full speed. When the motor reaches about 80% of full speed, the centrifugal switch clicks open and the machine continues to run on the magnetic flux created by the run winding only.

Split-phase motors are constant speed machines with a low starting torque and are used on light loads such as fans, pumps, refrigerators and washing machines. Reversal of rotation may be achieved by reversing the connections to the start or run windings, but not both.

SHADED POLE MOTORS

The shaded pole motor is a simple, robust single-phase motor, which is suitable for very small machines with a rating of less than about 50 watts. Figure 3.44 shows a shaded pole motor. It has a cage rotor and the moving field is produced by enclosing one side of each stator pole in a solid copper or brass ring, called a shading ring, which displaces the magnetic field and creates an artificial phase shift.

Shaded pole motors are constant speed machines with a very low starting torque and are used on very light

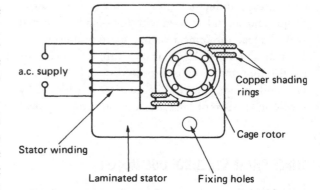

Fig. 3.44   Shaded pole motor.

loads such as oven fans, record turntable motors and electric fan heaters. Reversal of rotation is theoretically possible by moving the shading rings to the opposite side of the stator pole face. However, in practice this is often not a simple process, but the motors are symmetrical and it is sometimes easier to reverse the rotor by removing the fixing bolts and reversing the whole motor.

There are more motors operating from single-phase supplies than all other types of motor added together. Most of them operate as very small motors in domestic and business machines where single-phase supplies are most common.

Motor starters

The magnetic flux generated in the stator of an induction motor rotates immediately the supply is switched on, and therefore the machine is self-starting. The purpose of the motor starter is not to start the machine, as the name implies, but to reduce heavy starting currents and provide overload and no-volt protection in accordance with the requirements of Regulations 552.

Thermal overload protection is usually provided by means of a bimetal strip bending under overload conditions and breaking the starter contactor coil circuit. This de-energizes the coil and switches off the motor under fault conditions such as overloading or single phasing. Once the motor has automatically switched off under overload conditions or because a remote stop/start button has been operated, it is an important safety feature that the motor cannot restart without the operator going through the normal start-up procedure. Therefore, no-volt protection is provided by incorporating the safety devices into the motor starter control circuit which energizes the contactor coil.

Electronic thermistors (thermal transistors) provide an alternative method of sensing if a motor is overheating. These tiny heat-sensing transistors, about the size of a matchstick head, are embedded in the motor windings to sense the internal temperature, and the thermistor either trips out the contactor coil as described above or operates an alarm.

All electric motors with a rating above 0.37 kW must be supplied from a suitable motor starter and we will now consider the more common types.

DIRECT ON LINE (d.o.l.) STARTERS

The d.o.l. starter switches the main supply directly on to the motor. Since motor starting currents can be seven or eight times greater than the running current, the d.o.l. starter is only used for small motors of less than about 5 kW rating.

When the start button is pressed current will flow from the red phase through the control circuit and contactor coil to the blue phase which energizes the contactor coil and the contacts close, connecting the three-phase supply to the motor, as can be seen in Fig. 3.45. If the start button is released the control circuit is maintained by the hold on contact. If the stop button is pressed or the overload coils operate, the control circuit is broken and the contractor drops out, breaking the supply to the load. Once the supply is interrupted the supply to the motor can only be reconnected by pressing the start button.

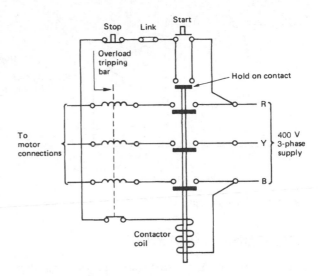

Fig. 3.45 Three-phase d.o.l. starter.

Therefore this type of arrangement also provides no-volt protection.

When large industrial motors have to be started, a way of reducing the excessive starting currents must be found. One method is to connect the motor to a star delta starter.

STAR DELTA STARTERS

When three loads, such as the three windings of a motor, are connected in star, the line current has only one-third of the value it has when the same load is connected in delta. A starter which can connect the motor windings in star during the initial starting period and then switch to delta connection will reduce the problems of an excessive starting current. This arrangement is shown in Fig. 3.46, where the six connections to the three stator phase windings are brought out to the starter. For starting, the motor windings are star-connected at the a–b–c end of the winding by the star making contacts. This reduces the phase voltage to about 58% of the running voltage which reduces the current and the motor's torque. Once the motor is running a double-throw switch makes the changeover from star starting to delta running, thereby achieving a minimum starting current and maximum running torque. The starter will incorporate overload and no-volt protection, but these are not shown in Fig. 3.46 in the interests of showing more clearly the principle of operation.

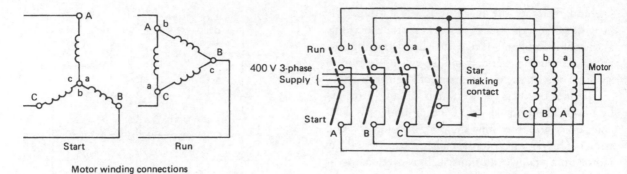

Motor winding connections

Fig. 3.46 Star delta starter.

AUTO-TRANSFORMER STARTER

An auto-transformer motor starter provides another method of reducing the starting current by reducing the voltage during the initial starting period. Since this also reduces the starting torque, the voltage is only reduced by a sufficient amount to reduce the starting current, being permanently connected to the tapping found to be most appropriate by the installing electrician. Switching the changeover switch to the start position connects the auto-transformer windings in series with the delta-connected motor starter winding. When sufficient speed has been achieved by the motor the changeover switch is moved to the run connections which connect the three-phase supply directly on to the motor as shown in Fig. 3.47.

This starting method has the advantage of only requiring three connection conductors between the motor starter and the motor. The starter will incorporate overload and no-volt protection in addition to some method of preventing the motor being switched to the run position while the motor is stopped. These protective devices are not shown in Fig. 3.47 in order to show more clearly the principle of operation.

ROTOR RESISTANCE STARTER

When starting a machine on load a wound rotor induction motor must generally be used since this allows an external resistance to be connected to the rotor winding through slip-rings and brushes, which increases the starting torque as shown in Fig. 3.42, curve (c).

When the motor is first switched on the external rotor resistance is at a maximum. As the motor speed increases the resistance is reduced until at full speed the external resistance is completely cut out and the

Fig. 3.47 Auto-transformer starting.

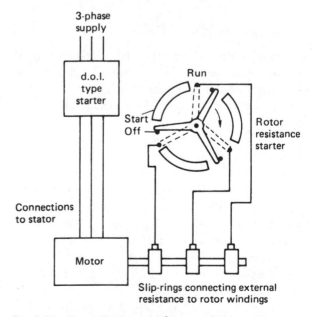

Fig. 3.48 Rotor resistance starter for a wound rotor machine.

machine runs as a cage induction motor. The starter is provided with overload and no-volt protection and an interlock to prevent the machine being switched on with no rotor resistance connected, but these are not shown in Fig. 3.48 since the purpose of the diagram is to show the principle of operation.

Remote control of motors

When it is required to have stop/start control of a motor at a position other than the starter position, additional start buttons may be connected in parallel and additional stop buttons in series, as shown in Fig. 3.49 for the d.o.l. starter. This is the diagram shown in Fig. 3.45 with the link removed and a remote stop/start button connected. Additional stop and start facilities are often provided for the safety and convenience of the machine operator.

Installation of motors

Electric motors vibrate when running and should be connected to the electrical installation through a flexible connection. This may also make final adjustments of the motor position easier. Where the final connection is made with flexible conduit, the tube must not be relied upon to provide a continuous earth path and a separate CPC must be run either inside or outside the flexible conduit (Regulation 543–02–01).

All motors over 0.37 kW rating must be connected to the source of supply through a suitable starter which incorporates overload protection and a device which prevents dangerous restarting of the motor following a mains failure (Regulations 552–01–02 and 03).

The cables supplying the motor must be capable of carrying at least the full load current of the motor (Regulation 552–01–01) and a local means of isolation

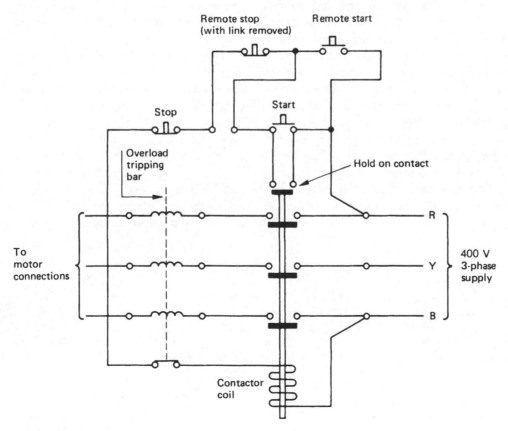

Fig. 3.49 Remote stop/start connections to d.o.l. starter.

must be provided to facilitate safe mechanical maintenance (Regulation 476–02–03).

At the supply end, the motor circuit will be protected by a fuse or MCB. The supply protection must be capable of withstanding the motor starting current while providing adequate overcurrent protection. There must also be discrimination so that the overcurrent device in the motor starter operates first in the event of an excessive motor current.

Most motors are 'continuously rated'. This is the load at which the motor may be operated continuously without overheating.

Many standard motors have class A insulation which is suitable for operating in ambient temperatures up to about 55°C. If a class A motor is to be operated in a higher ambient temperature, the continuous rating may need to be reduced to prevent damage to the motor. The motor and its enclosure must be suitable for the installed conditions and must additionally prevent anyone coming into contact with the internal live or moving parts. Many different enclosures are used, depending upon the atmosphere in which the motor is situated. Clean air, damp conditions, dust particles in the atmosphere, chemical or explosive vapours will determine the type of motor enclosure. In high ambient temperatures it may be necessary to provide additional ventilation to keep the motor cool and prevent the lubricating oil thinning. The following motor enclosures are examples of those to be found in industry.

Screen protected enclosures prevent access to the internal live and moving parts by covering openings in the motor casing with metal screens of perforated metal or wire mesh. Air flow for cooling is not restricted and is usually assisted by a fan mounted internally on the machine shaft. This type of enclosure is shown in Fig. 3.50.

A duct ventilated enclosure is used when the air in the room in which the motor is situated is unsuitable for passing through the motor for cooling – for example, when the atmosphere contains dust particles or chemical vapour. In these cases the air is drawn from a clean air zone outside the room in which the machine is installed, as shown in Fig. 3.50.

A totally enclosed enclosure is one in which the air inside the machine casing has no connection with the air in the room in which it is installed, but it is not necessarily airtight. A fan on the motor shaft inside the casing circulates the air through the windings and cooling is by conduction through motor casing. To increase the surface area and assist cooling, the casing is surrounded by fins, and an externally mounted fan can increase the flow of air over these fins. This type of enclosure is shown in Fig. 3.50.

A flameproof enclosure requires that further modifications be made to the totally enclosed casing to prevent inflammable gases coming into contact with sparks or arcing inside the motor. To ensure that the motor meets the stringent regulations for flameproof enclosures the shaft is usually enclosed in special bearings and the motor connections contained by a wide flange junction box.

When a motor is connected to a load, either by direct coupling or by a vee belt, it is important that the shafts or pulleys are exactly in line. This is usually best achieved by placing a straight edge or steel rule across the flange coupling of a direct drive or across the flat faces of a pair of pulleys, as shown in Fig. 3.51. Since pulley belts stretch in use it is also important to have some means of adjusting the tension of the vee belt. This is usually achieved by mounting the motor on a pair of slide rails as shown in Fig. 3.52. Adjustment is carried out by loosening the motor fixing bolts,

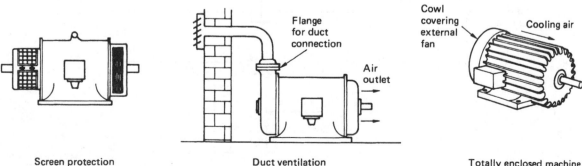

Screen protection　　　　　Duct ventilation　　　　　Totally enclosed machine

Fig. 3.50 Motor enclosures.

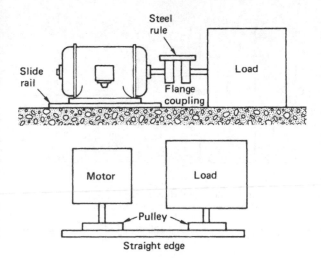

Fig. 3.51 Pulley and flange coupling arrangement.

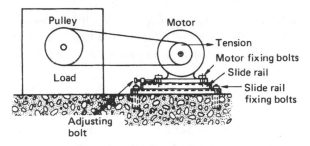

Fig. 3.52 Vee belt adjustment of slide rail mounted motor.

screwing in the adjusting bolts which push the motor back, and when the correct belt tension has been achieved the motor fixing bolts are tightened.

Motor maintenance

All rotating machines are subject to wear, simply because they rotate. Motor fans which provide cooling also pull dust particles from the surrounding air into the motor enclosure. Bearings dry out, drive belts stretch and lubricating oils and greases require replacement at regular intervals. Industrial electric motors are often operated in a hot, dirty, dusty or corrosive environment for many years. If they are to give good and reliable service they must be suitable for the task and the conditions in which they must operate. Maintenance at regular intervals is required, in the same way that a motor car is regularly serviced.

The solid construction of the cage rotor used in many a.c. machines makes them almost indestructible, and, since there are no external connections to the rotor, the need for slip-rings and brushes is eliminated. These characteristics give cage rotor a.c. machines maximum reliability with the minimum of maintenance and make the induction motor the most widely used in industry. Often the only maintenance required with an a.c. machine is to lubricate in accordance with the manufacturer's recommendations.

However, where high torque and variable speed characteristics are required d.c. machines are often used. These require a little more maintenance than a.c. machines because the carbon brushes, rubbing on the commutator, wear down and require replacing. New brushes must be of the correct grade and may require 'bedding in' or shaping with a piece of fine abrasive cloth to the curve of the commutator.

The commutator itself should be kept clean and any irregularities smoothed out with abrasive cloth. As the commutator wears, the mica insulation between the segments must be cut back with an undercutting tool or a hacksaw blade to keep the commutator surface smooth. If the commutator has become badly worn, and a groove is evident, the armature will need to be removed from the motor, and the commutator turned in a lathe.

Motors vibrate when operating and as a result fixing bolts and connections should be checked as part of the maintenance operation. Where a motor drives a load via a pulley belt, the motor should be adjusted on the slider rails until there is about 10 mm of play in the belt.

Planning maintenance work with forethought and keeping records of work done with dates can have the following advantages:

- the maintenance is carried out when it is most convenient;
- regular simple maintenance often results in less emergency maintenance;
- regular servicing and adjustment maintain the plant and machines at peak efficiency.

The result of planned maintenance is often that fewer breakdowns occur, which result in loss of production time. Therefore, a planned maintenance programme must be a sensible consideration for any commercial operator.

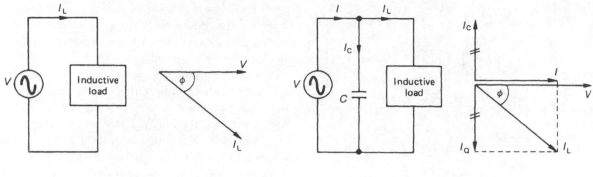

(a) Circuit and phasor diagram for an inductive load with low p.f.

(b) Circuit and phasor diagram for circuit (a) with capacitor correcting p.f. to unity

Fig. 3.53 Power-factor correction of inductive load: (a) circuit and phasor diagram for an inductive load with low p.f.; (b) circuit and phasor diagram for (a) with capacitor correcting p.f. to unity.

Power-factor correction

Most electrical installations have a low power factor because loads such as motors, transformers and discharge lighting circuits are inductive in nature and cause the current to lag behind the voltage. A capacitor has the opposite effect to an inductor, causing the current to lead the voltage. Therefore, by adding capacitance to an inductive circuit the bad power factor can be corrected. The load current I_L is made up of an in-phase component I and a quadrature component I_Q. The power factor can be corrected to unity when the capacitor current I_C is equal and opposite to the quadrature or reactive current I_Q of the inductive load. The quadrature or reactive current is responsible for setting up the magnetic field in an inductive circuit. Figure 3.53 shows the power factor corrected to unity, that is when $I_Q = I_C$.

A low power factor is considered a disadvantage because a given load takes more current at a low power factor than it does at a high power factor. Earlier in this chapter under the heading Power and Power Factor example 3 on page 133 we calculated that a 1.84 kW load at unity power factor took 8 A, but at a bad power factor of 0.4 a current of 20 A was required to supply the same load.

The supply authorities discourage industrial consumers from operating at a bad power factor because:

- larger cables and switchgear are necessary to supply a given load;
- larger currents give rise to greater copper losses in transmission cables and transformers;
- larger currents give rise to greater voltage drops in cables;
- larger cables may be required on the consumer's side of the electrical installation to carry the larger currents demanded by a load operating with a bad power factor.

Bad power factors are corrected by connecting a capacitor either across the individual piece of equipment or across the main busbars of the installation. When individual capacitors are used they are usually of the paper dielectric type of construction (see Fig. 2.21 in Chapter 2 of this book). This is the type of capacitor used for power-factor correction in a fluorescent luminaire. When large banks of capacitors are required to correct the p.f. of a whole installation, paper dielectric capacitors are immersed in an oil tank in a similar type construction to a transformer, and connected on to the main busbars of the electrical installation by suitably insulated and mechanically protected cables.

The current to be carried by the capacitor for p.f. correction and the value of the capacitor may be calculated as shown by the following example.

EXAMPLE

An 8 kW load with a power factor of 0.7 is connected across a 400 V, 50 Hz supply. Calculate:

(a) the current taken by this load
(b) the capacitor current required to raise the p.f. to unity
(c) the capacitance of the capacitor required to raise the p.f. to unity.

For (a), since $P = VI \cos \phi$ (W),

$$I = \frac{P}{V \cos \phi} \text{ (A)}$$

$$\therefore I = \frac{8000 \text{ W}}{400 \text{ V} \times 0.7} = 28.57 \text{ (A)}$$

This current lags the voltage by an angle of 45.6° (since $\cos^{-1} 0.7 = 45.6°$) and can therefore be drawn to scale as shown in Fig. 3.54 and represented by line AB.

For (b) at unity p.f. the current will be in phase with the voltage, represented by line AC in Fig. 3.54. To raise the load current to this value will require a capacitor current I_C which is equal and opposite to the value of the quadrature or reactive component I_Q. The value of I_Q is measured from the phasor diagram and found to be 20 A which is the value of the capacitor current required to raise the p.f. to unity and shown by line AD in Fig. 3.54.

For (c), since

$$I_C = \frac{V}{X_C} \text{ (A) and } X_C = \frac{V}{I_C} \text{ (}\Omega\text{)}$$

$$\therefore X_C = \frac{400 \text{ V}}{20 \text{ A}} = 20 \text{ }\Omega$$

Since

$$X_C = \frac{1}{2\pi fc} \text{ (}\Omega\text{) and } C = \frac{1}{2\pi fX_C} \text{ (F)}$$

$$\therefore C = \frac{1}{2 \times \pi \times 50 \text{ Hz} \times 20 \text{ }\Omega} = 159 \text{ }\mu\text{F}$$

A 159 μF capacitor connected in parallel with the 8 kW load would correct the power factor to unity.

Transformers

A transformer is an electrical machine which is used to change the value of an alternating voltage. They vary in size from miniature units used in electronics to huge power transformers used in power stations. A transformer will only work when an alternating voltage is connected. It will not normally work from a d.c. supply such as a battery.

A transformer, as shown in Fig. 3.55, consists of two coils, called the primary and secondary coils, or windings, which are insulated from each other and wound on to the same steel or iron core.

An alternating voltage applied to the primary winding produces an alternating current, which sets up an alternating magnetic flux throughout the core. This

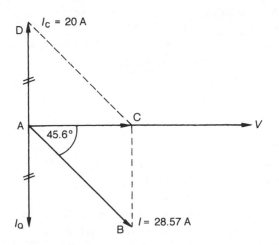

Fig. 3.54 Phasor diagram.

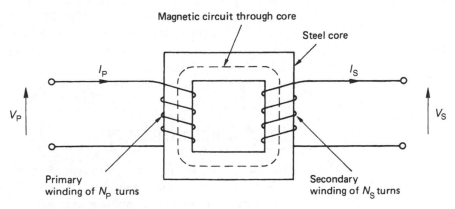

Fig. 3.55 A simple transformer.

magnetic flux induces an emf in the secondary winding, as described by Faraday's law, which says that when a conductor is cut by a magnetic field, an emf is induced in that conductor. Since both windings are linked by the same magnetic flux, the induced emf per turn will be the same for both windings. Therefore, the emf in both windings is proportional to the number of turns. In symbols,

$$\frac{V_P}{N_P} = \frac{V_S}{N_S} \qquad (1)$$

Most practical power transformers have a very high efficiency, and for an ideal transformer having 100% efficiency the primary power is equal to the secondary power:

$$\text{Primary power} = \text{Secondary power}$$

and, since

$$\text{Power} = \text{Voltage} \times \text{Current}$$

then

$$V_P \times I_P = V_S \times I_S \qquad (2)$$

Combining equations (1) and (2), we have

$$\frac{V_P}{V_S} = \frac{N_P}{N_S} = \frac{I_S}{I_P}$$

EXAMPLE

A 230 V to 12 V bell transformer is constructed with 800 turns on the primary winding. Calculate the number of secondary turns and the primary and secondary currents when the transformer supplies a 12 V 12 W alarm bell.

Collecting the information given in the question into a usable form, we have

$$V_P = 230\,V$$

$$V_S = 12\,V$$

$$N_P = 800$$

Power = 12 W
Information required: N_S, I_S and I_P
Secondary turns

$$N_S = \frac{N_P\,V_S}{V_P}$$

$$\therefore N_S = \frac{800 \times 12\,V}{230\,V} = 42\text{ turns}$$

Secondary current

$$I_S = \frac{\text{Power}}{V_S}$$

$$I_S = \frac{12\,W}{12\,V} = 1\,A$$

Primary current

$$I_P = \frac{I_S \times V_S}{V_P}$$

$$\therefore I_P = \frac{1\,A \times 12\,V}{230\,V} = 0.052\,A$$

Transformer losses

As they have no moving parts causing frictional losses, most transformers have a very high efficiency, usually better than 90%. However, the losses which do occur in a transformer can be grouped under two general headings: copper losses and iron losses.

Copper losses occur because of the small internal resistance of the windings. They are proportional to the load, increasing as the load increases because copper loss is an 'I^2R' loss.

Iron losses are made up of *hysteresis loss* and *eddy current loss*. The hysteresis loss depends upon the type of iron used to construct the core and consequently core materials are carefully chosen. Transformers will only operate on an alternating supply. Thus, the current which establishes the core flux is constantly changing from positive to negative. Each time there is a current reversal, the magnetic flux reverses and it is this build-up and collapse of magnetic flux in the core material which accounts for the hysteresis loss.

Eddy currents are circulating currents created in the core material by the changing magnetic flux. These are reduced by building up the core of thin slices or laminations of iron and insulating the separate laminations from each other. The iron loss is a constant loss consuming the same power from no load to full load.

Transformer efficiency

The efficiency of any machine is determined by the losses incurred by the machine in normal operation. The efficiency of rotating machines is usually in the region of 50–60% because they incur windage and friction losses; but the transformer has no moving parts so, therefore, these losses do not occur. However, the efficiency of a transformer can be calculated in the same way as for any other machine. The efficiency of a machine is generally given by:

$$\eta = \frac{\text{output power}}{\text{input power}}$$

(η is the Greek letter 'eta'). However, the input to the transformer must supply the output plus any losses which occur within the transformer. We can therefore say:

$$\text{input power} = \text{output power} + \text{losses}$$

Rewriting the basic formula, we have:

$$\eta = \frac{\text{output power}}{\text{output power} + \text{losses}}$$

EXAMPLE

A 100 kVA power transformer feeds a load operating at a power factor of 0.8. Find the efficiency of the transformer if the combined iron and copper loss at this load is 1 kW.

$$\text{Output power} = \text{kVA} \times \text{p.f.}$$
$$\therefore \text{Output power} = 100\,\text{kVA} \times 0.8$$
$$\text{Output power} = 80\,\text{kW}$$
$$\eta = \frac{\text{output power}}{\text{output power} + \text{losses}}$$
$$\eta = \frac{80\,\text{kW}}{80\,\text{kW} + 1\,\text{kW}} = 0.987$$

or, multiplying by 100 to give a percentage, the transformer has an efficiency of 98.7%.

Transformer construction

Transformers are constructed in a way which reduces the losses to a minimum. The core is usually made of silicon–iron laminations, because at fixed low frequencies silicon–iron has a small hysteresis loss and the laminations reduce the eddy current loss. The primary and secondary windings are wound close to each other on the same limb. If the windings are spread over two limbs, there will usually be half of each winding on each limb, as shown in Fig. 3.56.

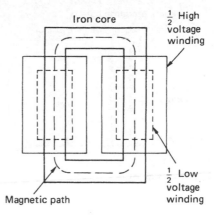

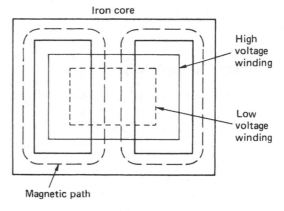

Fig. 3.56 Transformer construction.

AUTO-TRANSFORMERS

Transformers having a separate primary and secondary winding, as shown in Fig. 3.56, are called double-wound transformers, but it is possible to construct a transformer which has only one winding which is common to the primary and secondary circuits. The secondary voltage is supplied by means of a 'tapping' on the primary winding. An arrangement such as this is called an auto-transformer.

The auto-transformer is much cheaper and lighter than a double-wound transformer because less copper and iron are used in its construction. However, the primary and secondary windings are not electrically separate and a short circuit on the upper part of the winding shown in Fig. 3.57 would result in the primary voltage appearing across the secondary terminals. For this reason auto-transformers are mostly used where only a small difference is required between the primary and secondary voltages. When installing transformers,

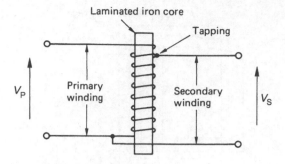

Fig. 3.57   An auto-transformer.

the regulations of Section 555 must be complied with, in addition to any other regulations relevant to the particular installation.

THREE-PHASE TRANSFORMERS

Most of the transformers used in industrial applications are designed for three-phase operation. In the double-wound type construction, as shown in Fig. 3.55, three separate single-phase transformers are wound on to a common laminated silicon–steel core to form the three-phase transformer. The primary and secondary windings may be either star or delta connected but in distribution transformers the primary is usually connected in delta and the secondary in star. This has the advantage of providing two secondary voltages, typically 400 V between phases and 230 V between phase and neutral from an 11 kV primary voltage. The coil arrangement is shown in Fig. 3.58.

The construction of the three-phase transformer is the same as the single-phase transformer, but because of the larger size the core is often cooled by oil.

OIL-IMMERSED CORE

As the rating of a transformer increases so does the problem of dissipating the heat generated in the core. In power distribution transformers the most common solution is to house the transformer in a steel casing containing insulating oil which completely covers the core and the windings. The oil is a coolant and an insulating medium for the core. On load the transformer heats up and establishes circulating convection currents in the oil which flows through the external tubes. Air passing over the tubes carries the heat away and cools the transformer. Figure 3.59 shows the construction of a typical oil-filled transformer.

TAP CHANGING

Under load conditions the secondary voltage of a transformer may fall and become less than that permitted by the Regulations. The tolerance permitted by the Regulations is the open-circuit voltage plus 10% or minus 6%. Because the voltage of a transformer is proportional to the number of turns, one solution is to vary the number of turns on either the primary or secondary winding, to achieve the desired voltage. This process is called tap changing, and most distribution transformers are fitted with a tap changing switch on the high-voltage (HV) winding so that

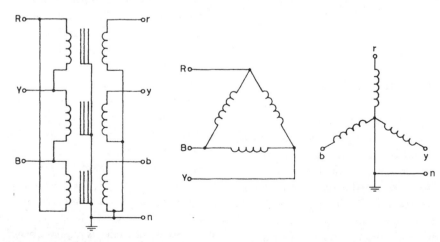

Fig. 3.58   Delta-star connected three-phase transformer.

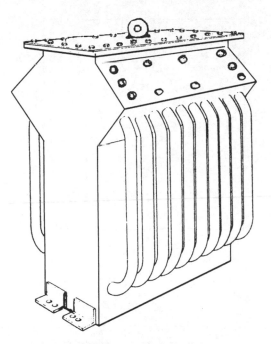

Fig. 3.59 Typical oil-filled power transformer.

the number of turns can be varied. These switches are always off-load devices and, therefore, the transformer must be isolated before the tap changing operation is carried out. The switch is usually padlocked to prevent unauthorized operation.

Instrument transformers

An instrument transformer works on the same principles as a power transformer but is designed to be used specifically in conjunction with an electrical measuring instrument to extend the range of an ammeter or voltmeter. The instrument transformer carries the current or voltage to be measured and the instrument is connected to the secondary winding of the transformer. In this way the instrument measures a small current or voltage which is proportional to the main current or voltage.

The advantages of using instrument transformers are as follows:

■ The secondary side of the instrument transformer is wound for low voltage which simplifies the insulation of the measuring instrument and makes it safe to handle.

■ The transformer isolates the instrument from the main circuit.
■ The measuring instrument can be read in a remote, convenient position connected by long leads to the instrument transformer.
■ The secondary voltage or current can be standardized (usually 110 V and 5 A), which simplifies instrument changes.

Voltage transformers

The construction of a voltage transformer (VT) is similar to the power transformer dealt with earlier in this chapter. The secondary winding of the VT is connected to a voltmeter as shown in Fig. 3.60. Voltage transformers are operated as step-down transformers, the secondary voltage usually being standardized at 110 V. A large number of turns are wound on the primary and a few on the secondary since

$$\frac{V_P}{V_S} = \frac{N_P}{N_S}$$

The voltmeter reading must be multiplied by the turns ratio to determine the load voltage.

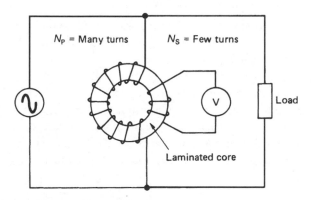

Fig. 3.60 A voltage transformer.

EXAMPLE 1

A voltmeter is connected to 50 turns on the secondary winding of a VT. The primary winding of 250 turns is connected to the main supply. Calculate the supply voltage if the voltmeter reading is 80 V.

$$\text{Primary voltage } V_P = \frac{N_P}{N_S} \times V_S$$

$$\therefore V_p = \frac{250}{50} \times 80 \text{ V}$$
$$V_p = 400 \text{ V}$$

As an alternative solution we could say the turns ratio is 250:50, that is 5:1, and therefore the supply voltage is $5 \times 80 = 400$ V.

EXAMPLE 2

An electrical contractor wishes to monitor a 660 V supply with a standard 110 V voltmeter. Determine the turns ratio of the VT to perform this task.

$$\frac{V_P}{V_S} = \frac{N_P}{N_S}$$
$$\frac{660 \text{ V}}{110 \text{ V}} = \frac{N_P}{N_S} = \frac{6}{1}$$

The turns ratio is 6:1. This means that the number of turns on the primary side of the VT must be six times as great as the number of turns on the secondary, which is connected to the 110 V voltmeter.

Current transformers

The operation of a current transformer (CT) is different to a power transformer although the transformer principle remains the same.

The secondary winding of the CT consists of a large number of turns connected to an ammeter as shown in Fig. 3.61(a). The ammeter is usually standardized at 1 A or 5 A and the transformer ratio chosen so that 1 A or 5 A flows when the main circuit carries full load current calculated from the transformer turns ratio

$$\frac{V_P}{V_S} = \frac{I_S}{I_P}$$

The primary winding is wound with only a few turns and when heavy currents are being measured one turn on the primary may be sufficient. In this case the conductor carrying the main current or the main busbar is passed through the centre of the CT as shown in Fig. 3.61(b). This is called a bar primary CT.

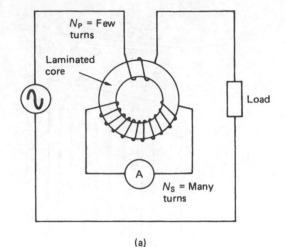

(a)

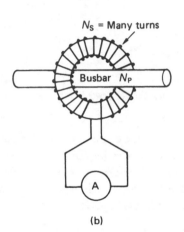

(b)

Fig. 3.61 Current transformers: (a) wound primary current transformer; (b) bar primary current transformer.

EXAMPLE 1

An ammeter having a full scale deflection of 5 A is used to measure a line current of 200 A. If the primary is wound with two turns, calculate the number of secondary turns required to give full scale deflection.

$$\frac{N_P}{N_S} = \frac{I_S}{I_P}$$
$$N_S = \frac{N_P \times I_P}{I_S}$$
$$N_S = \frac{2 \times 200 \text{ A}}{5 \text{ A}} = 80 \text{ turns}$$

With a power transformer a secondary load is necessary to cause a primary current to flow which maintains the magnetic flux in the core at a constant value. With a CT the primary current is the main circuit current and will flow whether the secondary is connected or not.

However, the secondary current through the ammeter is necessary to stabilize the magnetic flux in the core, and if the ammeter is removed the voltage across the secondary terminals could reach a dangerously high value and cause the insulation to break down or cause excessive heating of the core. The CT must never be operated with the secondary terminals open-circuited and overload protection should not be provided in the secondary circuit (Regulation 473–01–03). If the ammeter must be removed from the CT then the terminals must first be short-circuited. This will not damage the CT and will prevent a dangerous situation arising. The rating of an instrument transformer is measured in volt amperes and is called the burden. To reduce errors, the ammeter or voltmeter connected to the CT or VT should be operated at the rated burden.

EXAMPLE 2

To determine the power taken by a single-phase motor a wattmeter is connected to the circuit through a CT and VT. The test readings obtained were:

Wattmeter reading = 300 W
Voltage transformer turns ratio = 440/110 V
Current transformer turns ratio = 150/5 A.

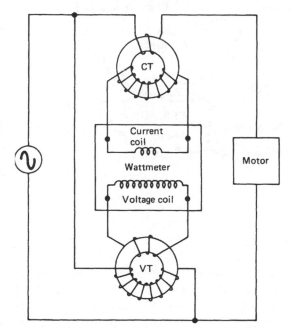

Fig. 3.62 Wattmeter connected through voltage transformer and current transformer.

Sketch the circuit arrangements and calculate the power taken by the motor.

The circuit arrangements are shown in Fig. 3.62.

Voltage transformer ratio = 440/110 = 4 : 1
Current transformer ratio = 150/5 = 30 : 1

$$\frac{\text{Team}}{\text{power}} = \frac{\text{Wattmeter}}{\text{reading}} \times \frac{\text{VT}}{\text{multiplier}} \times \frac{\text{CT}}{\text{multiplier}} \text{ (W)}$$

True power = 300 W × 4 × 30 = 36 kW
The power taken by the motor is therefore 36 kW.

Electricity generation, transmission and distribution

Following the denationalization of the electricity supply industry in March 1991, the generation of electrical energy in Great Britain has become the responsibility of three generating companies formed from the previously nationalized Central Electricity Generating Board. These are known as National Power, PowerGen and British Energy.

National Power was allocated 40 of the power stations previously operated by the CEGB. These have a generating capacity of 30 000 MW and provide about 48% of the electricity supply in England and Wales; 80% of them are coal-fired and 20% oil-fired. PowerGen was allocated 21 of the CEGB's power stations. These have a generating capacity of 19 000 MW and provide about 30% of the electricity supply in England and Wales; again, 80% of them are coal-fired and 20% oil-fired.

British Energy plc was formed in July 1996 to operate all of the nuclear power stations. They will provide between 20% and 25% of the daily base load power from 14 power stations.

The 12 electricity board areas have become the regional electricity companies which own the National Grid company and are responsible for the distribution of electricity to individual homes, offices and industry.

The North of Scotland Hydro-Electricity Board and the South of Scotland Electricity Board have been renamed Hydro-Electric and Scottish Power. These two electricity companies generate, transmit, distribute and sell electricity from the power station to the individual consumer.

Fig. 3.63 Regional electricity companies in the United Kingdom.

In England and Wales, the three generating companies generate the power, and the National Grid company transmits it to the 12 regional electricity companies which distribute and sell to the individual consumer. Figure 3.63 shows the areas covered by the Regional Electricity Companies.

Generation of electricity in most modern power stations is at 25 kV, and this voltage is then transformed to 400 kV for transmission. Virtually all the generators of electricity throughout the world are three-phase synchronous generators. The generator consists of a prime mover and a magnetic field excitor. The magnetic field is produced electrically by passing a direct current through a winding on an iron core, which rotates inside three phase windings on the stator of the machine. The magnetic field is rotated by means of a prime mover which may be a steam turbine, water turbine or gas turbine. Primary sources for electricity generation were discussed in Chapter 2 of this book.

The generators in modern power stations are rated between 500 MW and 100 MW. A 2000 MW station might contain four 500 MW sets, three 660 MW sets and a 20 MW gas turbine generator or two 1000 MW sets. Having a number of generator sets in a single power station provides the flexibility required for seasonal variations in the load and for maintenance of equipment. When generators are connected to a single system they must rotate at exactly the same speed, hence the term synchronous generator.

Very high voltages are used for transmission systems because, as a general principle, the higher the voltage the cheaper is the supply. Since power in an a.c. system is expressed as $P = VI \cos\theta$, it follows that an increase in voltage will reduce the current for a given amount of power. A lower current will result in reduced cable and switchgear size and the line power losses, given by the equation $P = I^2R$, will also be reduced.

The 132 kV grid and 400 kV Supergrid transmission lines are, for the most part, steel-cored aluminium conductors suspended on steel lattice towers, since this is about 16 times cheaper than the equivalent underground cable. Figure 3.64 shows a suspension tower on the Sizewell–Sundon 400 kV transmission line. The conductors are attached to porcelain insulator strings which are fixed to the cross-members of the tower as shown in Fig. 3.65. Three conductors comprise a single circuit of a three-phase system so that towers with six arms carry two separate circuits.

Primary distribution to consumers is from 11 kV substations, which for the most part are fed from 33 kV substations, but direct transformation between 132 kV and 11 kV is becoming common policy in city areas where over 100 MW can be economically distributed at 11 kV from one site. Figure 3.66 shows a block diagram indicating the voltages at the various stages of the transmission and distribution system.

Distribution systems at 11 kV may be ring or radial systems but a ring system offers a greater security of supply. The maintenance of a secure supply is an important consideration for any electrical engineer or

Fig. 3.64 Suspension tower on Sizewell–Sundon transmission line (by kind permission of the CEGB).

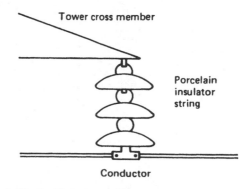

Fig. 3.65 Steel lattice tower cable supports.

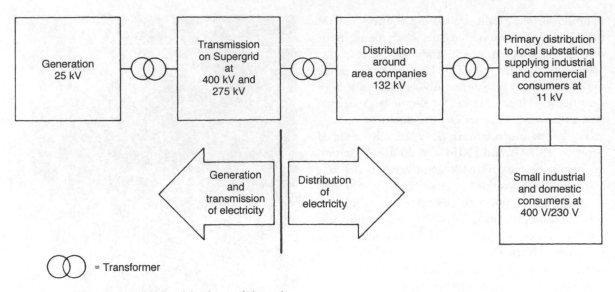

Fig. 3.66   Generation, transmission and distribution of electrical energy.

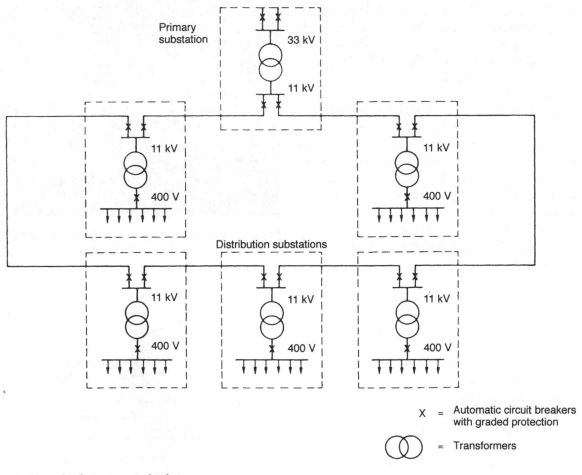

Fig. 3.67   High-voltage ring main distribution.

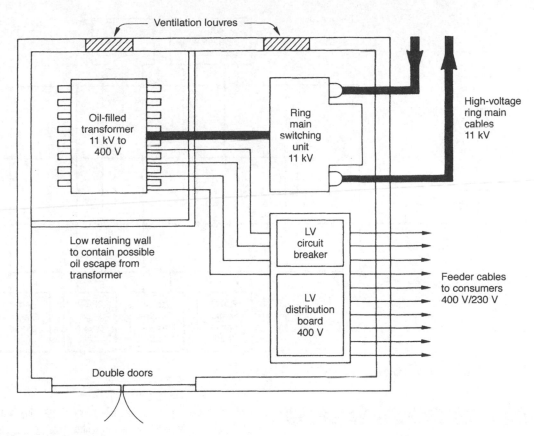

Fig. 3.68 Typical substation layout.

supply authority because electricity plays a vital part in an industrial society, and a loss of supply may cause inconvenience, financial loss or danger to the consumer or the public.

The principle employed with a ring system is that any consumer's substation is fed from two directions, and by carefully grading the overload and cable protection equipment a fault can be disconnected without loss of supply to other consumers.

High-voltage distribution to primary substations is used by the electricity boards to supply small industrial, commercial and domestic consumers. This distribution method is also suitable for large industrial consumers where 11 kV substations, as shown in Fig. 3.67, may be strategically placed at load centres around the factory site. Regulation 9 of the Electricity Supply Regulations and Regulation 31 of the Factories Act require that these substations be protected by 2.44 m high fences or enclosed in some other way so that no unauthorized person may gain access to the potentially dangerous

equipment required for 11 kV distribution. In towns and cities the substation equipment is usually enclosed in a brick building, as shown in Fig. 3.68.

The final connections to plant, distribution boards, commercial or domestic loads are usually by simple underground radial feeders at 400 V/230 V. These outgoing circuits are usually protected by circuit breakers in a distribution board.

The 400 V/230 V is derived from the 11 kV/400 V substation transformer by connecting the secondary winding in star as shown in Fig. 3.69. The star point is earthed to an earth electrode sunk into the ground below the substation, and from this point is taken the fourth conductor, the neutral. Loads connected between phases are fed at 400 V, and those fed between one phase and neutral at 230 V. A three-phase 400 V supply is used for supplying small industrial and commercial loads such as garages, schools and blocks of flats. A single-phase 230 V supply is usually provided for individual domestic consumers.

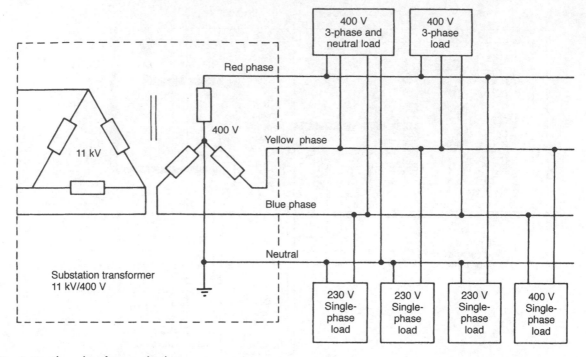

Fig. 3.69 Three-phase four-wire distribution.

EXAMPLE

Use a suitable diagram to show how a 400 V three-phase, four-wire supply may be obtained from an 11 kV delta-connected transformer. Assuming that the three-phase four-wire supply feeds a small factory, show how the following loads must be connected:

(a) a three-phase 400 V motor;
(b) a single-phase 400 V welder;
(c) a lighting load made up of discharge lamps arranged in a way which reduces the stroboscopic effect.
(d) State why 'balancing' of loads is desirable.
(e) State the advantages of using a three-phase four-wire supply to industrial premises instead of a single-phase supply.

3PH LOAD

Figure 3.69 shows the connections of the 11 kV to 400 V supply and the method of connecting a 400 V three-phase load such as a motor and a 400 V single-phase load such as a welder.

REDUCING STROBOSCOPIC EFFECT

The stroboscopic effect may be reduced by equally dividing the lighting load across the three phases of the supply. For example, if the lighting load were made up of 18 luminaires, then six luminaires should be

connected to the red phase and neutral, six to the blue phase and neutral and six to the yellow phase and neutral. (The stroboscopic effect and its elimination are discussed in some detail in Chapter 5 of this book.)

BALANCING 3PH LOADS

A three-phase load such as a motor has equally balanced phases since the resistance of each phase winding will be the same. Therefore the current taken by each phase will be equal. When connecting single-phase loads to a three-phase supply, care should be taken to distribute the single-phase loads equally across the three phases so that each phase carries approximately the same current. Equally distributing the single-phase loads across the three-phase supply is known as 'balancing' the load. A lighting load of 18 luminaires would be 'balanced' if six luminaires were connected to each of the three phases.

ADVANTAGES OF 3PH FOUR-WIRE SUPPLY

A three-phase four-wire supply gives a consumer the choice of a 400 V three-phase supply and a 230 V single-phase supply. Many industrial loads such as motors require a three-phase 400 V supply, while the lighting load in a factory, as in a house, will be 230 V. Industrial loads usually demand more power than a domestic load, and more power can be supplied by a 400 V three-phase supply than is possible with a 230 V single-phase supply for a given size of cable since Power = $VI \cos \theta$ (watts).

LOW-VOLTAGE DISTRIBUTION IN BUILDINGS

In domestic installations the final circuits for lights, sockets, cookers, immersion heating, etc. are connected to separate fuseways in the consumer's unit mounted at the service position, as shown in Fig. 2.61.

In commercial or industrial installations a three-phase 400 V supply must be distributed to appropriate equipment in addition to supplying single-phase 230 V loads such as lighting. It is now common practice to establish industrial estates speculatively, with the intention of encouraging local industry to use individual units. This presents the electrical contractor with an additional problem. The use and electrical demand of a single industrial unit are often unknown and the electrical supply equipment will need to be flexible in order to meet a changing demand due to expansion or change of use.

Busbar chambers incorporated into cubicle switchboards or on-site assemblies of switchboards are to be found at the incoming service position of commercial and industrial consumers, since this has proved to provide the flexibility required by these consumers. This is shown in Fig. 3.70.

Distribution fuse boards, which may incorporate circuit breakers, are wired by submain cables from the service position to load centres in other parts of the building, thereby keeping the length of cable to the final circuit as short as possible. This is shown in Fig. 3.71.

When high-rise buildings such as multi-storey flats have to be wired, it is usual to provide a three-phase four-wire rising main. This may comprise vertical busbars running from top to bottom at some central point in the building. Each floor or individual flat is then connected to the busbar to provide the consumer's supply. When individual dwellings receive a single-phase supply the electrical contractor must balance the load across the three phases. Fig. 3.72 shows a rising main system. The rising main must incorporate fire barriers to prevent the spread of fire throughout the building (Regulation 527–02).

OFF-PEAK SUPPLIES

Off-peak supplies are those made available by supply authorities to consumers during restricted periods, usually during the night. They attract a lower tariff than unrestricted supplies. (The off-peak charge is usually about half the normal supply charge.) Recent years have seen a growing demand for off-peak supplies by

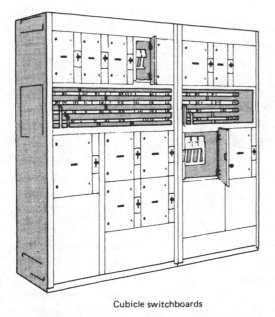

Cubicle switchboards

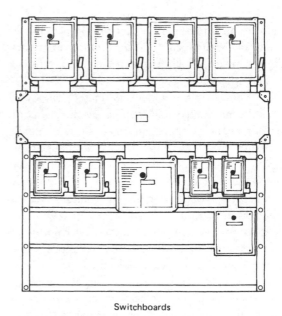

Switchboards

Fig. 3.70 Industrial consumer's service position equipment.

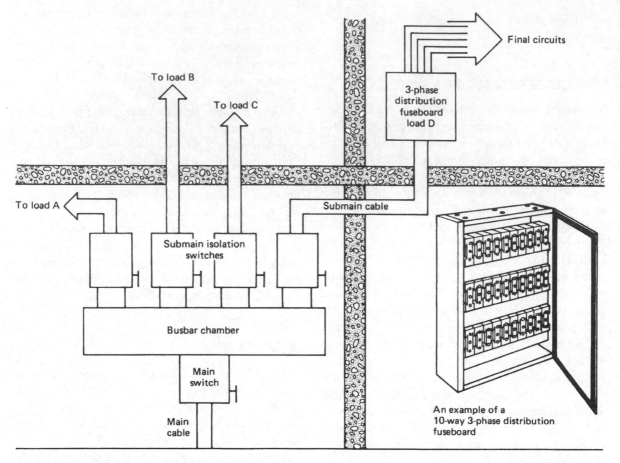

Fig. 3.71   Typical distribution in commercial or industrial building.

domestic consumers, partly as a result of the publicity given to the 'white meter' and the introduction of new materials making block storage heaters more efficient, slim and attractive.

The white meter is a two-rate meter. Units consumed during the day and evening are charged at the normal rate and all units consumed during the night are charged at a lower rate. To take advantage of an off-peak supply, the consumer must install storage heaters and heat the domestic hot water during the night, keeping it hot during the day by improved levels of thermal insulation.

The consumer's equipment at the service position is divided into two parts: an unrestricted supply feeding lighting, cooking and socket outlet final circuits; and a restricted supply, controlled by the supply authority's time switch, feeding space heating and water heating final circuits. The service position arrangements for a domestic consumer are shown in Fig. 3.73. The supply

authority's equipment and connector blocks are sealed to prevent unauthorized entry.

Protection of an electrical installation

The provision of protective devices in an electrical installation is fundamental to the whole concept of the safe use of electricity in buildings. The electrical installation as a whole must be protected against overload or short circuit and the people using the building must be protected against the risk of shock, fire or other risks arising from their own misuse of the installation or from a fault. The installation and maintenance of adequate and appropriate protective measures is a vital part of the safe use of electrical energy. I want to

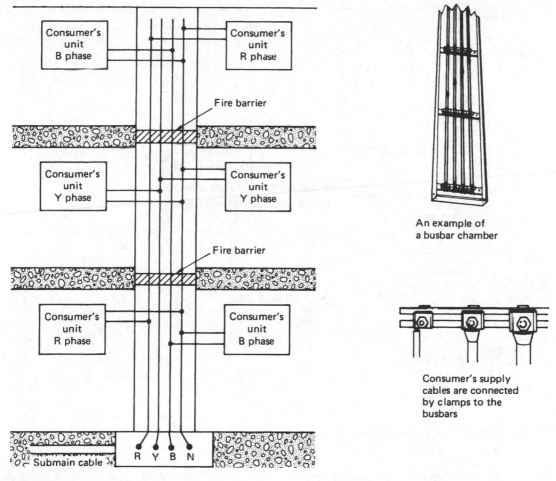

An example of
a busbar chamber

Consumer's supply
cables are connected
by clamps to the
busbars

Fig. 3.72 Busbar rising main system.

look at protection against an electric shock by both direct and indirect contact, at protection by equipotential bonding and automatic disconnection of the supply, and protection against excess current.

Let us first define some of the words we will be using. Chapter 54 of the IEE Regulations describes the earthing arrangements for an electrical installation. It gives the following definitions:

Earth – the conductive mass of the earth.

Bonding conductor – a protective conductor providing equipotential bonding.

Circuit protective conductor (CPC) – a protective conductor connecting exposed conductive parts of equipment to the main earthing terminal. This is the green and yellow insulated conductor in twin and earth cable.

Exposed conductive parts – the metalwork of an electrical appliance or the trunking and conduit of an electrical system which can be touched because they are not normally live, but which may become live under fault conditions.

Extraneous conductive parts – the structural steelwork of a building and other service pipes such as gas, water, radiators and sinks. They do not form a part of the electrical installation but may introduce a potential, generally earth potential, to the electrical installation.

DIRECT CONTACT PROTECTION

The human body's movements are controlled by the nervous system. Very tiny electrical signals travel between the central nervous system and the muscles, stimulating operation of the muscles, which enable us to walk, talk and run and remember that the heart is also a muscle.

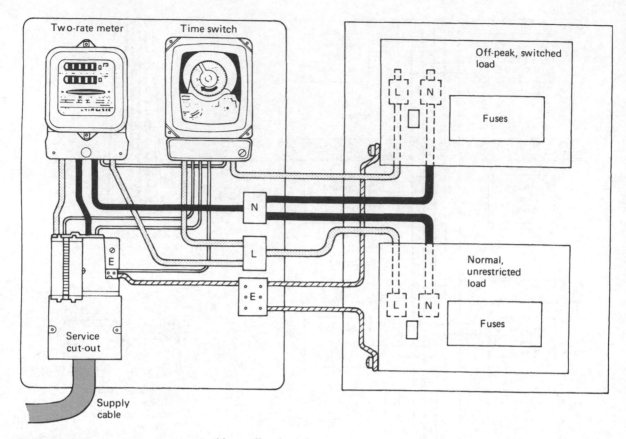

Fig. 3.73 Service position equipment required for an off-peak supply.

If the body becomes part of a more powerful external circuit, such as the electrical mains, and current flows through it, the body's normal electrical operations are disrupted. The shock current causes unnatural operation of the muscles and the result may be that the person is unable to release the live conductor causing the shock, or the person may be thrown across the room. The current which flows through the body is determined by the resistance of the human body and the surface resistance of the skin on the hands and feet.

This leads to the consideration of exceptional precautions where people with wet skin or wet surfaces are involved, and the need for special consideration in bathroom installations.

Two types of contact will result in a person receiving an electric shock. Direct contact with live parts which involves touching a terminal or phase conductor that is actually live. Indirect contact results from contact with an exposed conductive part such as the metal structure of a piece of equipment that has become live as a result of a fault.

In installations operating at normal mains voltage, the primary method of protections against direct contact is by insulation. All live parts are enclosed in insulating material such as rubber or plastic, which prevents contact with those parts. The insulating material must, of course, be suitable for the circumstances in which they will be used and the stresses to which they will be subjected.

Other methods of direct contact protection include the provision of barriers or enclosures which can only be opened by the use of a tool, or when the supply is first disconnected. Protection can also be provided by fixed obstacles such as a guard rail around an open switchboard or by placing live parts out of reach as with overhead lines.

EARTH FAULT PROTECTION

In Chapter 13 of the IEE Regulations we are told that where the metalwork of electrical equipment may become charged with electricity in such a manner as to cause danger, that metalwork will be connected with earth so as to discharge the electrical energy without danger.

There are five methods of protection against contact with metalwork which has become unintentionally live, that is, indirect contact with exposed conductive parts recognised by the IEE Regulations. These are:

1 earthed equipotential bonding coupled with automatic disconnection of the supply,
2 the use of Class II (double insulated) equipment,
3 the provision of a non-conducting location,
4 the use of earth free equipotential bonding,
5 electrical separation.

Methods 3 and 4 are limited to special situations under the effective supervision of trained personnel.

Method 5, electrical separation, is little used but does find an application in the domestic electric shaver supply unit which incorporates an isolating transformer.

Method 2, the use of Class II insulated equipment is limited to single pieces of equipment such as tools used on construction sites, because it relies upon effective supervision to ensure that no metallic equipment or extraneous earthed metalwork enters the area of the installation.

The method which is most universally used in the United Kingdom is, therefore, Method 1 – earthed equipotential bonding coupled with automatic disconnection of the supply.

This method relies upon all exposed metalwork being electrically connected together to an effective earth connection. Not only must all the metalwork associated with the electrical installation be so connected, that is conduits, trunking, metal switches and the metalwork of electrical appliances, but Regulation 413–02–02 tells us to connect the extraneous metalwork of water service pipes, gas and other service pipes and ducting, central heating and air conditioning systems, exposed metallic structural parts of the building and lightning protective systems to the main earthing terminal. In this way the possibility of a voltage appearing between two exposed metal parts is removed. Main equipotential bonding is shown in Fig. 2.64 in Chapter 2.

The second element of this protection method is the provision of a means of automatic disconnection of the supply in the event of a fault occurring that causes the exposed metalwork to become live.

The IEE Regulations recognise that the risk of an injurious shock is greater when the equipment concerned is portable and likely to be hand held, such as an electric drill, than when the equipment is fixed. The Regulations, therefore, specify that the disconnection must be effected within 0.4 seconds for circuits, which include socket outlets, but within 5.0 seconds for circuits connected to fixed equipment.

The achievement of these disconnection times is dependent upon the type of protective device used, fuse or circuit breaker, the circuit conductors to the fault and the provision of adequate equipotential bonding. The resistance, or we call it the impedance, of the earth fault loop must be less than the values given in Appendix 2 of the *On Site Guide* and Tables 41B1, 41B2 and 41D of the IEE Regulations. (Table 3.2 later in this Chapter shows the maximum value of the earth fault loop impedance for circuits protected by a semi-enclosed fuse to BS 3036.). We will look at this again later in this Chapter under the heading 'Earth Fault Loop Impedance Z_S'. Section 542 of the IEE Regulations gives details of the earthing arrangements to be incorporated in the supply system to meet these Regulations and these are described in the next Chapter of this book, Chapter 4, under the heading 'Low Voltage Supply Systems'.

RESIDUAL CURRENT PROTECTION

The IEE Regulations recognise the particular problems created when electrical equipment such as lawnmowers, hedge-trimmers, drills and lights are used outside buildings. In these circumstances the availability of an adequate earth return path is a matter of chance. The Regulations, therefore, require that any socket intended to be used to supply equipment outside a building shall have the additional protection of a residual current device (RCD) which has a rated operating current of not more than 30 milliamperes (mA).

An RCD is a type of circuit breaker that continuously compares the current in the phase and neutral conductors of the circuit. The currents in a healthy circuit will be equal, but in a circuit that develops a fault, some current will flow to earth and the phase and neutral currents will no longer balance. The RCD detects the imbalance and disconnects the circuit. Figure 2.71 in Chapter 2 shows an RCD.

ISOLATION AND SWITCHING

Part 4 of the IEE Regulations deals with the application of protective measures for safety and Chapter 53 with the regulations for switching devices or switchgear required for protection, isolation and switching of a consumer's installation.

The consumer's main switchgear must be readily accessible to the consumer and be able to:

- isolate the complete installation from the supply,
- protect against overcurrent,
- cut off the current in the event of a serious fault occurring.

The Regulations identify four separate types of switching: switching for isolation; switching for mechanical maintenance; emergency switching; and functional switching.

Isolation is defined as cutting off the electrical supply to a circuit or item of equipment in order to ensure the safety of those working on the equipment by making dead those parts which are live in normal service.

An isolator is a mechanical device which is operated manually and used to open or close a circuit off load. An isolator switch must be provided close to the supply point so that all equipment can be made safe for maintenance. Isolators for motor circuits must isolate the motor and the control equipment, and isolators for high-voltage discharge lighting luminaires must be an integral part of the luminaire so that it is isolated when the cover is removed (Regulations 461, 476–02 and 537–02). Devices which are suitable for isolation are isolation switches, fuse links, circuit breakers, plugs and socket outlets.

Isolation at the consumer's service position can be achieved by a double pole switch which opens or closes all conductors simultaneously. On three-phase supplies the switch need only break the live conductors with a solid link in the neutral, provided that the neutral link cannot be removed before opening the switch.

The switching for mechanical maintenance requirements is similar to those for isolation except that the control switch must be capable of switching the full load current of the circuit or piece of equipment. Switches for mechanical maintenance must not have exposed live parts when the appliance is opened, must be connected in the main electrical circuit and have a reliable on/off indication or visible contact gap (Regulations 462 and 537–03). Devices which are suitable for switching off for mechanical maintenance are switches, circuit breakers, plug and socket outlets.

Emergency switching involves the rapid disconnection of the electrical supply by a single action to remove or prevent danger. The device used for emergency switching must be immediately accessible and identifiable, and be capable of cutting off the full load current. A fireman's switch provides emergency switching for high-voltage signs, as described in Chapter 5.

Electrical machines must be provided with a means of emergency switching, and a person operating an electrically driven machine must have access to an emergency switch so that the machine can be stopped in an emergency. The remote stop/start arrangement shown in Fig. 3.49 could meet this requirement for an electrically driven machine (Regulations 463, 476–03 and 537–04). Devices which are suitable for emergency switching are switches, circuit breakers and contactors. Where contactors are operated by remote control they should *open* when the coil is de-energized, that is, fail safe. Push-buttons used for emergency switching must be coloured red and latch in the stop or off position. They should be installed where danger may arise and be clearly identified as emergency switches. Plugs and socket outlets cannot be considered appropriate for emergency disconnection of supplies.

Functional switching involves the switching on or off or varying the supply of electrically operated equipment in normal service. The device must be capable of interrupting the total steady current of the circuit or appliance. When the device controls a discharge lighting circuit it must have a current rating capable of switching an inductive load. Plug and socket outlets may be used as switching devices and recent years have seen an increase in the number of electronic dimmer switches being used for the control and functional switching of lighting circuits (Regulations 537–05–01).

Where more than one of these functions is performed by a common device, it must meet the individual requirements for each function (Regulation 476–01–01).

OVERCURRENT PROTECTION

The consumer's mains equipment must provide protection against overcurrent, that is a current exceeding the rated value (Regulation 431–01–01). Fuses provide overcurrent protection when situated in the live conductors; they must not be connected in the neutral conductor. Circuit breakers may be used in place of fuses, in which case the circuit breaker may also provide the means of isolation, although a further means of isolation is usually provided so that maintenance can be carried out on the circuit breakers themselves.

Overcurrent can be subdivided into overload current, and short-circuit current. An overload current can be defined as a current which exceeds the rated value in an otherwise healthy circuit. Overload currents usually occur because the circuit is abused or because it has been badly designed or modified. A short circuit is an overcurrent resulting from a fault of negligible impedance connected between conductors. Short circuits usually occur as a result of an accident which could not have been predicted before the event.

An overload may result in currents of two or three times the rated current flowing in the circuit. Short-circuit currents may be hundreds of times greater than the rated current. In both cases the basic requirements for protection are that the fault currents should be interrupted quickly and the circuit isolated safely before the fault current causes a temperature rise which might damage the insulation and terminations of the circuit conductors.

The selected protective device should have a current rating which is not less than the full load current of the circuit but which does not exceed the cable current rating. The cable is then fully protected against both overload and short-circuit faults (Regulation 433–02–01). Devices which provide overcurrent protection are:

- HBC fuses to BS 88. These are for industrial applications having a maximum fault capacity of 80 kA.
- Cartridge fuses to BS 1361. These are used for a.c. circuits on industrial and domestic installations having a fault capacity of about 30 kA.
- Cartridge fuses to BS 1362. These are used in 13 A plug tops and have a maximum fault capacity of about 6 kA.
- Semi-enclosed fuses to BS 3036. These were previously called rewirable fuses and are used mainly on domestic installations having a maximum fault capacity of about 4 kA.
- MCBs to BS 3871. These are miniature circuit breakers which may be used as an alternative to fuses for some installations. The British Standard

includes ratings up to 100 A and maximum fault capacities of 9 kA. They are graded according to their instantaneous tripping currents – that is, the current at which they will trip within 100 milli-seconds. This is less than the time taken to blink an eye.

MCB Type 1 to BS 3871 will trip instantly at between 2.7 and four times its rated current and is therefore more suitable on loads with minimal or no switching surges such as domestic or commercial installations.

MCB Type B to BS EN 60898 will trip instantly at between three and five times its rated current and is also suitable for domestic and commercial installations.

MCB Type 2 to BS 3871 will trip instantly at between four and seven times its rated current. It offers fast protection on small overloads combined with a slower operation on heavier faults, which reduces the possibility of nuisance tripping. Its characteristics are very similar to those of an HBC fuse, and this MCB is possibly best suited for general commercial and industrial use.

MCB Type C to BS EN 60898 will trip instantly at between five and ten times its rated current. It is more suitable for highly inductive commercial and industrial loads.

MCB Type 3 to BS 3871 will trip instantly at between seven and ten times its rated current. It is more suitable for protecting highly inductive circuits and is used on circuits supplying transformers, chokes and lighting banks.

MCB Type D to BS EN 69898 will trip instantly at between 10 and 25 times its rated current. It is suitable for welding and X-ray machines where large inrush currents may occur.

MCB Type 4 to BS 3871 will trip instantly between 10 and 50 times the rated current and is more suitable for special industrial applications such as welding equipment and X-ray machines.

The construction, advantages and disadvantages of the various protective devices are discussed in Chapter 2 of this book and shown in Figs 2.66 to 2.70.

POSITION OF PROTECTIVE DEVICES

The general principle to be followed is that a protective device must be placed at a point where a reduction occurs in the current carrying capacity of the circuit conductors. A reduction may occur because of a change in the size or type of conductor or because of a change in the method of installation or a change in the environmental conditions. The only exceptions to this rule are where an overload protective device opening a circuit might cause a greater danger than the overload itself – for example, a circuit feeding an overhead electro-magnet in a scrapyard.

DISCONNECTION TIME CALCULATIONS

The overcurrent protection device protecting socket outlet circuits and any fixed equipment in bathrooms must operate within 0.4 seconds. Those protecting fixed equipment circuits in rooms other than bathrooms must operate within 5 seconds (Regulation 413–02–08 to 13 and 601).

The reason for the more rapid disconnection of the socket outlet circuits is that portable equipment plugged into the socket outlet is considered a higher risk than fixed equipment since it is more likely to be firmly held by a person. The more rapid disconnection times for fixed equipment in bathrooms take account of a possibly reduced body resistance in the bathroom environment.

The IEE Regulations permit us to assume that where an overload protective device is also intended to provide short-circuit protection, and has a rated breaking capacity greater than the prospective short-circuit current at the point of its installation, the conductors on the load side of the protective device are considered to be adequately protected against short-circuit currents without further proof. This is because the cable rating and the overload rating of the device are compatible. However, if this condition is not met or if there is some doubt, it must be verified that fault currents will be interrupted quickly before they can cause a dangerously high temperature rise in the circuit conductors. Regulation 434–03–03 provides an equation for calculating the maximum operating time of the protective device to prevent the permitted conductor temperature rise being exceeded as follows:

$$t = \frac{k^2 S^2}{I^2} \text{ (seconds)}$$

where

t = duration time in seconds

S = cross-sectional area of conductor in square millimetres

I = short-circuit rms current in amperes

k = a constant dependent upon the conductor metal and type of insulation (see Table 43A of the IEE Regulations).

EXAMPLE

A 10 mm PVC insulated copper cable is short-circuited when connected to a 400 V supply. The impedance of the short-circuit path is 0.1 Ω. Calculate the maximum permissible disconnection time and show that a 50 A type 2 MCB to BS 3871 will meet this requirement.

$$I = \frac{V}{Z} \text{ (A)} \qquad I = \frac{400 \text{ V}}{0.1 \text{ } \Omega} = 4000 \text{ A}$$

For copper conductor and PVC insulation, Table 43A gives a value for k of 115. So,

$$t = \frac{k^2 \text{ } S^2}{I^2} \text{ (s)}$$

$$\therefore t = \frac{115^2 \times 10^2 \text{ mm}^2}{4000 \text{ A}} = 82.66 \times 10^{-3} \text{ S}$$

The maximum time that a 4000 A fault current can be applied to this 10 mm² cable without dangerously raising the conductor temperature is 82.66 milliseconds. Therefore, the protective device must disconnect the supply to the cable in less than 82.66 milliseconds under short-circuit conditions. Manufacturers' information and Appendix 3 of the IEE Regulations give the operating times of protective devices at various short-circuit currents in the form of graphs, similar to those shown in Figs 3.74 and 3.75.

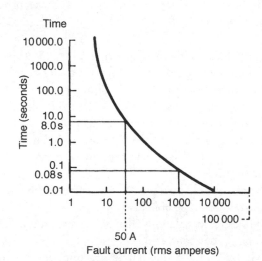

Fig. 3.74 Time/current characteristic of an overcurrent protective device.

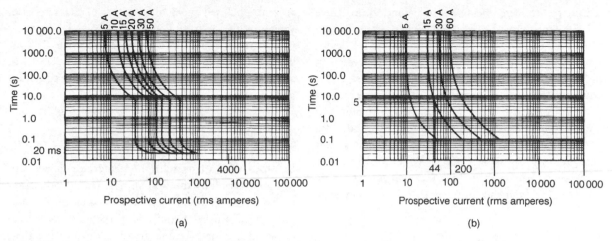

Fig. 3.75 Time/current characteristics of (a) a type 2 MCB to BS 3871; (b) semi-enclosed fuse to BS 3036.

TIME/CURRENT CHARACTERISTICS OF PROTECTIVE DEVICES

Disconnection times for various overcurrent devices are given in the form of a logarithmic graph. This means that each successive graduation of the axis represents a ten times change over the previous graduation.

These logarithmic scales are shown in the graphs of Figs 3.74 and 3.75. From Fig. 3.74 it can be seen that the particular protective device represented by this characteristic will take 8 seconds to disconnect a fault current of 50 A and 0.08 seconds to clear a fault current of 1000 A.

Figure 3.75 (a) shows the time/current characteristics for a type 2 MCB to BS 3871. This graph shows that a fault current of 4000 A will trip the protective device in 20 milliseconds. Since this is quicker than 82.66 milliseconds, the 50 A type 2 MCB will clear the fault current before the temperature of the cable is raised to a dangerous level.

Appendix 3 of the IEE Regulations gives the time/current characteristics and specific values of prospective short-circuit current for a number of protective devices.

These indicate the value of fault current which will cause the protective device to operate in the times indicated by Chapter 413 of the IEE Regulations, that is 0.4 and 5 seconds in the case of domestic socket outlet circuits and distribution circuits feeding fixed appliances.

Figures 1, 2 and 3 in Appendix 3 of the IEE Regulations deal with fuses and Figs 4 to 8 with MCBs.

It can be seen that the prospective fault current required to trip an MCB in the required time is a multiple of the current rating of the device. The multiple depends upon the characteristics of the particular devices. Thus:

- type 1 MCB to BS 3871 has a multiple of 4
- type 2 MCB to BS 3871 has a multiple of 7
- type 3 MCB to BS 3871 has a multiple of 10
- type B MCB to BS EN 60898 has a multiple of 5
- type C MCB to BS EN 60898 has a multiple of 10
- type D MCB to BS EN 60898 has a multiple of 20.

EXAMPLE

A 6 A type 1 MCB to BS 3871 used to protect a domestic lighting circuit will trip within 5 seconds when 6 A times a multiple of 4, that is 24 A, flows under fault conditions.

Therefore if the earth fault loop impedance is low enough to allow at least 24 A to flow in the circuit under fault conditions, the protective device will operate within the time required by Regulation 413–02–14.

The characteristics shown in Appendix 3 of the IEE Regulations give the specific values of prospective short-circuit current for all standard sizes of protective device.

DISCRIMINATION

In the event of a fault occurring on an electrical installation only the protective device nearest to the fault should operate, leaving other healthy circuits unaffected. A circuit designed in this way would be considered to have effective discrimination. Effective discrimination can be achieved by graded protection since the speed of operation of the protective device increases as the rating decreases. This can be seen in Fig. 3.75 (a). A fault current of 200 A will cause a 15 A semi-enclosed fuse to operate in about 0.1 seconds, a 30 A semi-enclosed fuse in about 0.45 seconds and a 60 A semi-enclosed fuse in about 5.4 seconds. If a circuit is arranged as shown in Fig. 3.76 and a fault occurs on the appliance, effective discrimination will be achieved because the 15 A fuse will operate more quickly than the other protective devices if they were

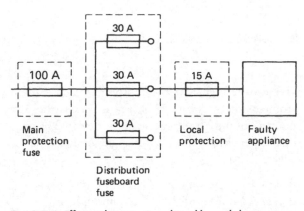

Fig. 3.76   Effective discrimination achieved by graded protection.

all semi-enclosed types fuses with the characteristics shown in Fig. 3.75 (b).

Security of supply, and therefore effective discrimination, is an important consideration for an electrical engineer and is also a requirement of Regulation 533–01–06.

EARTH FAULT LOOP IMPEDANCE Z_S

In order that an overcurrent protective device can operate successfully, meeting the required disconnection times, of less than 0.4 seconds for socket outlets and 5.0 seconds for fixed equipment, the earth fault loop impedance value measured in ohms must be less than those values given in Appendix 2 of the *On Site Guide* and Tables 41B1 and 41B2 of the IEE Regulations for socket outlet circuits and Tables 41B2 and 41D for circuits supplying fixed equipment. The value of the earth fault loop impedance may be verified by means of an earth fault loop impedance test as described in Chapter 4 of this book. The formula is:

$$Z_S = Z_E + (R_1 + R_2) \ (\Omega)$$

Here Z_E is the impedance of the supply side of the earth fault loop. The actual value will depend upon many

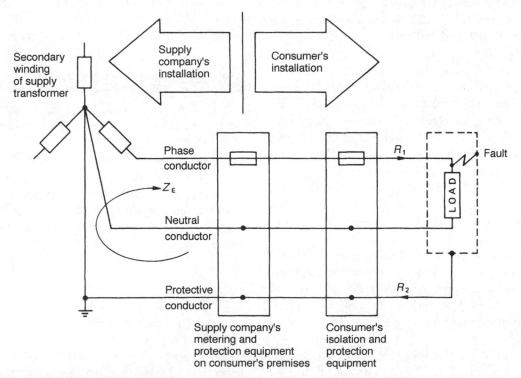

Fig. 3.77   Earth fault loop path for a TN-S system.

Table 3.1 Table 9A of the IEE *On Site* Guide

Table 9A   Value of resistance/metre for copper and aluminium conductors and of $R_1 + R_2$ per metre at 20°C in milliohms/metre

Cross-sectional area (mm²)		Resistance/metre or $(R_1 + R_2)$/metre (mΩ/m)	
Phase conductor	Protective conductor	Copper	Aluminium
1	–	18.10	
1	1	36.20	
1.5	–	12.10	
1.5	1	30.20	
1.5	1.5	24.20	
2.5	–	7.41	
2.5	1	25.51	
2.5	1.5	19.51	
2.5	2.5	14.82	
4	–	4.61	
4	1.5	16.71	
4	2.5	12.02	
4	4	9.22	
6	–	3.08	
6	2.5	10.49	
6	4	7.69	
6	6	6.16	
10	–	1.83	
10	4	6.44	
10	6	4.91	
10	10	3.66	
16	–	1.15	1.91
16	6	4.23	–
16	10	2.98	–
16	16	2.30	3.82
25	–	0.727	1.20
25	10	2.557	–
25	16	1.877	–
25	25	1.454	2.40
35	–	0.524	0.87
35	16	1.674	2.78
35	25	1.251	2.07
35	35	1.048	1.74
50	–	0.387	0.64
50	25	1.114	1.84
50	35	0.911	1.51
50	50	0.774	1.28

Reproduced from the IEE *On Site Guide* by kind permission of the Institution of Electrical Engineers

factors: the type of supply, the ground conditions, the distance from the transformer, etc. The value can be obtained from the area electricity companies, but typical values are 0.35 Ω for TN-C-S (PME) supplies and 0.8 Ω for TN-S (cable sheath earth) supplies. Also in the above formula, R_1 is the resistance of the phase conductor and R_2 is the resistance of the earth conductor. The complete earth fault loop path is shown in Fig. 3.77.

Values of $R_1 + R_2$ have been calculated for copper and aluminium conductors and are given in Table 9A of the *On Site Guide* as shown in Table 3.1.

EXAMPLE

A 20 A radial socket outlet circuit is wired in 2.5 mm² PVC cable incorporating a 1.5 mm² CPC. The cable length is 30 m installed in an ambient temperature of 20°C and the consumer's protection is by semi-enclosed fuse to BS 3036. The earth fault loop impedance of the supply is 0.5 Ω. Calculate the total earth fault loop impedance Z_S, and establish that the value is less than the maximum value permissible for this type of circuit.

We have

$$Z_S = Z_E + (R_1 + R_2) \ (\Omega)$$
$$Z_E = 0.5 \ \Omega \text{ (value given in the question).}$$

From the value given in Table 9A of the *On Site Guide* and reproduced in Table 3.1 a 2.5 mm phase conductor with a 1.5 mm protective conductor has an $(R_1 + R_2)$ value of $19.51 \times 10^{-3} \, \Omega/\text{m}$

$$(R_1 + R_2) = 19.51 \times 10^{-3} \, \Omega/\text{m} \times 30 \, \text{m} = 0.585 \, \Omega$$

However, under fault conditions, the temperature and therefore the cable resistance will increase. To take account of this, we must multiply the value of cable resistance by the factor given in Table 9C of the *On Site Guide*. In this case the factor is 1.20 and therefore the cable resistance under fault conditions will be:

$$0.585 \, \Omega \times 1.20 = 0.702 \, \Omega$$

The total earth fault loop impedance is therefore

$$Z_S = 0.5 \, \Omega + 0.702 \, \Omega = 1.202 \, \Omega$$

The maximum permitted value given in Table 2A of the *On Site Guide* for a 20 A fuse to BS 3036 protecting a socket outlet is 1.48 Ω as shown by Table 3.2. The circuit earth fault loop impedance is less than this value and therefore the protective device will operate within the required disconnection time of 0.4 s.

PROTECTIVE CONDUCTOR SIZE

The circuit protective conductor forms an integral part of the total earth fault loop impedance, so it is necessary to check that the cross-section of this conductor is adequate. If the cross-section of the circuit protective conductor complies with Table 54G of the IEE Regulations, there is no need to carry out further checks. Where phase and protective conductors are made from the same material, Table 54G tells us that:

- for phase conductors equal to or less than 16 mm², the protective conductor should equal the phase conductor;
- for phase conductors greater than 16 mm² but less than 35 mm², the protective conductor should have a cross-sectional area of 16 mm²;
- for phase conductors greater than 35 mm², the protective conductor should be half the size of the phase conductor.

However, where the conductor cross-section does not comply with this table, then the formula given in Regulation 543–01–03 must be used:

$$S = \frac{\sqrt{I^2 t}}{k} \, (\text{mm}^2)$$

Table 3.2   Table 2A of the IEE *On Site Guide*

Table 2A   Semi enclosed fuses. Maximum measured earth fault loop impedance (in ohms) when overcurrent protective device is a semi-enclosed fuse to BS 3036

Protective conductor (mm²)	Fuse rating (amperes)				
	5	15	20	30	45
(i) 0.4 second disconnection					
1.0	8.00	2.14	1.48	NP	NP
1.5	8.00	2.14	1.48	0.91	NP
2.5 to 16.0	8.00	2.14	1.48	0.91	0.50
(ii) 5 seconds disconnection					
1.0	14.80	4.46	2.79	NP	NP
1.5	14.80	4.46	3.20	2.08	NP
2.5	14.80	4.46	3.20	2.21	1.20
4.0 to 16.0	14.80	4.46	3.20	2.21	1.33

NP protective conductor, fuse combination NOT PERMITITTED.

Reproduced from the IEE *On Site Guide* by kind permission of the Institution of Electrical Engineers.

where

S = cross-sectional area in mm^2
I = value of maximum fault current in amperes
t = operating time of the protective device
k = a factor for the particular protective conductor (see Table 54B to 54F of the IEE Regulations).

EXAMPLE 1

A 230 V ring main circuit of socket outlets is wired in 2.5 mm single PVC copper cables in a plastic conduit with a separate 1.5 mm CPC. An earth fault loop impedance test identifies Z_s as 1.15 Ω. Verify that the 1.5 mm CPC meets the requirements of Regulation 543–01–03 when the protective device is a 30 A semi-enclosed fuse.

$$I = \text{Maximum fault current} = \frac{V}{Z_s} \text{ (A)}$$

$$\therefore I = \frac{230}{1.15} = 200 \text{ A}$$

t = Maximum operating time of the protective device for a socket outlet circuit is 0.4 seconds from Regulation 413–02–09. From Fig. 3.75(b) you can see that the time taken to clear a fault of 200 A is about 0.4 seconds.

k = 115 (from Table 54C).

$$S = \frac{\sqrt{I^2 t}}{k} \text{ (mm}^2)$$

$$S = \frac{\sqrt{(200 \text{ A})^2 \times 0.4 \text{ s}}}{115} = 1.10 \text{ mm}^2$$

A 1.5 mm^2 CPC is acceptable since this is the nearest standard-size conductor above the minimum cross-sectional area of 1.10 mm^2 found by calculation.

EXAMPLE 2

A domestic immersion heater is wired in 2.5 mm^2 PVC insulated copper cable and incorporates a 1.5 mm^2 CPC. The circuit is correctly protected with a 15 A semi-enclosed fuse to BS 3036. Establish by calculation that the CPC is of an adequate size to meet the requirements of Regulation 543–01–03. The characteristics of the protective device are given in Fig. 3.75(b).

For circuits feeding fixed appliances the maximum operating time of the protective device is 5 seconds. From Fig. 3.5(b) it can be seen that

a current of about 44 A will trip the 15 A fuse in 5 seconds. Alternatively Table 2A in Appendix 3 of the IEE Regulations gives a value of 43 A. Let us assume a value of 43 A.

$$\therefore I = 43 \text{ A}$$
$$t = 5 \text{ seconds for fixed appliances}$$
$$k = 115 \text{ (from Table 54C)}$$

$$S = \frac{\sqrt{I^2 t}}{k} \text{ (mm}^2) \text{ (from Regulation 543–01–03)}$$

$$S = \frac{\sqrt{(43 \text{ A})^2 \times 5 \text{ s}}}{115} = 0.836 \text{ mm}^2$$

The circuit protective conductor of the cable is greater than 0.836 mm^2 and is therefore suitable. If the protective conductor is a separate conductor, that is, it does not form part of a cable as in this example and is not enclosed in a wiring system as in Example 1, the cross-section of the protective conductor must be not less than 2.5 mm^2 where mechanical protection is provided or 4.0 mm^2 where mechanical protection is *not* provided in order to comply with Regulation 547–03–03.

Exercises

1 An a.c. series circuit has an inductive reactance of 4 Ω and a resistance of 3 Ω. The impedance of this circuit will be:
   (a) 5 Ω
   (b) 7 Ω
   (c) 12 Ω
   (d) 25 Ω.

2 An a.c. series circuit has a capacitive reactance of 12 Ω and a resistance of 9 Ω. The impedance of this circuit will be:
   (a) 3 Ω
   (b) 15 Ω
   (c) 20 Ω
   (d) 108 Ω.

3 A circuit whose resistance is 3 Ω, reactance 5 Ω and impedance 5.83 Ω will have a p.f. of:
   (a) 0.515
   (b) 0.600
   (c) 0.858
   (d) 1.666.

4 A circuit whose resistance is 5 Ω, capacitive reactance 12 Ω and inductive reactance 20 Ω will have an impedance of:
   (a) 9.434 Ω
   (b) 21.189 Ω
   (c) 23.853 Ω
   (d) 32.388 Ω.

5 The inductive reactance of a 100 mH coil when connected to 50 Hz will be:
   (a) $0.5\,\Omega$
   (b) $0.0318\,\Omega$
   (c) $5.0\,\Omega$
   (d) $31.416\,\Omega$.

6 The capacitive reactance of a 100 μF capacitor connected to a 50 Hz supply will be:
   (a) $0.5\,\Omega$
   (b) $5.0\,m\Omega$
   (c) $31.83\,\Omega$
   (d) $31415.93\,\Omega$.

7 Two a.c. voltages V_1 and V_2 have values of 20 and 30 V, respectively. If V_1 leads V_2 by 45° the resultant voltage will be:
   (a) 16 V at 24°
   (b) 45 V at 90°
   (c) 46 V at 18°
   (d) 50 V at 45°.

8 Two parallel branch currents I_1 and I_2 are found to have values of 9 A and 12 A, respectively. If I_1 leads I_2 by 90° the resultant current will be:
   (a) 3 A at 21°
   (b) 10.5 A at 90°
   (c) 15 A at 37°
   (d) 21 A at 45°.

9 The phasor diagram which shows V and I at a power factor of unity is:

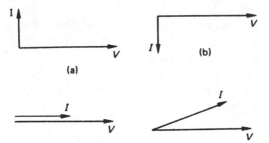

10 The phasor diagram for a lagging p.f. is given by:

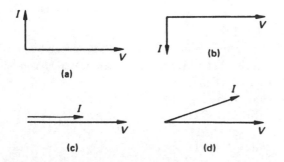

11 The phasor diagram for a purely capacitive circuit is given by:

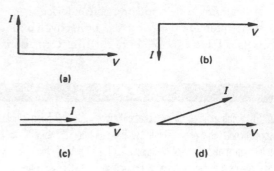

12 To avoid back injuries when manually lifting heavy weights from ground level a worker should:
   (a) bend both legs and back
   (b) bend legs but keep back straight
   (c) keep legs straight but bend back
   (d) keep both legs and back straight.

13 The angle of a ladder to the building upon which it is resting should be in the proportions of:
   (a) 1 up to 4 out
   (b) 4 up to 75 out
   (c) 4 up to 1 out
   (d) 75 up to 4 out.

14 The angle which a correctly erected ladder should make with level ground is:
   (a) 41°
   (b) 45°
   (c) 57°
   (d) 75°.

15 The working platform of a trestle scaffold which is 1.5 m above the ground should be:
   (a) at least two boards wide
   (b) fitted with toeboards
   (c) fitted with toeboards and guardrails
   (d) secured to the building to give stability.

16 For good stability mobile towers must have a base width to tower height ratio of:
   (a) 1 : 2
   (b) 1 : 3
   (c) 1 : 4
   (d) 1 : 5.

17 The most important far-reaching recent piece of safety legislation has been:
   (a) the IEE Regulations
   (b) the Electricity at Work Regulations 1989

(c) the Health and Safety at Work Act 1974

(d) The British and European Standards Mark.

18 An overload current may be defined as:

(a) a current in excess of at least 15 A

(b) a current which exceeds the rated value in an otherwise healthy circuit

(c) an overcurrent resulting from a fault between live and neutral conductors

(d) a current in excess of 60 A.

19 A short circuit may be defined as:

(a) a current in excess of at least 15 A

(b) a current which exceeds the rated value in an otherwise healthy circuit

(c) an overcurrent resulting from a fault between live and neutral conductors

(d) a current in excess of 60 A.

20 Hazard may be defined as:

(a) anything that can cause harm

(b) the chance, large or small, of harm actually being done

(c) someone who has the necessary training and expertise to safely carry out an activity

(d) the rules and regulations of the working environment.

21 Risk may be defined as:

(a) anything that can cause harm

(b) the chance, large or small, of harm actually being done

(c) someone who has the necessary training and expertise to safely carry out an activity

(d) the rules and regulations of the working environment.

22 A coil is made up of 30 m of conductor and is laid within and at right angles to a magnetic field of 4 T. The force exerted upon this coil when 5 A flows will be:

(a) 1.875 N

(b) 30 N

(c) 240 N

(d) 600 N.

23 An instrument coil of 20 mm diameter is wound with 100 turns and placed within and at right angles to a magnetic field of flux density 5 T. The force exerted on this coil when 15 mA flows in the coil conductors will be:

(a) 0.15 N

(b) 0.47 N

(c) 0.628 N

(d) 1.875 N.

24 The synchronous speed N_s of a four-pole machine connected to a 50 Hz supply will be:

(a) 200 rpm

(b) 750 rpm

(c) 1500 rpm

(d) 3000 rpm.

25 A four-pole induction motor running at 1425 rpm has a percentage slip of:

(a) 2%

(b) 5%

(c) 52.5%

(d) 75%.

26 A laminated cylinder of silicon steel with copper or aluminium bars slotted into holes around the circumference and short-circuited at each end of the cylinder, is one description of:

(a) a cage rotor

(b) an electromagnet

(c) a linear motor

(d) an induction motor.

27 All electric motors with a rating above 0.37 kW must be supplied with:

(a) protection by MCB

(b) protection by HBC fuses

(c) a motor starter

(d) remote stop/start switches.

28 A star delta starter:

(a) increases the initial starting torque of the motor

(b) reduces the initial starting current of the motor

(c) gives direct connection of the main voltage to the motor during starting

(d) requires only three connecting conductors between the motor and starter.

29 An auto-transformer starter:

(a) increases the initial starting torque of the motor

(b) increases the initial starting current of the motor

(c) gives direct connection of the mains voltage to the motor during starting

(d) requires only three connecting conductors between the motor and starter.

30 When an electric motor is to be connected to a load via a vee belt, it is recommended that the motor be mounted:

(a) firmly and secured by rawlbolts

(b) firmly and secured on slide rails

(c) loosely and connected by flexible conduit

(d) adjacent to the motor starter.

31 A series d.c. motor has the characteristic of:
(a) constant speed about 5% below synchronous speed
(b) start winding 90° out of phase with the run winding
(c) low starting torque but almost constant speed
(d) high starting torque and a speed which varies with load.

32 A shunt motor has the characteristic of:
(a) constant speed about 5% below synchronous speed
(b) start winding 90° out of phase with the run winding
(c) low starting torque but almost constant speed
(d) high starting torque and a speed which varies with load.

33 A three-phase induction motor has the characteristic of:
(a) constant speed about 5% below synchronous speed
(b) start winding 90° out of phase with the run winding
(c) high starting torque but almost constant speed
(d) high starting torque and a speed which varies with load.

34 A single-phase induction motor has:
(a) variable speed control operated by a centrifugal switch
(b) a start winding approximately 90° out of phase with the run winding
(c) high starting torque and almost constant speed
(d) high starting torque and a speed which varies with load.

35 One advantage of all d.c. machines is:
(a) that they are almost indestructible
(b) that starters are never required
(c) that they may be operated on a.c. or d.c. supplies
(d) the ease with which speed may be controlled.

36 One advantage of a cage rotor is:
(a) that it is almost indestructible
(b) that starters are never required
(c) that it may be operated on a.c. or d.c. supplies
(d) the ease with which speed may be controlled.

37 One advantage of a series d.c. motor is that:
(a) it is almost indestructible
(b) starters are never required
(c) it may be operated on a.c. or d.c. supplies
(d) speed is constant at all loads.

38 A shaded pole motor would normally be used for:
(a) an industrial process drive motor
(b) a portable electric drill motor
(c) a constant speed lathe motor
(d) a record turntable drive motor.

39 A d.c. shunt motor would normally be used for a:
(a) domestic oven fan motor
(b) portable electric drill motor
(c) constant speed lathe motor
(d) record turntable drive motor.

40 A d.c. series motor would normally be used for a:
(a) domestic oven fan motor
(b) portable electric drill motor
(c) constant speed lathe motor
(d) record turntable drive motor.

41 The core of a transformer is laminated to:
(a) reduce cost
(b) reduce copper losses
(c) reduce hysteresis loss
(d) reduce eddy current loss.

42 The transformation ratio of a step-down transformer is 2 0:1. If the primary voltage is 230 V the secondary voltage will be:
(a) 2.3 V
(b) 11.5 V
(c) 23 V
(d) 46 V.

43 Before an ammeter can be removed from the secondary terminals of a current transformer connected to a load, the transformer terminals must be:
(a) open-circuited
(b) short-circuited
(c) connected to the primary winding
(d) connected to earth.

44 The transmission of electricity is, for the most part, by overhead conductors suspended on steel towers because:
(a) this is environmentally more acceptable than running cables underground
(b) this is very many times cheaper than running an equivalent cable underground
(c) high-voltage electricity cables cannot be buried on agricultural land
(d) more power can be carried by overhead conductors than is possible by an underground cable.

45 A ring distribution system of electrical supply:
  (a) is cheaper than a radial distribution system
  (b) can use smaller supply cables than a radial distribution system
  (c) offers greater security of supply than a radial distribution system
  (d) is safer than a radial system.

46 An off-peak supply is available to a consumer:
  (a) for water heating purposes only
  (b) during the day time only
  (c) during the night time only
  (d) at all times provided that peak demand does not exceed 4 hours.

47 The Regulations define isolation switching as:
  (a) a mechanical switching device capable of making, carrying and breaking current under normal circuit conditions
  (b) cutting off an electrical installation or circuit from every source of electrical energy
  (c) the rapid disconnection of the electrical supply to remove or prevent danger
  (d) the switching of electrical equipment in normal service.

48 Functional switching may be defined as:
  (a) a mechanical switching device capable of making, carrying and breaking current under normal circuit conditions
  (b) cutting off an electrical installation or circuit from every source of electrical energy
  (c) the rapid disconnection of the electrical supply to remove or prevent danger
  (d) the switching of electrical equipment in normal service.

49 Emergency switching can be defined as:
  (a) a mechanical switching device capable of making, carrying and breaking current under normal circuit conditions
  (b) cutting off an electrical installation or circuit from every source of electrical energy
  (c) the rapid disconnection of the electrical supply to remove or prevent danger
  (d) the switching of electrical equipment in normal service.

50 The Regulations require that an overcurrent protective device interrupts a fault quickly and isolates the circuit before:
  (a) the voltage on any extraneous conductive parts reaches 50 V
  (b) the earth loop impedance reaches $0.4\,\Omega$ on circuits feeding 13 A socket outlets
  (c) the fault causes damage to the circuit isolating switches
  (d) the fault causes a temperature rise which might damage the insulation and terminations of the circuit conductors.

51 The maximum permissible value of the earth loop impedance of a circuit supplying fixed equipment and protected by a 30 A semi-enclosed fuse to BS 3036 and having a 2.5 mm protective conductor is found by reference to the tables in Appendix 2 of the *On Site Guide* (see also Table 3.2 of this book) to be:
  (a) $0.91\,\Omega$
  (b) $1.48\,\Omega$
  (c) $2.08\,\Omega$
  (d) $2.21\,\Omega$.

52 The earth fault loop impedance of a socket outlet circuit protected by a 30 A cartridge fuse to BS 1361 must not exceed (see IEE Table 41B1):
  (a) $0.4\,\Omega$
  (b) $1.14\,\Omega$
  (c) $1.20\,\Omega$
  (d) $2.0\,\Omega$.

53 The $(R_1 + R_2)$ resistance of 1000 m of PVC insulated copper cable having a $4.0\,mm^2$ phase conductor and $2.5\,mm^2$ protective conductor will be found from Table 9A of the *On Site Guide* to be (see also Table 3.1 in this book):
  (a) $4.61\,\Omega$
  (b) $9.22\,\Omega$
  (c) $12.02\,\Omega$
  (d) $16.71\,\Omega$.

54 The $(R_1 + R_2)$ resistance of 176 m of PVC insulated copper cable having a $2.5\,mm^2$ phase and protective conductor is (see Table 3.1 in this book):
  (a) $2.608\,\Omega$
  (b) $7.41\,\Omega$
  (c) $14.82\,\Omega$
  (d) $19.51\,\Omega$.

55 The value of the earth fault loop impedance Z_S of a circuit fed by 40 m of PVC insulated copper cable having a $2.5\,mm^2$ phase conductor and $1.5\,mm^2$ protective conductor connected to a supply having an impedance Z_E of $0.5\,\Omega$ under fault conditions will be
  (a) $1.436\,\Omega$
  (b) $9.755\,\Omega$

(c) $20.01 \, \Omega$

(d) $780.4 \, m\Omega$.

56 The time/current characteristics shown in Fig. 3.75(b) indicate that a fault current of 300 A will cause a 30 A semi-enclosed fuse to BS 3036 to operate in:

(a) $0.01 \, s$

(b) $0.1 \, s$

(c) $0.2 \, s$

(d) $2.0 \, s$.

57 The time/current characteristics shown in Fig. 3.75(a) indicate that a fault current of 30 A will cause a 10 A type 2 MCB to BS 3871 to operate in:

(a) $0.02 \, s$

(b) $8 \, s$

(c) $30 \, s$

(d) $200 \, s$.

58 'Under fault conditions the protective device nearest to the fault should operate leaving other healthy circuits unaffected'. This is one definition of:

(a) fusing factor

(b) effective discrimination

(c) a miniature circuit breaker

(d) a circuit protective conductor.

59 The overcurrent protective device protecting socket outlet circuits and any fixed equipment in bathrooms must operate within:

(a) $0.02 \, s$

(b) $0.4 \, s$

(c) $5 \, s$

(d) $45 \, s$.

60 The overcurrent protective device protecting fixed equipment in rooms other than bathrooms must operate within:

(a) $0.02 \, s$

(b) $0.4 \, s$

(c) $5 \, s$

(d) $45 \, s$.

61 Sketch an 11 kV ring distribution system suitable for supplying four load centres in a factory from a 33 kV primary substation. Show how the supply can be secured to all four load centres even though a fault occurs on a cable between two of the distribution substations.

62 Use a block diagram to describe the electrical equipment which should be installed in an 11 kV/400 V substation of brick construction.

63 A 400 V three-phase and neutral busbar rising main is to be used to provide a 230 V supply to each of eight individual flats on four floors of a building.

(a) Sketch the arrangement and describe how each flat's supply must be connected to the rising main if the total load is to be balanced.

(b) Describe the method used to prevent the spread of fire.

64 State the advantages and disadvantages of transmitting electricity:

(a) at very high voltage

(b) by overhead lines suspended on steel towers.

65 (a) State the meaning of an off-peak supply.

(b) An installation is made up of a lighting load, a cooking load, a space heating load and a water heating load. Which of these loads are suitable for connection to an off-peak supply? Sketch the arrangements required at the service position to make the connections.

66 Explain why the maximum values of earth fault loop impedance Z_S specified by the IEE Regulations and given in Tables 41B1 and 41B2 should not be exceeded.

67 A 50 mm² PVC insulated cable with copper conductors is subjected to a short circuit when connected to a 400 V supply. The impedance of the fault path is 83 mΩ. Calculate the maximum operating time of the protective device to prevent damage to the cable.

68 By referring to the table in the IEE Regulations (Tables 41B1, 42B2 and 41D), determine the maximum permitted earth fault loop impedance Z_S for the following circuits:

(a) a ring main of 13 A socket outlets protected by a 30 A semi-enclosed fuse to BS 3036

(b) a ring main of 13 A socket outlets protected by a 30 A cartridge fuse to BS 1361

(c) a single socket outlet protected by a 15 A type 1 MCB to BS 3871

(d) a water heating circuit protected by a 15 A semi-enclosed fuse to BS 3036

(e) a lighting circuit protected by a 6 A HBC fuse to BS 88 Part 2

(f) a lighting circuit protected by a 5 A semi-enclosed fuse to BS 3036.

69 10 mm² cables with PVC insulated copper conductors feed a commercial cooker connected to a 400 V supply. An earth loop impedance test indicates that Z_S has a value of 1.5 Ω. Calculate the minimum size of the protective conductor.

70 It is proposed to protect the commercial cooker circuit described in Exercise 35 with 30 A
(a) semi-enclosed fuses to BS 3036
(b) type 2 MCBs to BS 3871.
 Determine the time taken for each protection device to clear an earth fault on this circuit by referring to the characteristics of Fig. 3.20.

71 A 2.5 mm^2 PVC insulated and sheathed cable is used to feed a single 13 A socket outlet from a 15 A semi-enclosed fuse in a consumer's unit connected to a 230 V supply. Calculate the minimum size of the protective conductor to comply with the Regulations, given that the value of Z_S was 0.9 Ω.

72 Describe, with the aid of a circuit diagram, how speed control may be achieved with:
(a) a d.c. series motor
(b) a d.c. shunt motor.

73 Describe what is meant by the back emf of an electric motor and describe the construction and operation of a d.c. motor starter.

74 With the aid of a sketch describe the construction of a cage rotor. State the advantages and disadvantages of a rotor constructed in this way.

75 With the aid of a sketch describe how the turning forces are established which rotate the cage rotor of an induction motor.

76 Describe how a rotating magnetic flux produces a turning force on the rotor conductors of an induction motor.

77 Describe, with sketches, the construction of a wound rotor and cage rotor. State one advantage and one disadvantage for each type of construction.

78 The Regulations require that motor starters incorporate overload protection and no-volt protection. Describe what is meant by overload protection and no-volt protection.

79 Sketch a three-phase direct on line motor starter and describe its operation.

80 Sketch the wiring diagram for a three-phase direct on line motor starter which incorporates remote stop/start buttons.

81 Sketch the wiring diagram for a star delta motor starter and describe its operation.

82 Sketch the wiring diagram for an auto-transformer motor starter and describe its operation.

83 Sketch the wiring diagram for a rotor resistance motor starter and describe its operation.

84 Use sketches to explain how an electric motor should be installed and connected to a load via a vee belt.

85 Use a block diagram to explain the sequence of control for an electric motor of about 5 kW.

86 Describe what is meant by the 'continuous rating' of a motor.

87 Describe three types of motor enclosure and state one typical application for each type.

88 Use a phasor diagram to explain the meaning of a 'bad power factor'. Describe two methods of correcting the bad power factor due to a number of industrial motors.

89 A 7.5 kW motor with a p.f. of 0.866 is connected to a 400 V 50 Hz supply. Calculate:
(a) the current taken by the motor
(b) the value of the capacitor required to correct the p.f. to unity.

90 With the aid of neat sketches, describe the construction of:
(a) a double-wound transformer
(b) an auto-transformer
and state the losses which occur in a transformer.

91 Describe the construction and use of a voltage transformer.

92 Describe the construction and use of a bar primary current transformer.

93 Draw a circuit diagram to show an ammeter and current transformer connected to measure the current in a single-phase a.c. circuit. Explain why the secondary winding must not be open-circuited when the transformer is connected to the supply.

94 Draw a circuit diagram showing a voltmeter and voltage transformer connected to measure the voltage in a single-phase load.

95 Describe the construction of an oil-immersed transformer.

96 Describe what is meant by a tap changing transformer.

97 A 10 Ω resistor is connected in series with an inductor of reactance 15 Ω across a 230 V a.c. supply. Calculate
(a) the impedance
(b) the current
(c) the voltage across each component
(d) the power factor.
Draw to scale the phasor diagram.

98 A 9 Ω resistor is connected in series with a capacitor of 265.25 μF across a 230 V 50 Hz supply. Calculate
   (a) the impedance
   (b) the current
   (c) the voltage across each component
   (d) the power factor.
   Draw to scale the phasor diagram.

99 A pure inductor of 100 mH is connected in parallel with a 15 Ω resistor across a 230 V 50 Hz supply. Calculate the current in each branch and the total current and power factor. Sketch the phasor diagram.

100 A 60 μF capacitor is connected in parallel with a 20 Ω resistor across a 230 V 50 Hz supply. Calculate the current in each branch and the total current and power factor. Sketch the phasor diagram.

4

INSTALLATION (BUILDING AND STRUCTURES)

Regulations and responsibilities

Electricity generation as we know it today began when Michael Faraday conducted the famous ring experiment in 1831. This experiment, together with many other experiments of the time, made it possible for Lord Kelvin and Sebastian de Ferranti to patent in 1882 the designs for an electrical machine called the Ferranti–Thompson dynamo, which enabled the generation of electricity on a commercial scale.

In 1887 the London Electric Supply Corporation was formed with Ferranti as chief engineer. This was one of many privately owned electricity generating stations supplying the electrical needs of the UK. As the demand for electricity grew, more privately owned generating stations were built until eventually the government realized that electricity was a national asset which would benefit from nationalization.

In 1926 the Electricity Supply Act placed the responsibility for generation in the hands of the Central Electricity Board. In England and Wales the Central Electricity Generating Board (CEGB) had the responsibility for the generation and transmission of electricity on the Supergrid. In Scotland, generation was the joint responsibility of the North of Scotland Hydro-Electricity Board and the South of Scotland Electricity Board. In Northern Ireland electricity generation was the responsibility of the Northern Ireland Electricity Service.

In 1988 Cecil Parkinson, the Secretary of State for Energy in the Conservative government, proposed the denationalization of the electricity supply industry; this became law in March 1991, thereby returning the responsibility for generation, transmission and distribution to the private sector. It is anticipated that this action, together with new legislation over the security of supplies, will lead to a guaranteed quality of provision, with increased competition leading eventually to cheaper electricity.

During the period of development of the electricity services, particularly in the early days, poor design and installation led to many buildings being damaged by fire and the electrocution of human beings and livestock. It was the insurance companies which originally drew up a set of rules and guidelines of good practice in the interest of reducing the number of claims made upon them. The first rules were made by the American Board of Fire Underwriters and were quickly followed by the Phoenix Rules of 1882. In the same year the first edition of the Rules and Regulations for the Prevention of Fire Risk arising from Electrical Lighting was issued by the Institution of Electrical Engineers.

The current edition of these regulations is called the Requirements for Electrical Installations, IEE Wiring Regulations (BS 7671: 2001), and since January 1991 we have been using the 16th edition. All the rules have been revised, updated and amended at regular intervals to take account of modern developments, and the 16th edition brought the UK Regulations into harmony with those of the rest of Europe. The electrotechnical industry is now controlled by many rules, regulations and standards.

In Chapter 1 of this book we looked at a number of Statutory and Non-Statutory Regulations as they apply to the electrotechnical industry. In this section we will look at our responsibilities in our working environment.

The Health and Safety at Work Act 1974

The Health and Safety at Work Act provides a legal framework for stimulating and encouraging high standards of health and safety for everyone at work and the public at large from risks arising from work activities. The Act was a result of recommendations made by a Royal Commission in 1970 which looked at the health and safety of employees at work, and concluded that the main cause of accidents was apathy on the part of employer and employee. The new Act places the responsibility for safety at work on both the employer and the employee.

The employer has a duty to care for the health and safety of employees (Section 2 of the Act). To do this he must ensure that

- the working conditions and standard of hygiene are appropriate;
- the plant, tools and equipment are properly maintained;
- the necessary safety equipment – such as personal protective equipment, dust and fume extractors and machine guards – is available and properly used;
- the workers are trained to use equipment and plant safely.

Employees have a duty to care for their own health and safety and that of others who may be affected by their actions (Section 7 of the Act). To do this they must:

- take reasonable care to avoid injury to themselves or others as a result of their work activity;
- co-operate with their employer, helping him or her to comply with the requirements of the Act;
- not interfere with or misuse anything provided to protect their health and safety.

Failure to comply with the Health and Safety at Work Act is a criminal offence and any infringement of the law can result in heavy fines, a prison sentence or both.

ENFORCEMENT

Laws and rules must be enforced if they are to be effective. The system of control under the Health and Safety at Work Act comes from the Health and Safety Executive (HSE) which is charged with enforcing the law. The HSE is divided into a number of specialist inspectorates or sections which operate from local offices throughout the UK. From the local offices the inspectors visit individual places of work.

The HSE inspectors have been given wide-ranging powers to assist them in the enforcement of the law. They can:

1 enter premises unannounced and carry out investigations, take measurements or photographs;
2 take statements from individuals;
3 check the records and documents required by legislation;
4 give information and advice to an employee or employer about safety in the workplace;
5 demand the dismantling or destruction of any equipment, material or substance likely to cause immediate serious injury;
6 issue an improvement notice which will require an employer to put right, within a specified period of time, a minor infringement of the legislation;
7 issue a prohibition notice which will require an employer to stop immediately any activity likely to result in serious injury, and which will be enforced until the situation is corrected;
8 prosecute all persons who fail to comply with their safety duties, including employers, employees, designers, manufacturers, suppliers and the self-employed.

SAFETY DOCUMENTATION

Under the Health and Safety at Work Act, the employer is responsible for ensuring that adequate instruction and information is given to employees to make them safety-conscious. Part 1, section 3 of the Act instructs all employers to prepare a written health and safety policy statement and to bring this to the notice of all employees.

To promote adequate health and safety measures the employer must consult with the employees' safety representatives. All actions of the safety representatives should be documented and recorded as evidence that the company takes seriously its health and safety policy.

The Act provides for criminal proceedings to be instituted against those who do not satisfy the requirements of the Regulations.

Under the general protective umbrella of the Health and Safety at Work Act, other pieces of legislation also affect those working in the electrotechnical industry.

The Electricity at Work Regulations 1989 (EWR)

This legislation came into force in 1990 and replaced earlier regulations such as the Electricity (Factories Act) Special Regulations 1944. The Regulations are made under the Health and Safety at Work Act 1974, and enforced by the Health and Safety Executive. The purpose of the Regulations is to 'require precautions to be taken against the risk of death or personal injury from electricity in work activities'.

Section 4 of the EWR tells us that 'all systems must be constructed so as to prevent danger ..., and be properly maintained. ... Every work activity shall be carried out in a manner which does not give rise to danger. ... In the case of work of an electrical nature, it is preferable that the conductors be made dead before work commences'.

The EWR do not tell us specifically how to carry out our work activities and ensure compliance, but if proceedings were brought against an individual for breaking the EWR, the only acceptable defence would be 'to prove that all reasonable steps were taken and all diligence exercised to avoid the offence' (Regulation 29).

An electrical contractor could reasonably be expected to have 'exercised all diligence' if the installation was wired according to the IEE Wiring Regulations (see below).

ABSOLUTE DUTY OF CARE

The Health and Safety at Work Act and the Electricity at Work Regulations make numerous references to employer and employees having a 'duty of care' for the health and safety of others in the work environment. In this context the Electricity at Work Regulations refer to a person as a 'duty holder'. This phrase recognises the level of responsibility which electricians are expected to take on as a part of their job in order to control electrical safety in the work environment.

Everyone has a duty of care but not everyone is a duty holder. The Regulations recognise the amount of control that an individual might exercise over the whole electrical installation. The person who exercises 'control over the whole systems, equipment and conductors' and is the Electrical Company's representative on site, is the duty holder. He might be a supervisor or manager, but he will have a duty of care on behalf of his employer for the electrical, health, safety and environmental issues on that site.

Duties referred to in the Regulations may have the qualifying terms 'reasonably practicable' or 'absolute'. If the requirement of the regulation is absolute, then that regulation must be met regardless of cost or any other consideration. If the regulation is to be met 'so far as is reasonably practicable', then risks, cost, time, trouble and difficulty can be considered.

Often there is a cost effective way to reduce a particular risk and prevent an accident occurring. For example, placing a fire-guard in front of the fire at home when there are young children in the family is a reasonably practicable way of reducing the risk of a child being burned.

If a regulation is not qualified with 'so far as is reasonably practicable', then it must be assumed that the regulation is absolute. In the context of the Electricity at Work Regulations, where the risk is very often death by electrocution, the level of duty to prevent danger more often approaches that of an absolute duty of care.

The IEE Wiring Regulations (BS 7671: 2001)

The Institution of Electrical Engineers Requirements for Electrical Installations (the IEE Regulations) are non-statutory regulations. They relate principally to the design, selection, erection, inspection and testing of electrical installations, whether permanent or temporary, in and about buildings generally and to agricultural and horticultural premises, construction sites and caravans and their sites. Paragraph 7 of the introduction to the EWR says: 'the IEE Wiring Regulations is a code of practice which is widely recognised and accepted in the United Kingdom and compliance with them is likely to achieve compliance with all relevant aspects of the Electricity At Work Regulations'. The IEE Wiring Regulations only apply to installations operating at a voltage up to 1000 V a.c. They do not apply to electrical installations in mines and quarries where special regulations apply because of the adverse conditions experienced there.

The current edition of the IEE Wiring Regulations is the 16th edition incorporating amendment number 1: 2002 and 2: 2004. The main reason for incorporating the IEE Wiring Regulations into British Standard BS 7671 was to create harmonization with European Standards.

To assist electricians in their understanding of the Regulations a number of guidance notes have been published. The guidance notes, to which I will frequently make reference in this book, are those contained in the *On Site Guide*. Seven other guidance note booklets are also currently available. These are:

- *Selection and Erection;*
- *Isolation and Switching;*
- *Inspection and Testing;*
- *Protection against Fire;*
- *Protection against Electric Shock;*
- *Protection against Overcurrent;*
- *Special Locations.*

The IEE *On Site Guide* is prepared by the Institution of Electrical Engineers to simplify some aspects of the IEE Regulations BS 7671: 2001.

The Guide is intended to be used by skilled persons (electricians) carrying out limited applications of BS 7671 in:

(a) domestic installations generally, including off-peak supplies and supplies to associated garages, outbuildings and the like;
(b) industrial and commercial single and three phase installations where the local distribution fuse-boards or consumer unit is located at or near the supplier's cut-out.

These guidance notes are intended to be read in conjunction with the Regulations.

The IEE Wiring Regulations are the electrician's bible and provide an authoritative framework for all the work activities which we undertake as electricians.

British and European Standards

Goods manufactured to the exacting specifications laid down by the British Standards Institution (BSI) are suitable for the purpose for which they were made. There seems to be a British Standard for practically everything made today, and compliance with the relevant British Standard is, in most cases, voluntary. However, when specifying or installing equipment, the electrical designer or contractor needs to be sure that the materials are suitable for their purpose and offer a degree of safety, and should only use equipment which carries the appropriate British Standards number.

The BSI has created two important marks of safety, the BSI kite mark and the BSI safety mark, which are shown in Fig. 4.1.

The BSI kite mark is an assurance that the product carrying the label has been produced under a system of supervision, control and testing and can only be used by manufacturers who have been granted a licence under the scheme. It does not necessarily cover safety unless the appropriate British Standard specifies a safety requirement.

The BSI safety mark is a guarantee of the product's electrical, mechanical and thermal safety. It does not guarantee the product's performance.

BSI kite mark

BSI safety mark

Fig. 4.1 BSI kite and safety marks.

Fig. 4.2 European Commission safety mark.

The CE mark (Fig. 4.2) is not a quality mark but an indication given by the manufacturer or importer that the product or system meets the legal safety requirements of the European Commission and can therefore be presumed safe to use. The mark is applied by the manufacturer after carrying out the appropriate tests to ensure compliance with the relevant safety standards. The CE mark gives the manufacturer the right to sell the product in all the countries of the European Economic Area. All electrical products used by electrical contractors after 1 January 1997 must bear the CE mark.

INDEX OF PROTECTION (IP) BS EN 60529

IEE Regulation 713–07–01 tells us that where barriers and enclosures have been installed to prevent direct contact with live parts they must comply with IP2X and IP4X, but what does this mean?

The Index of Protection is a code which gives us a means of specifying the suitability of equipment for the environmental conditions in which it will be used. The tests to be carried out for the various degrees of protection are given in the British and European Standard BS EN 60529.

The code is written as IP (Index of Protection) followed by two numbers XX. The first number gives the degree of protection against the penetration of solid objects into the enclosure. The second number gives the degree of protection against water penetration. For example, a piece of equipment classified as IP45 will have barriers installed which prevent a 1 mm diameter rigid steel bar from making contact with live parts and be protected against the ingress of water from jets of water applied from any direction. Where a degree of protection is not specified, the number is replaced by an 'X' which simply means that the degree of protection is not specified although some protection may be afforded. The 'X' is used instead of '0' since '0' would indicate that no protection was given. The index of protection code is shown in Fig. 4.3.

FLAMEPROOF STANDARDS

The British Standards concerned with hazardous areas were first published in the 1920s and were concerned with the connection of electrical apparatus in the mining industry. Since those early days many national and international standards, as well as codes of practice, have been published to inform the manufacture, installation and maintenance of electrical equipment in all hazardous areas.

The main committee charged with harmonizing standards in Europe is the Centre Européen des Normes Électriques (CENELEC). Members of CENELEC are drawn from the national electrotechnical committees of the individual EU countries plus Norway and Switzerland.

In the UK the CENELEC Standards are drafted into English under the auspices of the Health and Safety Executive; they are also published by the British Standards Institution with BS numbers.

Once the certifying authority has assessed a piece of flameproof equipment it issues a certificate. All equipment must be marked with the testing house mark and information which shows where the equipment can be used. All equipment used in a hazardous area must be appropriately marked to maintain the integrity of the system. Hazardous areas are discussed later in this chapter and flameproof marks are shown in Fig. 4.44.

The IEE Regulations make specific reference to many British Standard specifications and British Standard codes of practice in the 16th edition of the Regulations.

The Electricity Supply Regulations forbid electricity supply authorities to connect electric lines and apparatus to the supply system unless their insulation is capable of withstanding the tests prescribed by the appropriate British Standards.

It is clearly in the interests of the electrical contractor to be aware of, and to comply with, any regulations which are relevant to the particular installation. IEE Regulation 130–01 states that good workmanship and the use of proper materials are essential for compliance with the Regulations.

In order to try to ensure that all electrical installation work is carried out to a minimum standard, the National Inspection Council for Electrical Installation Contracting (NICEIC) was established in 1956. NICEIC is supported by all sections of the electrical

First number (DEGREE OF PROTECTION AGAINST SOLID OBJECT PENETRATION)		Second number (DEGREE OF PROTECTION AGAINST WATER PENETRATION)	
0	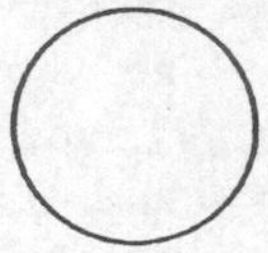 Non-protected.	0	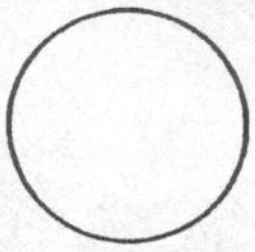 Non-protected.
1	 Protected against a solid object greater than 50 mm, such as a hand.	1	 Protected against water dripping vertically, such as condensation.
2	 Protected against a solid object greater than 12 mm, such as a finger.	2	 Protected against dripping water when tilted up to 15°.
3	 Protected against a solid object greater than 2.5 mm, such as a tool or wire.	3	 Protected against water spraying at an angle of up to 60°.
4	 Protected against a solid object greater than 1.0 mm, such as thin wire or strips.	4	 Protected against water splashing from any direction.
5	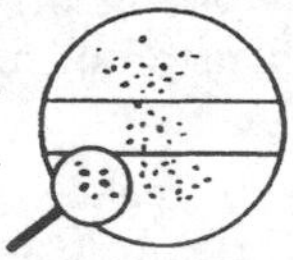 Dust protected. Prevents ingress of dust sufficient to cause harm.	5	 Protected against jets of water from any direction.
6	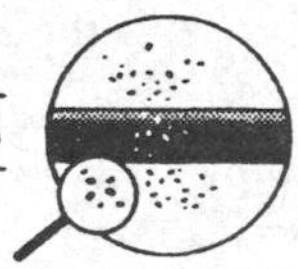 Dust tight. No dust ingress.	6	 Protected against heavy seas or powerful jets of water. Prevents ingress sufficient to cause harm.
		7	 Protected against harmful ingress of water when immersed to a depth of between 150 mm and 1 m.
		8	 Protected against submersion. Suitable for continuous immersion in water.

Fig. 4.3   Index of protection codes.

industry and its aims are to provide consumers with protection against faulty, unsafe or otherwise defective electrical installations. It maintains an approved roll of members who regularly have their premises, equipment and installations inspected by NICEIC engineers. Through this inspectorate the council is able to ensure a minimum standard of workmanship among its members. Electricians employed by an NICEIC-approved contractor are also, by association with their employer, accepted as being competent to carry out electrical installation work to an approved standard.

Safe working environment

In Chapter 1 we looked at some of the laws and regulations that affect our working environment. We looked at Safety Signs and PPE and how to recognise and use different types of fire extinguishers. The structure of companies within the electrotechnical industry and the ways in which they communicate information by drawings, symbols and standard forms was also discussed in Chapter 1.

We began to look at safe electrical isolation procedures in Chapter 1 and then discussed this topic further in Chapter 3, together with safe manual handling techniques and safe procedures for working above ground.

In Chapter 3, under the heading Health and Safety Applications, we looked at the common causes of accidents at work and how to control the risks associated with various hazards.

If your career in the electrotechnical industry is to be a long, happy and safe one, you must always behave responsibly and sensibly in order to maintain a safe working environment. Before starting work, make a safety assessment. What is going to be hazardous, will you require PPE, do you need any special access equipment. Carry out safe isolation procedures before beginning any work. You do not necessarily have to do these things formally, such as carrying out a risk assessment as described in Chapter 3, but just get into the habit of always working safely and being aware of the potential hazards around you when you are working. Finally, when the job is finished, clean up and dispose of all waste material responsibly as described in Chapter 3 under the heading Disposing of Waste.

Low-voltage supply systems

The British government agreed on 1 January 1995 that the electricity supplies in the UK would be harmonized with those of the rest of Europe. Thus the voltages used previously in low-voltage supply systems of 415 V and 240 V have become 400 V for three-phase supplies and 230 V for single-phase supplies. The Electricity Supply Regulations 1988 have also been amended to permit a range of variation from the new declared nominal voltage. From January 1995 the permitted tolerance is the nominal voltage +10% or −6%. Previously it was ±6%. This gives a voltage range of 216 V to 253 V for a nominal voltage of 230 V and 376 V to 440 V for a nominal voltage of 400 V.

The next change comes in 2005 when the tolerance levels will be adjusted to ±10% of the declared nominal voltage. All EU countries will adjust their voltages to comply with a nominal voltage of 230 V single-phase and 400 V three-phase.

The low-voltage supply to a domestic, commercial or small industrial consumer's installation is usually protected at the incoming service cable position with a 100 A high breaking capacity (HBC) fuse. Other items of equipment at this position are the energy meter and the consumer's distribution unit, providing the protection for the final circuits and the earthing arrangements for the installation.

An efficient and effective earthing system is essential to allow protective devices to operate. The limiting values of earth fault loop impedance are given in Tables 41, 604 and 605 of the IEE Regulations, and Section 542 gives details of the earthing arrangements to be incorporated in the supply system to meet the requirements of the Regulations. Five systems are described in the definitions but only the TN-S, TN-C-S and TT systems are suitable for public supplies.

A system consists of an electrical installation connected to a supply. Systems are classified by a capital letter designation.

The supply earthing

Arrangements are indicated by the first letter, where T means one or more points of the supply are directly connected to earth and I means the supply is not earthed or one point is earthed through a fault-limiting impedance.

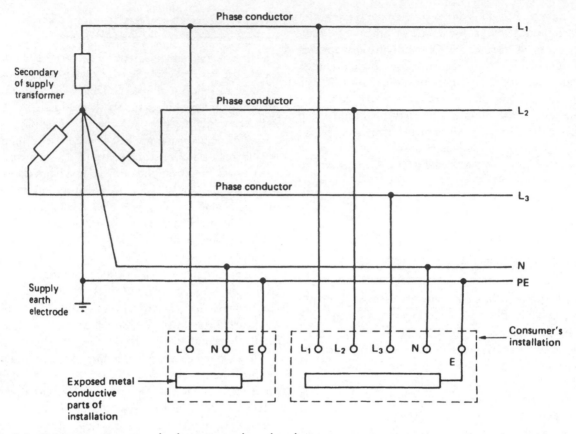

Fig. 4.4 TN-S system: separate neutral and protective conductor throughout.

The installation earthing

Arrangements are indicated by the second letter, where T means the exposed conductive parts are connected directly to earth and N means the exposed conductive parts are connected directly to the earthed point of the source of the electrical supply.

The earthed supply conductor

Arrangements are indicated by the third letter, where S means a separate neutral and protective conductor and C means that the neutral and protective conductors are combined in a single conductor.

TN-S SYSTEM

This is one of the commonest types of supply system to be found in the UK where the electricity companies' supply is provided by underground cables. The neutral and protective conductor are separate

throughout the system. The protective earth conductor (PE) is the metal sheath and armour of the underground cable, and this is connected to the consumer's main earthing terminal. All extraneous conductive parts of the installation, gas pipes, water pipes and any lightning protective system are connected to the protective conductor via the main earthing terminal of the installation. The arrangement is shown in Fig. 4.4.

TN-C-S SYSTEM

This type of underground supply is becoming increasingly popular to supply new installations in the UK. It is more commonly referred to as protective multiple earthing (PME). The supply cable uses a combined protective earth and neutral conductor (PEN conductor). At the supply intake point a consumer's main earthing terminal is formed by connecting the earthing terminal to the neutral conductor. All extraneous conductive parts of the installation, gas pipes, water pipes and any lightning protective system are then

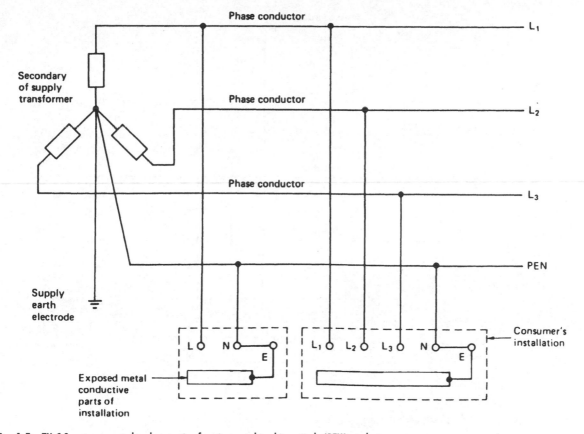

Fig. 4.5 TN-C-S system: neutral and protective functions combined in a single (PEN) conductor.

connected to the main earthing terminals. Thus phase to earth faults are effectively converted into phase to neutral faults. The arrangement is shown in Fig. 4.5.

TT SYSTEM

This is the type of supply more often found when the installation is fed from overhead cables. The supply authorities do not provide an earth terminal and the installation's circuit protective conductors must be connected to earth via an earth electrode provided by the consumer. An effective earth connection is sometimes difficult to obtain and in most cases a residual current device is provided when this type of supply is used. The arrangement is shown in Fig. 4.6.

The layout of a typical domestic service position for these three supply systems is shown in Fig. 4.7. There are two other systems of supply, the TN-C and IT systems but they do not comply with the supply regulations and therefore cannot be used for public supplies. Their use is restricted to private generating plants. For this reason I shall not include them here but they can be seen in Part 2 of the IEE Regulations.

Wiring circuits

LIGHTING CIRCUITS

Table 1A in Appendix 1 of the *On Site Guide* deals with the assumed current demand of points, and states that for lighting outlets we should assume a current equivalent to a minimum of 100 W per lampholder. This means that for a domestic lighting circuit rated at 5 A, a maximum of 11 lighting outlets could be connected to each circuit. In practice, it is usual to divide the fixed lighting outlets into two or more circuits of seven or eight outlets each. In this way the whole installation is not plunged into darkness if one lighting circuit fuses.

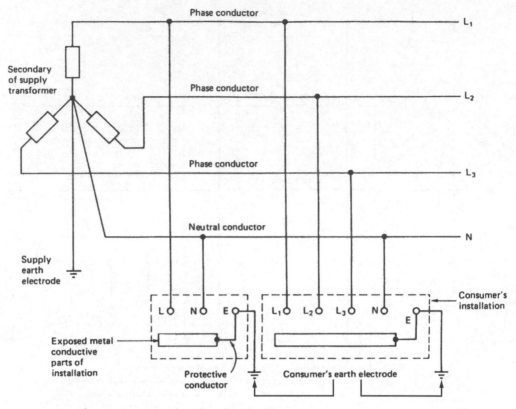

Figure 4.6 TT systems: earthing arrangements independent of supply cable.

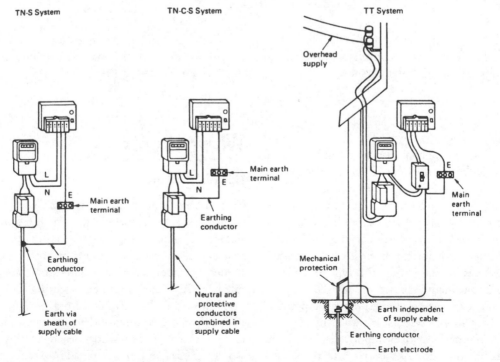

Fig. 4.7 Service arrangements for TN-S, TN-C-S and TT systems of supply.

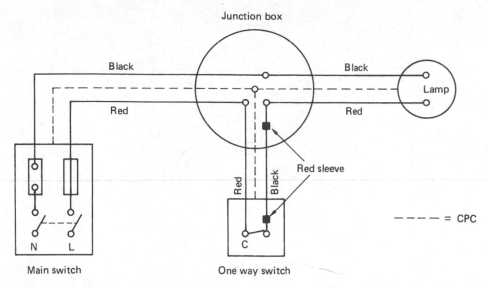

Fig. 4.8 Wiring diagram of one-way switch control.

Lighting circuits are usually wired in 1.0 mm or 1.5 mm cable using either a loop-in or joint-box method of installation. The loop-in method is universally employed with conduit installations or when access from above or below is prohibited after installation, as is the case with some industrial installations or blocks of flats. In this method the only joints are at the switches or lighting points, the live conductors being looped from switch to switch and the neutrals from one lighting point to another.

The use of junction boxes with fixed brass terminals is the method often adopted in domestic installations, since the joint boxes can be made accessible but are out of site in the loft area and under floorboards.

All switches and ceiling roses must contain an earth connection (Regulation 471–09–02) and the live conductors must be broken at the switch position in order to comply with the polarity regulations (713–09–01). A ceiling rose may only be connected to installations operating at 250 V maximum and must only accommodate one flexible cord unless it is specially designed to take more than one (553–04–02). Lampholders must comply with Regulation 553–03–02 and be suspended from flexible cords capable of suspending the mass of the luminaire fixed to the lampholder (554–01–01).

The type of circuit used will depend upon the installation conditions and the customer's requirements. One light controlled by one switch is called one-way switch control (see Fig. 4.8). A room with two access doors might benefit from a two-way switch control (see Fig. 4.9) so that the lights may be switched on or off at either position. A long staircase with more than two switches controlling the same lights would require intermediate switching (see Fig. 4.10).

One-way, two-way or intermediate switches can be obtained as plate switches for wall mounting or ceiling mounted cord switches. Cord switches can provide a convenient method of control in bedrooms or bathrooms and for independently controlling an office luminaire.

To convert an existing one-way switch control into a two-way switch control, a three-core and earth cable is run from the existing switch position to the proposed second switch position. The existing one-way switch is replaced by a two-way switch and connected as shown in Fig. 4.11.

SOCKET OUTLET CIRCUITS

A plug top is connected to an appliance by a flexible cord which should normally be no longer than 2 m (Regulation 553–01–07). Pressing the plug top into a socket outlet connects the appliance to the source of supply. Socket outlets therefore provide an easy and convenient method of connecting portable electrical appliances to a source of supply.

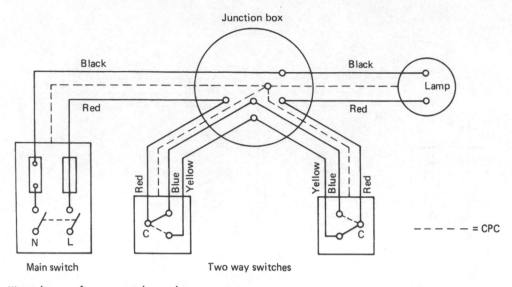

Fig. 4.9 Wiring diagram of two-way switch control.

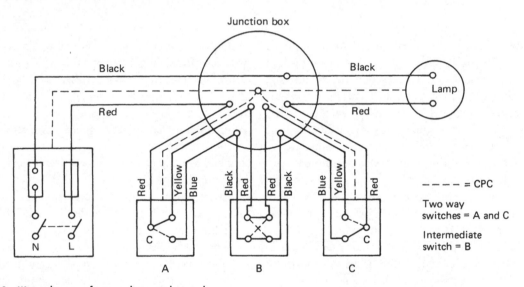

Fig. 4.10 Wiring diagram of intermediate switch control.

Socket outlets can be obtained in 15, 13, 5 and 2 A ratings, but the 13 A flat pin type complying with BS 1363 is the most popular for domestic installations in the United Kingdom. Each 13 A plug top contains a cartridge fuse to give maximum potential protection to the flexible cord and the appliances which it serves.

Socket outlets may be wired on a ring or radial circuit and in order that every appliance can be fed from an adjacent and convenient socket outlet, the number of sockets is unlimited provided that the floor area covered by the circuit does not exceed that given in Table 8A, Appendix 8 of the *On Site Guide*.

RADIAL CIRCUITS

In a radial circuit each socket outlet is fed from the previous one. Live is connected to live, neutral to neutral and earth to earth at each socket outlet. The fuse and cable sizes are given in Table 8A of Appendix 8 but circuits may also be expressed with a block diagram, as shown in Fig. 4.12. The number of permitted

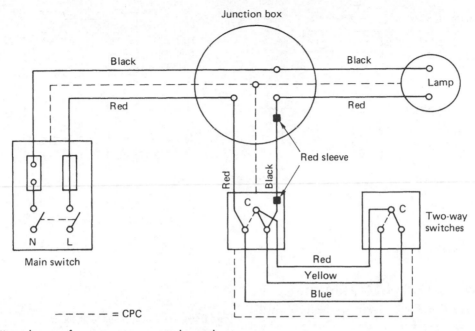

Fig. 4.11 Wiring diagram of one-way to two-way switch control.

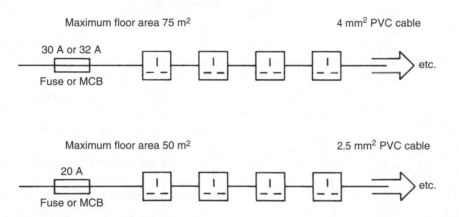

Fig. 4.12 Block diagram of radial circuits.

socket outlets is unlimited but each radial circuit must not exceed the floor area stated and the known or estimated load.

Where two or more circuits are installed in the same premises, the socket outlets and permanently connected equipment should be reasonably shared out among the circuits, so that the total load is balanced.

When designing ring or radial circuits special consideration should be given to the loading in kitchens which may require separate circuits. This is because the maximum demand of current-using equipment in kitchens may exceed the rating of the circuit cable and protection devices.

Ring and radial circuits may be used for domestic or other premises where the maximum demand of the current using equipment is estimated not to exceed the rating of the protective devices for the chosen circuit.

RING CIRCUITS

Ring circuits are very similar to radial circuits in that each socket outlet is fed from the previous one, but in

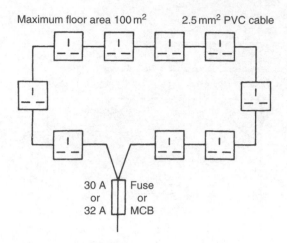

Fig. 4.13 Block diagram of ring circuits.

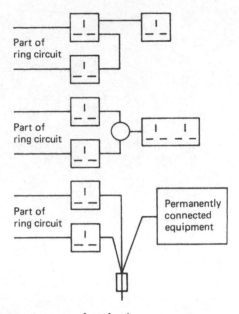

Fig. 4.14 Connection of non-fused spurs.

ring circuits the last socket is wired back to the source of supply. Each ring final circuit conductor must be looped into every socket outlet or joint box which forms the ring and must be electrically continuous throughout its length. The number of permitted socket outlets is unlimited but each ring circuit must not cover more than 100 m² of floor area.

The circuit details are given in Table 8A, Appendix 8 of the *On Site Guide* but may also be expressed by the block diagram given in Fig. 4.13.

Spurs to ring circuits

A spur is defined in Part 2 of the Regulations as a branch cable from a ring final circuit.

Non-fused spurs

The total number of non-fused spurs must not exceed the total number of socket outlets and pieces of stationary equipment connected directly in the circuit. The cable used for non-fused spurs must not be less than that of the ring circuit. The requirements concerning spurs are given in Appendix 8 of the *On Site Guide* but the various circuit arrangements may be expressed by the block diagrams of Fig. 4.14.

A non-fused spur may only feed one single or one twin socket or one permanently connected piece of equipment.

Non-fused spurs may be connected into the ring circuit at the terminals of socket outlets or at joint boxes or at the origin of the circuit.

Fused spurs

The total number of fused spurs is unlimited. A fused spur is connected to the circuit through a fused connection unit, the rating of which should be suitable for the conductor forming the spur but should not exceed 13 A. The requirements for fused spurs are also given in Appendix 8 but the various circuit arrangements may be expressed by the block diagrams of Fig. 4.15.

The general arrangement shown in Fig. 4.16 shows 11 socket outlets connected to the ring, three non-fused spur connections and two fused spur connections.

WATER HEATING CIRCUITS

A small, single-point over-sink type water heater may be considered as a permanently connected appliance and so may be connected to a ring circuit through a fused connection unit. A water heater of the immersion type is usually rated at a maximum of 3 kW, and could be considered as a permanently connected appliance, fed from a fused connection unit. However, many immersion heating systems are connected into storage vessels of about 150 litres in domestic installations, and Appendix 8 of the *On Site Guide* states that immersion heaters fitted to vessels in excess of 15 litres should be supplied by their own circuit.

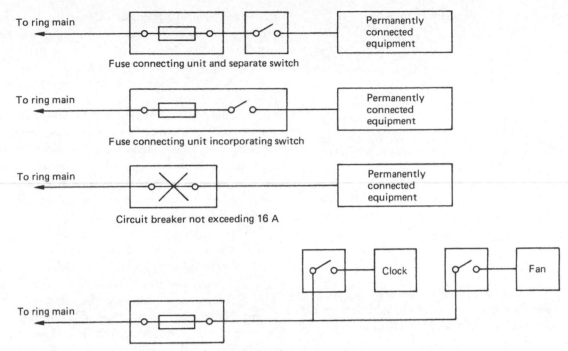

Fig. 4.15 Connection of fused spurs.

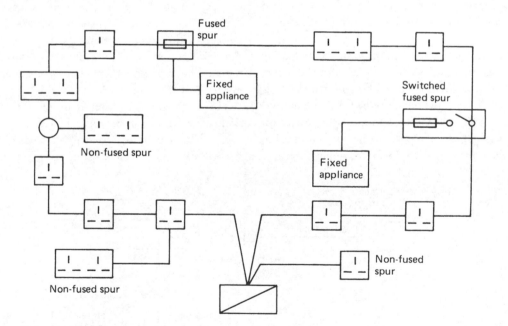

Fig. 4.16 Typical ring circuit with spurs.

Therefore, immersion heaters must be wired on a separate radial circuit when they are connected to water vessels which hold more than 15 litres. Figure 4.17 shows the wiring arrangements for an immersion heater. The hot and cold water connections must be connected to an earth connection in order to meet the supplementary bonding requirements of Regulation 413–05–02. Every switch must be a double-pole (DP)

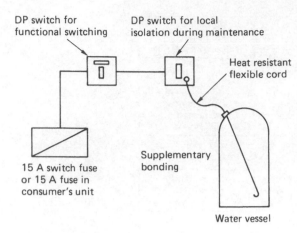

Fig. 4.17 Immersion heater wiring.

switch and out of reach of anyone using a fixed bath or shower (Regulation 601–08–01) when the immersion heater is fitted to a vessel in a bathroom.

ELECTRIC SPACE HEATING CIRCUITS

Electrical heating systems can be broadly divided into two categories: unrestricted local heating and off-peak heating.

Unrestricted local heating may be provided by portable electric radiators which plug into the socket outlets of the installation. Fixed heaters that are wall-mounted or inset must be connected through a fused connection and incorporate a local switch, either on the heater itself or as a part of the fuse connecting unit, as shown in Fig. 4.15. Heating appliances where the heating element can be touched must have a double-pole switch which disconnects all conductors. This requirement includes radiators which have an element inside a silica-glass sheath (601–12–01).

Off-peak heating systems may provide central heating from storage radiators, ducted warm air or under-floor heating elements. All three systems use the thermal storage principle, whereby a large mass of heat-retaining material is heated during the off-peak period and allowed to emit the stored heat throughout the day. The final circuits of all off-peak heating installations must be fed from a separate supply controlled by an electricity board time clock.

When calculating the size of cable required to supply a single storage radiator, it is good practice to assume a current demand equal to 3.4 kW at each point. This will allow the radiator to be changed at a future time with the minimum disturbance to the installation. Each radiator must have a 20 A double-pole means of isolation adjacent to the heater and the final connection should be via a flex outlet. See Fig. 4.18 for wiring arrangements.

Ducted warm air systems have a centrally sited thermal storage heater with a high storage capacity. The unit is charged during the off-peak period, and a fan drives the stored heat in the form of warm air through large air ducts to outlet grilles in the various rooms. The wiring arrangements for this type of heating are shown in Fig. 4.19.

The single storage heater is heated by an electric element embedded in bricks and rated between 6 kW and 15 kW depending upon its thermal capacity. A radiator of this capacity must be supplied on its own

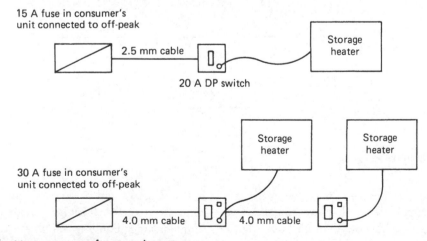

Fig. 4.18 Possible wiring arrangements for storage heaters.

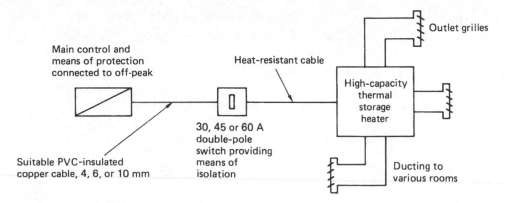

Fig. 4.19 Ducted warm air heating system.

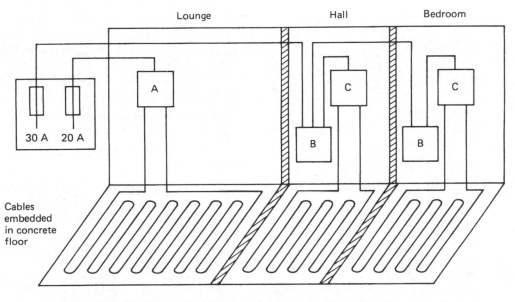

A = Thermostat incorporating DP switch fed by 2.5 mm PVC/copper
B = DP switch fuse fed by 4.0 mm PVC/copper
C = Thermostat fed by 2.5 mm PVC/copper

Fig. 4.20 Floor warming installations.

circuit, in cable capable of carrying the maximum current demand and protected by a fuse or MCB of 30 A, 45 A or 60 A as appropriate. At the heater position, a double-pole switch must be installed to terminate the fixed heater wiring. The flexible cables used for the final connection to the heaters must be of the heat-resistant type.

Floor warming installations use the thermal storage properties of concrete. Special cables are embedded in the concrete floor screed during construction. When current is passed through the cables they become heated, the concrete absorbs this heat and radiates it into the room. The wiring arrangements are shown in Fig. 4.20. Once heated, the concrete will give off heat for a long time after the supply is switched off and is, therefore, suitable for connection to an off-peak supply.

Underfloor heating cables installed in bathrooms or shower rooms must incorporate an earthed metallic sheath or be covered by an earthed metallic grid connected to the supplementary bonding (Regulation 601–12–02).

COOKER CIRCUIT

A cooker with a rating above 3 kW must be supplied on its own circuit but since it is unlikely that in normal use every heating element will be switched on at the same time, a diversity factor may be applied in calculating the cable size, as detailed in Table 1A in Appendix 1 of the *On Site Guide*.

Consider, as an example, a cooker with the following elements fed from a cooker control unit incorporating a 13 A socket:

$$4 \times 2 \text{ kW fast boiling rings} = 8000 \text{ W}$$
$$1 \times 2 \text{ kW grill} = 2000 \text{ W}$$
$$1 \times 2 \text{ kW oven} = 2000 \text{ W}$$
$$\text{Total loading} = 12\,000 \text{ W}$$

When connected to 250 V

$$\text{Current rating} = \frac{12\,000}{250} = 48 \text{ A}.$$

Applying the diversity factor of Table 1A,

$$\text{Total current rating} = 48 \text{ A}$$
$$\text{First 10 amperes} = 10 \text{ A}$$
$$30\% \text{ of } 38 \text{ A} = 11.4 \text{ A}$$
$$\text{Socket outlet} = 5 \text{ A}$$

Assessed current demand $= 10 + 11.4 + 5 = 26.4$ A

Therefore, a cable capable of carrying 26.4 A may be used safely rather than a 48 A cable.

A cooking appliance must be controlled by a switch separate from the cooker but in a readily accessible position (Regulation 476–03–04). Where two cooking appliances are installed in one room, such as split-level cookers, one switch may be used to control both appliances provided that neither appliance is more than 2 m from the switch (*On Site Guide*, Appendix 8).

CONDUCTOR SIZE CALCULATIONS

The size of a cable to be used for an installation depends upon:

■ the current rating of the cable under defined installation conditions and
■ the maximum permitted drop in voltage as defined by Regulation 525–01.

The factors which influence the current rating are:

1 the design current – the cable must carry the full load current;
2 the type of cable – PVC, MICC, copper conductors or aluminium conductors;
3 the installed conditions – clipped to a surface or installed with other cables in a trunking;
4 the surrounding temperature – cable resistance increases as temperature increases and insulation may melt if the temperature is too high;
5 the type of protection – for how long will the cable have to carry a fault current?

Regulation 525–01 states that the drop in voltage from the supply terminals to the fixed current-using equipment must not exceed 4% of the mains voltage. That is a maximum of 9.2 V on a 230 V installation. The volt drop for a particular cable may be found from

$$VD = \text{Factor} \times \text{Design current} \times \text{Length of run}$$

The factor is given in the tables of Appendix 4 of the IEE Regulations and Appendix 6 of the *On Site Guide*. See Table 4.3.

The cable rating, denoted I_t, may be determined as follows:

$$I_t = \frac{\text{Current rating of protective device}}{\text{Any applicable correction factors}}$$

The cable rating must be chosen to comply with Regulation 433–02–01. The correction factors which may need applying are given below as:

Ca   the ambient or surrounding temperature correction factor, which is given in Tables 4C1 and 4C2 of Appendix 4 of the IEE Regulations and 6A1 and 6A2 of the *On Site Guide* and shown in Table 4.1.

Cg   the grouping correction factor given in Tables 4B1, 4B2 and 4B3, of the IEE Regulations and 6C of the *On Site Guide*.

Cr   the 0.725 correction factor to be applied when semi-enclosed fuses protect the circuit as described in item 6.2 of the preface to Appendix 4 of the IEE Regulations.

Ci   the correction factor to be used when cables are enclosed in thermal insulation. Regulation 523–04 gives us three possible correction values:

■ Where one side of the cable is in contact with thermal insulation we must read the current rating from the column in the table which relates to reference method 4. See Table 4.2.

Table 4.1 Ambient temperature correction factors. Reproduced from the IEE *On Site Guide* by kind permission of the Institution of Electrical Engineers

Table 6A1   Ambient temperature factors

Correction factors for ambient temperature where protection is against short-circuit and overload

Type of insulation	Operating temperature	Ambient temperature °C								
		25	30	35	40	45	50	55	60	65
Thermoplastic (general purpose pvc)	70°C	1.03	1.0	0.94	0.87	0.79	0.71	0.61	0.50	0.35

Table 6A2   Ambient temperature factors

Correction factors for ambient temperature where the overload protective device is a semi-enclosed fuse to BS 3036

Type of insulation	Operating temperature	Ambient temperature °C								
		25	30	35	40	45	50	55	60	65
Thermoplastic (general purpose pvc)	70°C	1.03	1.0	0.97	0.94	0.91	0.87	0.84	0.69	0.48

Table 4.2 Current carrying capacity of cables. Reproduced from the IEE *On Site Guide* by kind permission of the Institution of Electrical Engineers

Table 6E1   Multicore cables having thermoplastic (pvc) or thermosetting insulation (note 1), non-armoured, (COPPER CONDUCTORS)

Ambient temperature: 30°C. Conductor operating temperature: 70°C
CURRENT-CARRYING CAPACITY (Amperes): BS 6004, BS 7629

Conductor cross-sectional area	Reference Method 4 (enclosed in an insulated wall, etc.)		Reference Method 3 (enclosed in conduit on a wall or ceiling, or in trunking)		Reference Method 1 (clipped direct)		Reference Method 11 (on a perforated cable tray) or Reference Method 13 (free air)	
	1 two-core cable*, single-phase a.c. or d.c.	1 three-core cable* or 1 four-core cable, three-phase a.c.	1 two-core cable*, single-phase a.c. or d.c.	1 three-core cable* or 1 four-core cable, three-phase a.c.	1 two-core cable*, single-phase a.c. or d.c.	1 three-core cable* or 1 four-core cable, three-phase a.c.	1 two-core cable*, single-phase a.c. or d.c.	1 three-core cable* or 1 four-core cable, three-phase a.c.
1	2	3	4	5	6	7	8	9
mm^2	A	A	A	A	A	A	A	A
1	11	10	13	11.5	15	13.5	17	14.5
1.5	14	13	16.5	15	19.5	17.5	22	18.5
2.5	18.5	17.5	23	20	27	24	30	25
4	25	23	30	27	36	32	40	34
6	32	29	38	34	46	41	51	43
10	43	39	52	46	63	57	70	60
16	57	52	69	62	85	76	94	80
25	75	68	90	80	112	96	119	101
35	92	83	111	99	138	119	148	126
50	110	99	133	118	168	144	180	153
70	139	125	168	149	213	184	232	196
95	167	150	201	179	258	223	282	238

See Notes overleaf. For a fuller treatment see Appendix 4 of BS 7671 Table 4D2A.

- Where the cable is *totally* surrounded over a length greater than 0.5 m we must apply a factor of 0.5.
- Where the cable is *totally* surrounded over a short length, the appropriate factor given in Table 52A of the IEE Regulations or Table 6B of the *On Site Guide* should be applied.

Having calculated the cable rating, the smallest cable should be chosen from the appropriate table which will carry that current. This cable must also meet the voltage drop Regulation 525–01 and this should be calculated as described earlier. When the calculated value is less than 4% of the mains voltage the cable may be considered suitable. If the calculated value is greater than the 4% value, the next larger cable size must be tested until a cable is found which meets both the current rating and voltage drop criteria.

EXAMPLE

A house extension has a total load of 6 kW installed some 18 m away from the mains consumer unit. A PVC insulated and sheathed twin and earth cable will provide a submain to this load and be clipped to the side of the ceiling joists over much of its length in a roof space which is anticipated to reach 35°C in the summer and where insulation is installed up to the top of the joists. Calculate the minimum cable size if the circuit is to be protected (a) by a semi-enclosed fuse to BS 3036 and (b) by a type 2 MCB to BS 3871. Assume a TN-S supply, that is, a supply having a separate neutral and protective conductor throughout.

Let us solve this question using only the tables given in the *On Site Guide*. The tables in the Regulations will give the same values, but this will simplify the problem. Refer to Tables 4.1, 4.2 and 4.3 in this book.

$$\text{Design current } I_b = \frac{\text{Power}}{\text{Volts}} = \frac{6000\,\text{W}}{240\,\text{V}} = 26.09\,\text{A}.$$

Nominal current setting of the protection for this load $I_n = 30\,\text{A}$.

For (a) the correction factors to be included in this calculation are:

Ca ambient temperature; from Table 6A2 and shown in Table 4.1 the correction factor for 35°C is 0.97.

Cg the grouping correction factor is not applied since the cable is to be clipped direct to a surface and not in contact with other cables.

Cr the protection is by a semi-enclosed fuse and, therefore, a factor of 0.725 must be applied.

Ci thermal insulation is in contact with one side of the cable and we must therefore assume installed method 4. See Table 4.2.

The cable rating, I_t is given by

$$I_t = \frac{\text{Current rating of protective device}}{\text{The product of the correction factors}}$$
$$= \frac{30\,\text{A}}{0.97 \times 0.725} = 42.66\,\text{A}$$

From column 2 of Table 6E1, shown in Table 4.2, a 10 mm cable having a rating of 43 A is required to carry this current.

Now test for volt drop: The maximum permissible volt drop is $4\% \times 230\,\text{V} = 9.2\,\text{V}$. From Table 6E2 shown in Table 4.3 the volt drop per ampere per metre for a 10 mm cable is 4.4 mV.

Therefore, the volt drop for this cable length and load is equal to

$$4.4 \times 10^{-3}\,\text{V}/(\text{A m}) \times 26.09\,\text{A} \times 18\,\text{m} = 2.07\,\text{V}.$$

Since this is less than the maximum permissible value of 9.2 V, a 10 mm cable satisfies the current and drop in voltage requirements and is therefore the chosen cable when semi-enclosed fuse protection is used.

For (b) the correction factors to be included in this calculation are:

Ca ambient temperature; from Table 6A1 shown in Table 4.1 the correction factor for 35°C is 0.94.

Cg grouping factors need not be applied.

Cr since protection is by MCB no factor need be applied.

Ci thermal insulation once more demands that we assume installed method 4. See Table 4.2.

The design current is still 26.09 A and we will therefore choose a 30 A MCB for the nominal current setting of the protective device, I_n.

$$\text{Cable rating} = I_t = \frac{30}{0.94} = 31.9\,\text{A}$$

From column 2 of Table 6E1 and shown in Table 4.2, a 6 mm cable, having a rating of 32 A, is required to carry this current.

Now test for volt drop: from Table 6E2 and shown in Table 4.3 the volt drop per ampere per metre for a 6 mm cable is 7.3 mV. So the volt drop for this cable length and load is equal to

$$7.3 \times 10^{-3}\,\text{V}/(\text{A m}) \times 26.09\,\text{A} \times 18\,\text{m} = 3.43\,\text{V}.$$

Since this is less than the maximum permissible value of 9.2 V, a 6 mm cable satisfies the current and drop in voltage requirements when the circuit is protected by an MCB. From the above calculations it is clear that better protection can reduce the cable size. Even though an MCB is

Table 4.3 Voltage Drop in Cables Factor. Reproduced from the IEE *On site Guide* by kind permission of the Institution of Electrical Engineers.

Table 6E2

Voltage drop: (per ampere per metre):			Conductor operating temperature: 70 °C
Conductor cross-sectional area	Two-core cable, d.c.	Two-core cable, single-phase a.c.	Three- or four-core cable, three-phase
1	2	3	4
mm²	mV/A/m	mV/A/m	mV/A/m
1	44	44	38
1.5	29	29	25
2.5	18	18	15
4	11	11	9.5
6	7.3	7.3	6.4
10	4.4	4.4	3.8
16	2.8	2.8	2.4
		r	r
25	1.75	1.75	1.50
35	1.25	1.25	1.10
50	0.93	0.93	0.80
70	0.63	0.63	0.55
95	0.46	0.47	0.41

Note: For a fuller treatment see Appendix 4 of BS 7671 Table 4D2B.

more expensive than a semi-enclosed fuse, the installation of a 6 mm cable with an MCB may be less expensive than 10 mm cable protected by a semi-enclosed fuse. These are some of the decisions which the electrical contractor must make when designing an installation which meets the requirements of the customer and the IEE Regulations.

If you are unsure of the standard fuse and MCB rating of protective devices, you can refer to Tables 2A, 2B, 2C and 2D of the *On Site Guide*.

Wiring systems

The final choice of a wiring system must rest with those designing the installation and those ordering the work, but whatever system is employed, good workmanship and the use of proper materials is essential for compliance with the Regulations (IEE Regulation 130–02–01). The necessary skills can be acquired by an electrical trainee who has the correct attitude and dedication to his craft.

PVC INSULATED AND SHEATHED CABLE INSTALLATIONS

PVC insulated and sheathed wiring systems are used extensively for lighting and socket installations in domestic dwellings. Mechanical damage to the cable caused by impact, abrasion, penetration, compression or tension must be minimized during installation (Regulation 522–06–01). The cables are generally fixed using plastic clips incorporating a masonry nail, which means the cables can be fixed to wood, plaster or brick with almost equal ease. Cables should be run horizontally or vertically, not diagonally, down a wall. All kinds should be removed so that the cable is run straight and neatly between clips fixed at equal distances providing adequate support for the cable so that it does not become damaged by its own weight (Regulation 522–08–04 and Table 4A of the *On Site Guide*). Table 4A of the IEE *On Site Guide* is shown in Table 4.4 of this chapter. Where cables are bent, the radius of the bend should not cause the conductors to be damaged (Regulation 522–08–03 and Table 4E of the *On Site Guide*).

Terminations or joints in the cable may be made in ceiling roses, junction boxes, or behind sockets or switches, provided that they are enclosed in a non-ignitable material, are properly insulated and are mechanically and electrically secure (IEE Regulation 526). All joints must be accessible for inspection testing and maintenance when the installation is completed.

Where PVC insulated and sheathed cables are concealed in walls, floors or partitions, they must be provided with a box incorporating an earth terminal at each outlet position. PVC cables do not react chemically with plaster, as do some cables, and consequently PVC cables may be buried under plaster. Further protection by channel or conduit is only necessary if mechanical protection from nails or screws is required. However, Regulation 522–06–06 now tells us that where PVC cables are to be embedded in wet plaster, they should be run along one of the permitted routes shown in Fig. 4.22 or be covered in capping to protect them from the plasterer's trowel and penetration by nails and screws. Figure 4.21 shows a typical PVC installation. To identify the most probable cable routes, Regulation 522–06–06 tells us that outside a zone formed by a 150 mm border all around a wall edge, cables can only be run horizontally or vertically

Table 4.4 Spacing of cable supports. Reproduced from the IEE *On Site Guide* by kind permission of the Institution of Electrical Engineers

Table 4A   Spacings of supports for cables in accessible positions

Overall diameter of cable*	Maximum spacings of clips							
	Non-armoured thermosetting, thermoplastic or lead sheathed cables				Armoured cables		Mineral insulated copper sheathed or aluminium sheathed cables	
	Generally		In caravans					
	Horizontal[†] 2	Vertical[†] 3	Horizontal[†] 4	Vertical[†] 5	Horizontal[†] 6	Vertical[†] 7	Horizontal[†] 8	Vertical[†] 9
mm	mm	mm	mm	mm	mm	mm	mm	mm
Not exceeding 9	250	400			–	–	600	800
Exceeding 9 and not exceeding 15	300	400	250 (for all sizes)	400 (for all sizes)	350	450	900	1200
Exceeding 15 and not exceeding 20	350	450			400	550	1500	2000
Exceeding 20 and not exceeding 40	400	550			450	600	–	–

Note: For the spacing of supports for cables having an overall diameter exceeding 40 mm, and for single-core cables having conductors of cross-sectional area 300 mm^2 and larger, the manufacturer's recommendations should be observed.

* For flat cables taken as the dimension of the major axis.

† The spacings stated for horizontal runs may be applied also to runs at an angle of more than 30 from the vertical. For runs at an angle of 30° or less from the vertical, the vertical spacings are applicable.

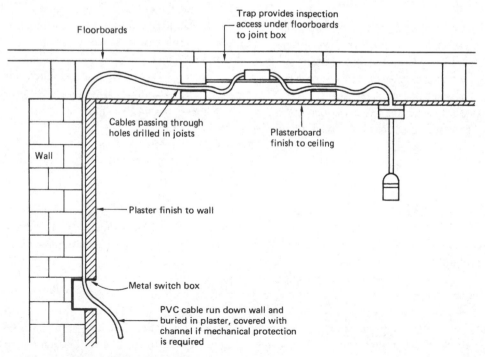

Fig. 4.21   A concealed PVC sheathed wiring system.

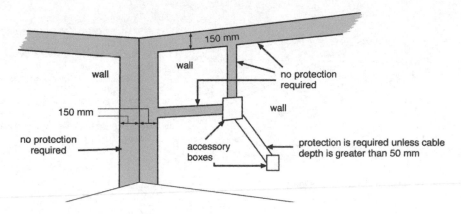

Figure. 4.22 Permitted cable routes.

to a point or accessory unless they are contained in a substantial earthed enclosure, such as a conduit, which can withstand nail penetration, as shown in Fig. 4.22.

Where cables pass through walls, floors and ceilings the hole should be made good with incombustible material such as mortar or plaster to prevent the spread of fire (Regulation 527–02–01). Cables passing through metal boxes should be bushed with a rubber grommet to prevent abrasion of the cable. Holes drilled in floor joists through which cables are run should be 50 mm below the top or 50 mm above the bottom of the joist to prevent damage to the cable by nail penetration (Regulation 522–06–05), as shown in Fig. 4.23. PVC cables should not be installed when the surrounding temperature is below 0°C or when the cable temperature has been below 0°C for the previous 24 hours because the insulation becomes brittle at low temperatures and may be damaged during installation.

CONDUIT INSTALLATIONS

A conduit is a tube, channel or pipe in which insulated conductors are contained. The conduit, in effect, replaces the PVC outer sheath of a cable, providing mechanical protection for the insulated conductors. A conduit installation can be rewired easily or altered at any time, and this flexibility, coupled with mechanical protection, makes conduit installations popular for commercial and industrial applications.

There are three types of conduit used in electrical installation work: steel, PVC and flexible.

Steel conduit

Steel conduits are made to a specification defined by BS 4568 and are either heavy gauge welded or solid drawn. Heavy gauge is made from a sheet of steel welded along the seam to form a tube and is used for most installation work. Solid drawn conduit is a seamless tube which is much more expensive and only used for special gas-tight, explosion-proof or flameproof installations.

Conduit is supplied in 3.75 m lengths and typical sizes are 16, 20, 25 and 32 mm. Conduit tubing and fittings are supplied in a black enamel finish for internal use or hot galvanized finish for use on external or damp installations. A wide range of fittings are available and the conduit is fixed using saddles or pipe hooks, as shown in Fig. 4.24.

Metal conduits are threaded with stocks and dies and bent using special bending machines. The metal conduit is also utilized as the circuit protective conductor and, therefore, all connections must be screwed up tightly and all burrs removed so that cables will not be damaged as they are drawn into the conduit. Metal conduits containing a.c. circuits must contain phase and neutral conductors in the same conduit to prevent eddy currents flowing, which would result in the metal conduit becoming hot (Regulation 521–02–01).

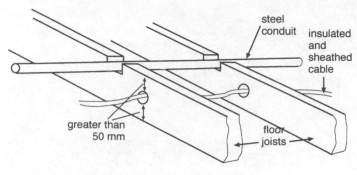

Notes:

1. Maximum diameter of hole should be 0.25 x joist depth.
2. Holes on centre line in a zone between 0.25 and 0.4 x span.
3. Maximum depth of notch should be 0.125 x joist depth.
4. Notches on top in a zone between 0.1 and 0.25 x span.
5. Holes in the same joist should be at least 3 diameters apart.

Fig. 4.23 Correct installation of conductors in floor joists.

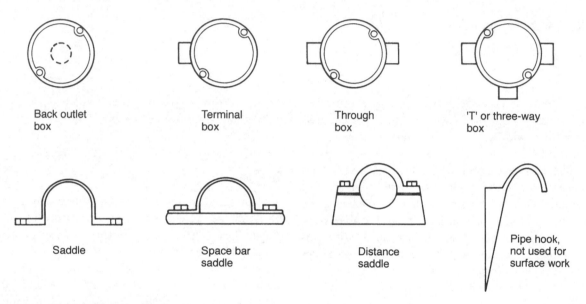

Fig. 4.24 Conduit fittings and saddles.

PVC conduit

PVC conduit used on typical electrical installations is heavy gauge standard impact tube manufactured to BS 4607. The conduit size and range of fittings are the same as those available for metal conduit. PVC conduit is most often joined by placing the end of the conduit into the appropriate fitting and fixing with a PVC solvent adhesive. PVC conduit can be bent by hand using a bending spring of the same diameter as the inside of the conduit. The spring is pushed into the conduit to the point of the intended bend and the conduit then bent over the knee. The spring ensures that the conduit keeps its circular shape. In cold weather, a little warmth applied to the point of the intended bend often helps to achieve a more successful bend.

The advantages of a PVC conduit system are that it may be installed much more quickly than steel conduit and is non-corrosive, but it does not have the mechanical strength of steel conduit. Since PVC conduit is an insulator it cannot be used as the CPC and a separate earth conductor must be run to every outlet. It is not suitable for installations subjected to temperatures below $-5°C$ or above $60°C$. Where luminaires are suspended from PVC conduit boxes, precautions must be taken to ensure that the lamp does not raise the box temperature or that the mass of the luminaire supported by each box does not exceed the maximum recommended by the manufacturer (IEE Regulation 522–01). PVC conduit also expands much more than metal conduit and so long runs require an expansion coupling to allow for conduit movement and help to prevent distortion during temperature changes.

All conduit installations must be erected first before any wiring is installed (IEE Regulation 522–080–02). The radius of all bends in conduit must not cause the cables to suffer damage, and therefore the minimum radius of bends given in Table 4E of the *On Site Guide* applies (IEE Regulation 522–08–03). All conduits should terminate in a box or fitting and meet the boxes or fittings at right angles, as shown in Fig. 4.25. Any unused conduit box entries should be blanked off and all boxes covered with a box lid, fitting or accessory to provide complete enclosure of the conduit system. Conduit runs should be separate from other services, unless intentionally bonded, to prevent arcing occurring from a faulty circuit within the conduit, which might cause the pipe of another service to become punctured.

When drawing cables into conduit they must first be *run off* the cable drum. That is, the drum must be rotated as shown in Fig. 4.26 and not allowed to *spiral off*, which will cause the cable to twist.

Cables should be fed into the conduit in a manner which prevents any cable crossing over and becoming twisted inside the conduit. The cable insulation must not be damaged on the metal edges of the draw-in box. Cables can be pulled in on a draw wire if the run is a long one. The draw wire itself may be drawn in on a fish tape, which is a thin spring steel or plastic tape.

A limit must be placed on the number of bends between boxes in a conduit run and the number of cables which may be drawn into a conduit to prevent the cables being strained during wiring. Appendix 5 of the *On Site Guide* gives a guide to the cable capacities of conduits and trunking.

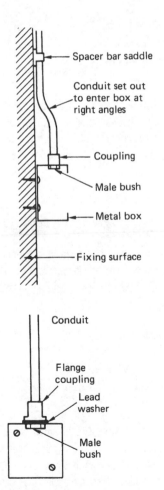

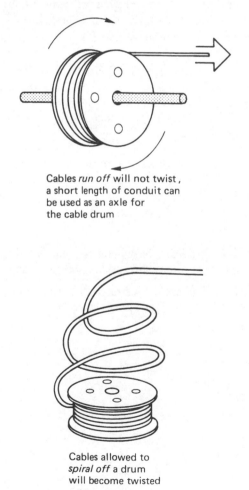

Fig. 4.25 Terminating conduits.

Fig. 4.26 Running off cable from a drum.

Flexible conduit

Flexible conduit is made of interlinked metal spirals often covered with a PVC sleeving. The tubing must not be relied upon to provide a continuous earth path and, consequently, a separate CPC must be run either inside or outside the flexible tube (Regulation 543–02–01).

Flexible conduit is used for the final connection to motors so that the vibrations of the motor are not transmitted throughout the electrical installation and to allow for modifications to be made to the final motor position and drive belt adjustments.

Conduit capacities

Single PVC insulated conductors are usually drawn into the installed conduit to complete the installation. Having decided upon the type, size and number of cables required for a final circuit, it is then necessary to select the appropriate size of conduit to accommodate those cables.

The tables in Appendix 5 of the *On Site Guide* describe a 'factor system' for determining the size of conduit required to enclose a number of conductors. The tables are shown in Tables 4.5 and 4.6 of this chapter. The method is as follows:

- Identify the cable factor for the particular size of conductor. (This is given in Table 5A for straight conduit runs and Table 5C for cables run in conduits which incorporate bends, see Table 4.5 of this chapter.)

Table 4.5 Conduit cable factors. Reproduced from the IEE *On Site Guide* by kind permission of the Institution of Electrical Engineers

Table 5C   Cable factors for use in conduit in long straight runs over 3 m, or runs of any length incorporating bends

Type of conductor	Conductor cross-sectional area mm^2	Cable factor
Solid	1	16
or	1.5	22
stranded	2.5	30
	4	43
	6	58
	10	105
	16	145
	25	217

The inner radius of a conduit bend should be not less than 2.5 times the outside diameter of the conduit.

- Multiply the cable factor by the number of conductors, to give the sum of the cable factors.
- Identify the appropriate part of the conduit factor table given by the length of run and number of bends. (For straight runs of conduit less than 3 m in length, the conduit factors are given in Table 5B. For conduit runs in excess of 3 m or incorporating bends, the conduit factors are given in Table 5D, see Table 4.6 of this chapter.)
- The correct size of conduit to accommodate the cables is that conduit which has a factor equal to or greater than the sum of the cable factors.

EXAMPLE 1

Six 2.5 mm^2 PVC insulated cables are to be run in a conduit containing two bends between boxes 10 m apart. Determine the minimum size of conduit to contain these cables.

From Table 5C, shown in Table 4.5
The factor for one 2.5 mm^2 cable $= 30$
The sum of the cable factors $\quad = 6 \times 30$
$\qquad\qquad\qquad\qquad\qquad = 180$

From Table 5D shown in Table 4.6, a 25 mm conduit, 10 m long and containing two bends, has a factor of 260. A 20 mm conduit containing two bends only has a factor of 141 which is less than 180, the sum of the cable factors and, therefore, 25 mm conduit is the minimum size to contain these cables.

EXAMPLE 2

Ten 1.0 mm^2 PVC insulated cables are to be drawn into a plastic conduit which is 6 m long between boxes and contains one bend. A 4.0 mm PVC insulated CPC is also included. Determine the minimum size of conduit to contain these conductors.

From Table 5C, and shown in Table 4.5
The factor for one 1.0 mm cable $= 16$
The factor for one 4.0 mm cable $= 43$
The sum of the cable factors $\quad = (10 \times 16) + (1 \times 43)$
$\qquad\qquad\qquad\qquad\qquad = 203$

From Table 5D shown in Table 4.6, a 20 mm conduit, 6 m long and containing one bend, has a factor of 233. A 16 mm conduit containing one bend only has a factor of 143 which is less than 203, the sum of the cable factors and, therefore, 20 mm conduit is the minimum size to contain these cables.

Table 4.6 Conduit cable factors. Reproduced from the IEE *On Site Guide* by kind permission of the Institution of Electrical Engineers

Table 5D Cable factors for runs incorporating bends and long straight runs

Length of run m	Conduit diameter, mm																			
	16	20	25	32	16	20	25	32	16	20	25	32	16	20	25	32	16	20	25	32
	Straight				One bend				Two bends				Three bends				Four bends			
1	Covered by				188	303	543	947	177	286	514	900	158	256	463	818	130	213	388	692
1.5					182	294	528	923	167	270	487	857	143	233	422	750	111	182	333	600
2	Tables				177	286	514	900	158	256	463	818	130	213	388	692	97	159	292	529
2.5					171	278	500	878	150	244	442	783	120	196	358	643	86	141	260	474
3	A and B				167	270	487	857	143	233	422	750	111	182	333	600				
3.5	179	290	521	911	162	263	475	837	136	222	404	720	103	169	311	563				
4	177	286	514	900	158	256	463	818	130	213	388	692	97	159	292	529				
4.5	174	282	507	889	154	250	452	800	125	204	373	667	91	149	275	500				
5	171	278	500	878	150	244	442	783	120	196	358	643	86	141	260	474				
6	167	270	487	857	143	233	422	750	111	182	333	600								
7	162	263	475	837	136	222	404	720	103	169	311	563								
8	158	256	463	818	130	213	388	692	97	159	292	529								
9	154	250	452	800	125	204	373	667	91	149	275	500								
10	150	244	442	783	120	196	358	643	86	141	260	474								

Additional factors: For 38 mm diameter use ...... $1.4 \times$ (32 mm factor)

For 50 mm diameter use ...... $2.6 \times$ (32 mm factor)

For 63 mm diameter use ...... $4.2 \times$ (32 mm factor)

TRUNKING INSTALLATIONS

A trunking is an enclosure provided for the protection of cables which is normally square or rectangular in cross-section, having one removable side. Trunking may be thought of as a more accessible conduit system and for industrial and commercial installations it is replacing the larger conduit sizes. A trunking system can have great flexibility when used in conjunction with conduit; the trunking forms the background or framework for the installation, with conduits running from the trunking to the point controlling the current using apparatus. When an alteration or extension is required it is easy to drill a hole in the side of the trunking and run a conduit to the new point. The new wiring can then be drawn through the new conduit and the existing trunking to the supply point.

Trunking is supplied in 3 m lengths and various cross-sections measured in millimetres from 50×50 up to 300×150. Most trunking is available in either steel or plastic.

Metallic trunking

Metallic trunking is formed from mild steel sheet, coated with grey or silver enamel paint for internal use or a hot-dipped galvanized coating where damp conditions might be encountered. A wide range of accessories are available, such as 45° bends, 90° bends, tee and four-way junctions, for speedy on-site assembly. Alternatively, bends may be fabricated in lengths of trunking, as shown in Fig. 4.27. This may be necessary or more convenient if a bend or set is non-standard, but it does take more time to fabricate bends than merely to bolt on standard accessories.

When fabricating bends the trunking should be supported with wooden blocks for sawing and filing, in order to prevent the sheet steel vibrating or becoming deformed. Fish plates must be made and riveted or bolted to the trunking to form a solid and secure bend. When manufactured bends are used, the continuity of the earth path must be ensured across the joint by making all fixing screw connections very tight, or fitting a separate copper strap between the trunking

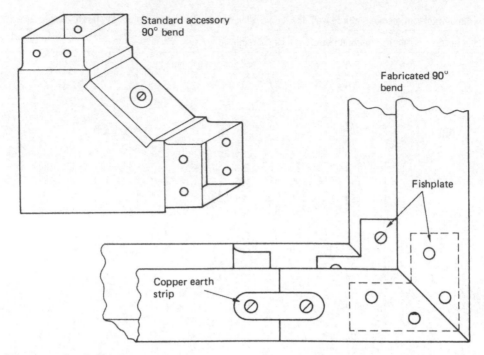

Fig. 4.27 Alternative trunking bends.

and the standard bend. If an earth continuity test on the trunking is found to be unsatisfactory, an insulated CPC must be installed inside the trunking. The size of the protective conductor will be determined by the largest cable contained in the trunking, as described by Table 54G of the IEE Regulations.

Non-metallic trunking

Trunking and trunking accessories are also available in high-impact PVC. The accessories are usually secured to the lengths of trunking with a PVC solvent adhesive. PVC trunking, like PVC conduit, is easy to install and is non-corrosive. A separate CPC will need to be installed and non-metallic trunking may required more frequent fixings because it is less rigid than metallic trunking. All trunking fixings should use round-headed screws to prevent damage to cables since the thin sheet construction makes it impossible to countersink screw heads.

Mini-trunking

Mini-trunking is very small PVC trunking, ideal for surface wiring in domestic and commercial installations such as offices. The trunking has a cross-section of 16 mm × 16 mm, 25 mm × 16 mm, 38 mm ×

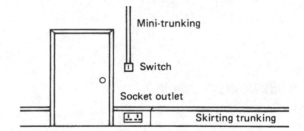

Fig. 4.28 Typical installation of skirting trunking and mini-trunking.

16 mm or 38 mm × 25 mm and is ideal for switch drops or for housing auxiliary circuits such as telephone or audio equipment wiring. The modern square look in switches and sockets is complemented by the mini-trunking which is very easy to install (see Fig. 4.28).

Skirting trunking

A trunking manufactured from PVC or steel and in the shape of a skirting board is frequently used in commercial buildings such as hospitals, laboratories and offices. The trunking is fitted around the walls of a room and contains the wiring for socket outlets and telephone points which are mounted on the lid, as shown in Fig. 4.28.

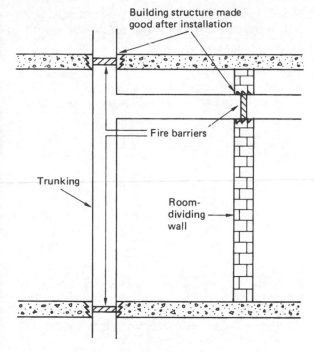

(a) Fire barriers in trunking

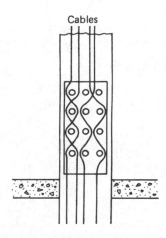

(b) Cable supports in vertical trunking

Fig. 4.29 Installation of trunking.

Where any trunking passes through walls, partitions, ceilings or floors, short lengths of lid should be fitted so that the remainder of the lid may be removed later without difficulty. Any damage to the structure of the buildings must be made good with mortar, plaster or concrete in order to prevent the spread of fire. Fire barriers must be fitted inside the trunking every 5 m,

or at every floor level or room dividing wall, if this is a shorter distance, as shown in Fig. 4.29(a).

Where trunking is installed vertically, the installed conductors must be supported so that the maximum unsupported length of non-sheathed cable does not exceed 5 m. Figure 4.29(b) shows cables woven through insulated pin supports, which is one method of supporting vertical cables.

PVC insulated cables are usually drawn into an erected conduit installation or laid into an erected trunking installation. Table 5D of the *On Site Guide* only gives factors for conduits up to 32 mm in diameter, which would indicate that conduits larger than this are not in frequent or common use. Where a cable enclosure greater than 32 mm is required because of the number or size of the conductors, it is generally more economical and convenient to use trunking.

Trunking capacities

The ratio of the space occupied by all the cables in a conduit or trunking to the whole space enclosed by the conduit or trunking is known as the *space factor*. Where sizes and types of cable and trunking are not covered by the tables in Appendix 5 of the *On Site Guide* a space factor of 45% must not be exceeded. This means that the cables must not fill more than 45% of the space enclosed by the trunking. The tables of Appendix 5 take this factor into account.

To calculate the size of trunking required to enclose a number of cables:

- Identify the cable factor for the particular size of conductor (Table 5E). See Table 4.7 of this chapter.
- Multiply the cable factor by the number of conductors to give the sum of the cable factors.
- Consider the factors for trunking (Table 5F) and shown in Table 4.8 of this chapter. The correct size of trunking to accommodate the cables is that trunking which has a factor equal to or greater than the sum of the cable factors.

EXAMPLE

Calculate the minimum size of trunking required to accommodate the following single-core PVC cables:

20 × 1.5 mm   solid conductors
20 × 2.5 mm   solid conductors

21 × 4.0 mm   stranded conductors
16 × 6.0 mm   stranded conductors

From Table 5E shown in Table 4.7, the cable factors are:

for 1.5 mm solid cable   –   8.0
for 2.5 mm solid cable   –   11.9
for 4.0 mm stranded cable   –   16.6
for 6.0 mm stranded cable   –   21.2

The sum of the cable terms is:
$(20 \times 8.0) + (20 \times 11.9) + (21 \times 16.6) + (16 \times 21.2) =$
1085.8. From Table 5F shown in Table 4.8, 75 mm × 38 mm trunking
has a factor of 1146 and, therefore, the minimum size of trunking to
accommodate these cables is 75 mm × 38 mm, although a larger size,
say 75 mm × 50 mm would be equally acceptable if this was more
readily available as a standard stock item.

SEGREGATION OF CIRCUITS

Where an installation comprises a mixture of low-
voltage and very low-voltage circuits such as mains
lighting and power, fire alarm and telecommunication
circuits, they must be separated or *segregated* to prevent
electrical contact (IEE Regulation 528–01–01).

For the purpose of these regulations various circuits
are identified by one of two bands and defined by Part
2 of the Regulations as follows:

Band I   telephone, radio, bell, call and intruder alarm
circuits, emergency circuits for fire alarm
and emergency lighting.
Band II   mains voltage circuits.

When Band I circuits are insulated to the same voltage as Band II circuits, they may be drawn into the same
compartment.

When trunking contains rigidly fixed metal barriers
along its length, the same trunking may be used to
enclose cables of the separate Bands without further
precautions, provided that each Band is separated by a
barrier, as shown in Fig. 4.30.

Multi-compartment PVC trunking cannot provide
band segregations since there is no metal screen
between the Bands. This can only be provided in
PVC trunking if screened cables are drawn into the
trunking.

Table 4.7 Trunking cable factors. Reproduced from the IEE *On Site Guide* by kind permission of the Institution of Electrical Engineers

Table 5E   Cable factors for trunking

Type of conductor	Conductor cross-sectional area mm^2	PVC BS 6004 Cable factor	Thermosetting BS 7211 Cable factor
Solid	1.5	8.0	8.6
	2.5	11.9	11.9
Stranded	1.5	8.6	9.6
	2.5	12.6	13.9
	4	16.6	18.1
	6	21.2	22.9
	10	35.3	36.3
	16	47.8	50.3
	25	73.9	75.4

Note:

(i) These factors are for metal trunking and may be optimistic for plastic trunking where the cross-sectional area available may be significantly reduced from the nominal by the thickness of the wall material.

(ii) The provision of spare space is advisable; however, any circuits added at a later date must take into account grouping. Appendix 4, BS 7671.

Table 4.8 Trunking cable factors. Reproduced from the IEE *On Site Guide* by kind permission of the Institution of Electrical Engineers

Table 5F   Factors for trunking

Dimensions of trunking mm × mm	Factor	Dimensions of trunking mm × mm	Factor
50 × 38	767	200 × 100	8572
50 × 50	1037	200 × 150	13001
75 × 25	738	200 × 200	17429
75 × 38	1146	225 × 38	3474
75 × 50	1555	225 × 50	4671
75 × 75	2371	225 × 75	7167
100 × 25	993	225 × 100	9662
100 × 38	1542	225 × 150	14652
100 × 50	2091	225 × 200	19643
100 × 75	3189	225 × 225	22138
100 × 100	4252	300 × 38	4648
150 × 38	2999	300 × 50	6251
150 × 50	3091	300 × 75	9590
150 × 75	4743	300 × 100	12929
150 × 100	6394	300 × 150	19607
150 × 150	9697	300 × 200	26285
200 × 38	3082	300 × 225	29624
200 × 50	4145	300 × 300	39428
200 × 75	6359		

Space factor – 45% with trunking thickness taken into account.

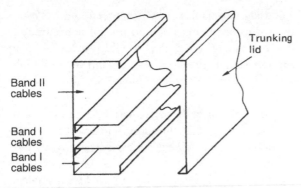

Fig. 4.30 *Segregation of cables in trunking.*

CABLE TRAY INSTALLATIONS

Cable tray is a sheet-steel channel with multiple holes. The most common finish is hot-dipped galvanized but PVC-coated tray is also available. It is used extensively on large industrial and commercial installations for supporting MI and SWA cables which are laid on the cable tray and secured with cable ties through the tray holes.

Cable tray should be adequately supported during installation by brackets which are appropriate for the particular installation. The tray should be bolted to the brackets with round-headed bolts and nuts, with the round head inside the tray so that cables drawn along the tray are not damaged.

The tray is supplied in standard widths from 50 mm to 900 mm, and a wide range of bends, tees and reducers are available. Figure 4.31 shows a factory-made 90° bend at B. The tray can also be bent using a cable tray bending machine to create bends such as that shown at A in Fig. 4.31. The installed tray should be securely bolted with round-headed bolts where lengths or accessories are attached, so that there is a continuous earth path which may be bonded to an electrical earth. The whole tray should provide a firm support for the cables and therefore the tray fixings must be capable of supporting the weight of both the tray and cables.

PVC/SWA CABLE INSTALLATIONS

Steel wire armoured PVC insulated cables are now extensively used on industrial installations and often laid on cable tray. This type of installation has the advantage of flexibility, allowing modifications to be made speedily as the need arises. The cable has a steel wire armouring giving mechanical protection and permitting it to be laid directly in the ground or inducts, or it may be fixed directly or laid on a cable tray.

It should be remembered that when several cables are grouped together the current rating will be reduced according to the correction factors given in Table 4B1 of the IEE Regulations and Table 6C of the *On Site Guide*.

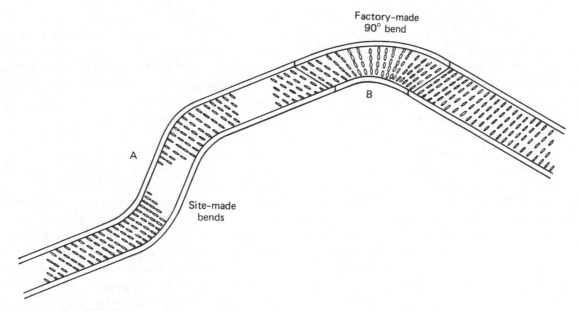

Fig. 4.31 *Cable tray with bends.*

The cable is easy to handle during installation, is pliable and may be bent to a radius of eight times the cable diameter. The PVC insulation would be damaged if installed in ambient temperatures over 70°C or below 0°C, but once installed the cable can operate at low temperatures.

The cable is terminated with a simple gland which compresses a compression ring on to the steel wire armouring to provide the earth continuity between the switchgear and the cable.

MI CABLE INSTALLATIONS

Mineral insulated cables are available for general wiring as:

- light-duty MI cables for voltages up to 600 V and sizes from $1.0 \, mm^2$ to $10 \, mm^2$,
- heavy-duty MI cables for voltages up to 1000 V and sizes from $1.0 \, mm^2$ to $150 \, mm^2$.

The cables are available with bare sheaths or with a PVC oversheath. The cable sheath provides sufficient mechanical protection for all but the most severe situations, where it may be necessary to fit a steel sheath or conduit over the cable to give extra protection, particularly near floor level in some industrial situations.

The cable may be laid directly in the ground, inducts, on cable tray or clipped directly to a structure. It is not affected by water, oil or the cutting fluids used in engineering and can withstand very high temperature or even fire. The cable diameter is small in relation to its current carrying capacity and it should last indefinitely if correctly installed because it is made from inorganic materials. These characteristics make the cable ideal for Band I emergency circuits, boilerhouses, furnaces, petrol stations and chemical plant installations.

The cable is supplied in coils and should be run off during installation and not spiralled off, as described in Fig. 4.26 for conduit. The cable can be work-hardened if over-handled or over-manipulated. This makes the copper outer sheath stiff and may result in fracture. The outer sheath of the cable must not be penetrated, otherwise moisture will enter the magnesium oxide insulation and lower its resistance. To reduce the risk of damage to the outer sheath during installation, cables should be straightened and formed by hammering with a hide hammer or a block of wood and a steel hammer. When bending MI cables the radius of the bend should not cause the cable to become damaged and clips should provide adequate support (Regulations 522–08–03 and 04 and Tables 4A and 4E of the *On Site Guide*). See also Table 4.4 earlier in this chapter.

The cable must be prepared for termination by removing the outer copper sheath to reveal the copper conductors. This can be achieved by using a rotary stripper tool or, if only a few cables are to be terminated, the outer sheath can be removed with side cutters, peeling off the cable in a similar way to peeling the skin from a piece of fruit with a knife. When enough conductor has been revealed, the outer sheath must be cut off square to facilitate the fitting of the sealing pot, and this can be done with a ringing tool. All excess magnesium oxide powder must be wiped from the conductors with a clean cloth. This is to prevent moisture from penetrating the seal by capillary action.

Cable ends must be terminated with a special seal to prevent the entry of moisture. Figure 2.55 shows a brass screw-on seal and gland assembly, which allows termination of the MI cables to standard switchgear and conduit fittings. The sealing pot is filled with a sealing compound, which is pressed in from one side only to prevent air pockets forming, and the pot closed by crimping home the sealing disc. Such an assembly is suitable for working temperatures up to 105°C. Other compounds or powdered glass can increase the working temperature up to 250°C.

The conductors are not identified during the manufacturing process and so it is necessary to identify them after the ends have been sealed. A simple continuity or polarity test, as described later in this chapter, can identify the conductors which are then sleeved or identified with coloured markers.

Connection of MI cables can be made directly to motors, but to absorb the vibrations a 360° loop should be made in the cable just before the termination. If excessive vibration is to be expected the MI cable should be terminated in a conduit through-box and the final connection made by flexible conduit.

Copper MI cables may develop a green incrustation or patina on the surface, even when exposed to normal atmospheres. This is not harmful and should not be removed. However, if the cable is exposed to an environment which might encourage corrosion, an MI cable with an overall PVC sheath should be used.

Support and fixing methods for electrical equipment

Individual conductors may be installed in trunking or conduit and individual cables may be clipped directly to a surface or laid on a tray using the wiring system which is most appropriate for the particular installation. The installation method chosen will depend upon the contract specification, the fabric of the building and the type of installation – domestic, commercial or industrial.

It is important that the wiring systems and fixing methods are appropriate for the particular type of installation and compatible with the structural materials used in the building construction. The electrical installation must be compatible with the installed conditions, must not damage the fabric of the building or weaken load-bearing girders or joists.

The installation designer must ask himself the following questions:

- Does this wiring system meet the contract specification?
- Is the wiring system compatible with this particular installation?
- Do I need to consider any special regulations such as those required by agricultural and horticultural installations, swimming pools, or flameproof installations?
- Will this type of electrical installation be aesthetically acceptable and compatible with the other structural materials?

The installation electrician must ask himself the following questions:

- Am I using materials and equipment which meet the relevant British Standards and the contract specification?
- Am I using an appropriate fixing method for this wiring system or piece of equipment?
- Will the structural material carry the extra load that my conduits and cables will place upon it?
- Will my fixings and fittings weaken the existing fabric of the building?
- Will the electrical installation interfere with other supplies and services?
- Will all terminations and joints be accessible upon completion of the erection period?

- Will the materials being used for the electrical installation be compatible with the intended use of the building?
- Am I working safely and efficiently and in accordance with the IEE Regulations (BS 7671)?

A domestic installation usually calls for a PVC insulated and sheathed wiring system. These cables are generally fixed using plastic clips incorporating a masonry nail which means that the cables can be fixed to wood, plaster or brick with almost equal ease.

Cables must be run straight and neatly between clips fixed at equal distances and providing adequate support for the cable so that it does not become damaged by its own weight (IEE Regulation 522–08–04 and Table 4A of the *On Site Guide*) and shown in Table 4.4 earlier in this chapter.

A commercial or industrial installation might call for a conduit or trunking wiring system. A conduit is a tube, channel or pipe in which insulated conductors are contained. The conduit, in effect, replaces the PVC outer sheath of a cable, providing mechanical protection for the insulated conductors. A conduit installation can be rewired easily or altered at any time and this flexibility, coupled with mechanical protection, makes conduit installations popular for commercial and industrial applications. Steel conduits and trunking are, however, much heavier than single cables and, therefore, need substantial and firm fixings and supports. A wide range of support brackets are available for fixing conduit, trunking and tray installations to the fabric of a commercial or industrial installation. Some of these are shown in Fig. 4.32.

When a heavier or more robust fixing is required to support cabling or equipment a nut and bolt or screw fixing is called for. Wood screws may be screwed directly into wood but when fixing to stone, brick or concrete it is first necessary to drill a hole in the masonry material which is then plugged with a material (usually plastic) to which a screw can be secured.

For the most robust fixing to masonry materials an expansion bolt such as that made by Rawlbolt should be used.

For lightweight fixings to hollow partitions or plasterboard a spring toggle can be used. Plasterboard cannot support a screw fixing directly into itself but the spring toggle spreads the load over a larger area, making the fixing suitable for light loads.

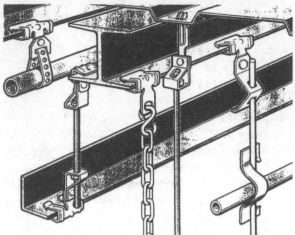
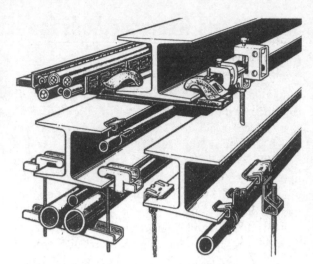

Fig. 4.32 Some manufactured supports for electrical equipment.

Let us look in a little more detail at individual joining, support and fixing methods.

Joining materials

Plastic can be joined with an appropriate solvent. Metals may be welded, brazed or soldered, but the most popular method of on-site joining of metals on electrical installations is by nuts and bolts or rivets.

A nut and bolt joint may be considered a temporary fastening since the parts can easily be separated if required by unscrewing the nut and removing the bolt. A rivet is a permanent fastening since the parts riveted together cannot be easily separated.

Two pieces of metal joined by a bolt and nut and by a machine screw and nut are shown in Fig. 4.33. The nut is tightened to secure the joint. When joining trunking or cable trays, a round head machine screw

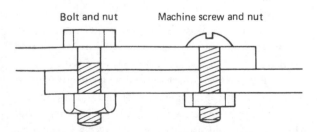

Fig. 4.33 Joining of metals.

should be used with the head inside to reduce the risk of damage to cables being drawn into the trunking or tray.

Thin sheet material such as trunking is often joined using a pop riveter. Special rivets are used with a hand-tool, as shown in Fig. 4.34. Where possible, the parts to be riveted should be clamped and drilled together with a clearance hole for the rivet. The stem of the rivet is pushed into the nose bush of the riveter until the alloy sleeve of the rivet is flush with the nose bush (a). The rivet is then placed in the hole and the handles squeezed together (b). The alloy sleeve is compressed and the rivet stem will break off when the rivet is set and the joint complete (c). To release the broken-off stem piece, the nose bush is turned upwards and the handles opened sharply. The stem will fall out and is discarded (d).

BRACKET SUPPORTS

Conduit and trunking may be fixed directly to a surface such as a brick wall or concrete ceiling, but where cable runs are across girders or other steel framework, spring steel clips may be used but support brackets or clips often require manufacturing.

The brackets are usually made from flat iron, which is painted after manufacturing to prevent corrosion. They may be made on-site by the electrician or, if many brackets are required, the electrical contractor may make a working sketch with dimensions and have the items manufactured by a blacksmith or metal fabricator.

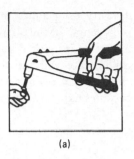
(a)

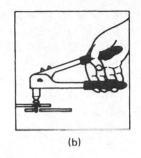
(b)

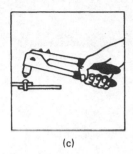
(c)

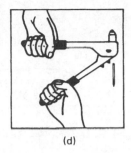
(d)

Fig. 4.34 Metal joining with pop rivets.

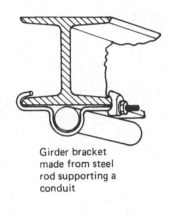

Girder bracket
made from steel
rod supporting a
conduit

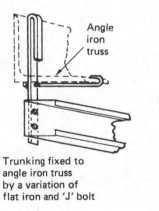

Girder bracket
made from flat
iron supporting
a trunking

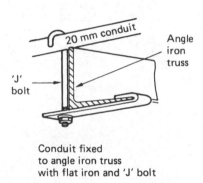

20 mm conduit

Angle
iron
truss

'J'
bolt

Conduit fixed
to angle iron truss
with flat iron and 'J' bolt

Angle
iron
truss

Trunking fixed to
angle iron truss
by a variation of
flat iron and 'J' bolt

Fig. 4.35 Bracket supports for conduits and trunking.

The type of bracket required will be determined by the installation, but Fig. 4.35 gives some examples of brackets which may be modified to suit particular circumstances.

Fixing methods

PVC insulated and sheathed wiring systems are usually fixed with PVC clips in order to comply with IEE Regulation 522–08 and Table 4A of the *On Site Guide* shown earlier in this chapter at Table 4.4. The clips are supplied in various sizes to hold the cable firmly, and the fixing nail is a hardened masonry nail. Figure 4.36 shows a cable clip of this type. The use of a masonry nail means that fixings to wood, plaster, brick or stone can be made with equal ease.

When heavier cables, trunking, conduit or luminaires have to be fixed a screw fixing is often needed. Wood screws may be screwed directly into wood but when fixing to brick, stone, plaster or concrete it is

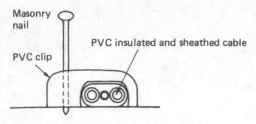

Fig. 4.36 PVC insulated and sheathed cable clip.

necessary to drill a hole in the masonry material, which is then plugged with a material to which the screw can be secured.

Plastic plugs

A plastic plug is made of a hollow plastic tube split up to half its length to allow for expansion. Each size of plastic plug is colour-coded to match a wood screw size.

A hole is drilled into the masonry, using a masonry drill of the same diameter, to the length of the plastic plug (see Fig. 4.37). The plastic plug is inserted into the hole and tapped home until it is level with the surface of the masonry. Finally, the fixing screw is driven into the plastic plug until it becomes tight and the fixture is secure.

Expansion bolts

The most common expansion bolt is made by Rawlbolt and consists of a split iron shell held together at one end by a steel ferrule and a spring wire clip at the other end. Tightening the bolt draws up an expanding bolt inside the split iron shell, forcing the iron to expand and grip the masonry. Rawlbolts are for heavy-duty masonry fixings (see Fig. 4.38).

A hole is drilled in the masonry to take the iron shell and ferrule. The iron shell is inserted with the

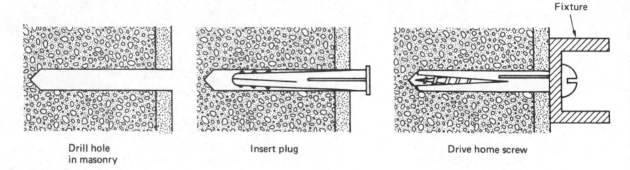

Drill hole in masonry

Insert plug

Drive home screw

Fig. 4.37 Screw fixing to plastic plug.

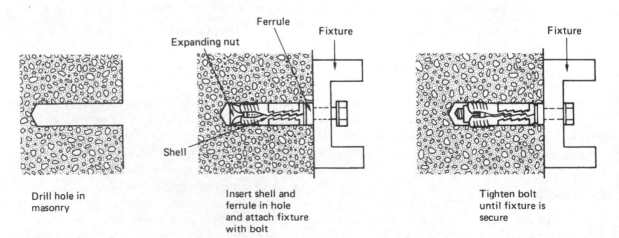

Drill hole in masonry

Insert shell and ferrule in hole and attach fixture with bolt

Tighten bolt until fixture is secure

Fig. 4.38 Expansion bolt fixing.

spring wire clip end first so that the ferrule is at the outer surface. The bolt is passed through the fixture, located in the expanding nut and tightened until the fixing becomes secure.

Spring toggle bolts

A spring toggle bolt provides one method of fixing to hollow partition walls which are usually faced with plasterboard and a plaster skimming. Plasterboard and plaster wall or ceiling surfaces are not strong enough to support a load fixed directly into the plasterboard, but the spring toggle spreads the load over a larger area, making the fixing suitable for light loads (see Fig. 4.39).

A hole is drilled through the plasterboard and into the cavity. The toggle bolt is passed through the fixture and the toggle wings screwed into the bolt. The toggle wings are compressed and passed through the hole in the plasterboard and into the cavity where they spring apart and rest on the cavity side of the plasterboard. The bolt is tightened until the fixing becomes firm. The bolt of the spring toggle cannot be removed after fixing without loosening the toggle wings. If it becomes necessary to remove and refix the fixture a new toggle bolt will have to be used.

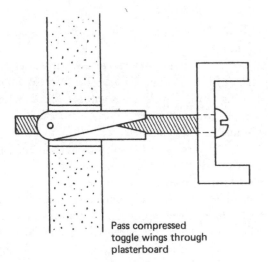

Pass compressed toggle wings through plasterboard

Tighten bolt until fixture is secure

Fig. 4.39 Spring toggle bolt fixing.

Special installations

All electrical installations and installed equipment must be safe to use and free from the dangers of electric shock, but some installations require special consideration because of the inherent dangers of the installed conditions. The danger may arise because of the corrosive or explosive nature of the atmosphere, because the installation must be used in damp or low-temperature conditions or because there is a need to provide additional mechanical protection for the electrical system. In this section we will consider some of the installations which require special consideration.

Temporary installations

Temporary electrical supplies provided on construction sites can save many man-hours of labour by providing the energy required for fixed and portable tools and lighting which speeds up the completion of a project. However, construction sites are dangerous places and the temporary electrical supply which is installed to assist the construction process must comply with all of the relevant wiring regulations for permanent installations (Regulation 110–01–01). All equipment must be of a robust construction in order to fulfil the on-site electrical requirements while being exposed to rough handling, vehicular nudging, the wind, rain and sun. All socket outlets, plugs and couplers must be of the industrial type to BS 4343 and specified by Regulation 604–12–02 as shown in Fig. 4.40.

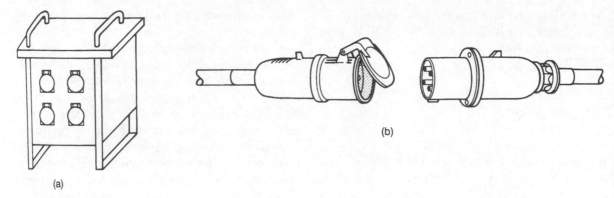

Fig. 4.40 110 V distribution unit and cable connector, suitable for construction site electrical supplies: (a) reduced-voltage distribution unit incorporating industrial sockets to BS 4343; (b) industrial plug and connector.

Where an electrician is not permanently on site, MCBs are preferred so that overcurrent protection devices can be safely reset by an unskilled person. The British Standards Code of Practice 1017, *The Distribution of Electricity on Construction and Building Sites*, advises that protection against earth faults may be obtained by first providing a low impedance path, so that overcurrent devices can operate quickly as described in Chapter 3, and secondly by fitting a residual current device (RCD) in addition to the overcurrent protection device. The 16th edition of the IEE Regulations considers construction sites very special locations, devoting the whole of Section 604 to their requirements. They have their own set of tables for disconnection times and maximum earth fault loop impedances (Regulations 604–04–03 and 04). A construction site installation should be tested and inspected in accordance with Part 7 of the Wiring Regulations every three months throughout the construction period.

The source of supply for the temporary installation may be from a petrol or diesel generating set or from the local supply company. When the local electricity company provides the supply, the incoming cable must be terminated in a waterproof and locked enclosure to prevent unauthorized access and provide metering arrangements.

IEE Regulation 604–02–02 recommends the following voltages for the distribution of electrical supplies to plant and equipment on construction sites:

■ 400 V three-phase for supplies to major items of plant having a rating above 3.75 kW such as cranes and lifts. These supplies must be wired in armoured cables.

■ 230 V single-phase for supplies to items of equipment which are robustly installed such as floodlighting towers, small hoists and site offices. These supplies must be wired in armoured cable unless run inside the site offices.

■ 110 V single-phase for supplies to all portable hand tools and all portable lighting equipment. The supply is usually provided by a reduced voltage distribution unit which incorporates splashproof sockets fed from a centre-tapped 110 V transformer. This arrangement limits the voltage to earth to 55 V, which is recognized as safe in most locations. A 110 V distribution unit is shown in Fig. 4.40. Edison screw lamps are used for 110 V lighting supplies so that they are not interchangeable with 230 V site office lamps.

There are occasions when even a 110 V supply from a centre-tapped transformer is too high, for example, supplies to inspection lamps for use inside damp or confined places. In these circumstances a safety extra-low voltage (SELV) supply would be required.

Industrial plugs have a keyway which prevents a tool from one voltage being connected to the socket outlet of a different voltage. They are also colour-coded for easy identification as follows:

440 V – red
230 V – blue
110 V – yellow
 50 V – white
 25 V – violet.

Agricultural and horticultural installations

Especially adverse installation conditions are to be encountered on agricultural and horticultural installations because of the presence of livestock, vermin, dampness, corrosive substances and mechanical damage. The 16th edition of the IEE Wiring Regulations considers these installations very special locations and has devoted the whole of Section 605 to their requirements. They have their own set of tables for disconnection times and maximum earth fault loop impedances (Regulations 605–05–03 and 04). In situations accessible to livestock the electrical equipment should be of a type which is appropriate for the external influences likely to occur, and should have at least protection IP44, that is, protection against solid objects and water splashing from any direction (Regulation 605–11–01).

In buildings intended for livestock, all fixed wiring systems must be inaccessible to the livestock and cables liable to be attacked by vermin must be suitably protected.

Horses and cattle have a very low body resistance, which makes them susceptible to an electric shock at voltages lower than 25 V rms, and so where protection is afforded by an overcurrent protective device the values of earth fault loop impedance are reduced as given in Tables 605B1 and 605B2. Similarly, where SELV is used, protection against direct contact must be provided by either barriers or enclosures giving a minimum protection against the standard finger 80 mm long and 12 mm diameter (IP2X), or by complete protection against a finger entering the enclosure (IPXXB), or by insulation that will withstand 500 V d.c. for 1 minute (Regulation 605–02–02).

PVC cables enclosed in heavy-duty PVC conduit are suitable for installations in most agricultural buildings. All exposed metalwork must be provided with supplementary equipotential bonding in areas where livestock is kept (Regulation 605–08–02). In many situations, waterproof socket outlets to BS 196 must be installed. Except for SELV circuits, all socket outlet circuits must be protected by an RCD complying with the appropriate British Standard. The operating current must not exceed 30 mA, and have a maximum operating time of 40 ms with a residual current of 150 mA (Regulation 605–03–01).

Cables buried on agricultural or horticultural land should be buried at a depth not less than 450 mm, or 600 mm where the ground may be cultivated, and the cable must have an armour sheath and be further protected by cable tiles. Overhead cables must be installed so that they are clear of farm machinery or placed at a minimum height of 5.2 m to comply with Regulation 522–08–01 and Table 4B of the *On Site Guide*.

The sensitivity of farm animals to electric shock means that they can be contained by an electric fence. An animal touching the fence receives a short pulse of electricity which passes through the animal to the general mass of earth and back to an earth electrode sunk near the controller, as shown in Fig. 4.41. The pulses are generated by a capacitor–resistor circuit inside the controller which may be mains- or battery-operated (capacitor–resistor circuits are discussed in *Advanced Electrical Installation Work*). There must be no risk to any human coming into contact with the controller, which should be manufactured to BS 2632. The output voltage of the controller must not exceed 10 kV and the energy must not be greater than 5 J. The duration of the pulse must not be greater than 1.5 milliseconds and the pulse must never have a frequency greater than one pulse per second. This shock level is very similar to that which can be experienced by touching a spark plug lead on a motor car. The energy levels are very low at 5 J. There are 3.6 million joules of energy in 1 kWh.

Earth electrodes connected to the earth terminal of an electric fence controller must be separate from the earthing system of any other circuit and should be situated outside the resistance area of any electrode used for protective earthing. The electric fence controller and the fence wire must be installed so that they do

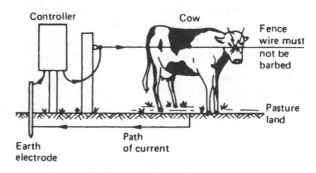

Fig. 4.41 Farm animal control by electric fence.

not come into contact with any power, telephone or radio systems, including poles (Regulation 605–14). Agricultural and horticultural installations should be tested and inspected in accordance with Part 7 of the Wiring Regulations every 3 years.

Caravans and caravan sites

The electrical installations on caravan sites, and within caravans, must comply in all respects with the wiring regulations for buildings. All the dangers which exist in buildings are present in and around caravans, including the added dangers associated with repeated connection and disconnection of the supply and the flexing of the caravan installation in a moving vehicle. The 16th edition of the Regulations has devoted Section 608 to the electrical installation in caravans, motor caravans and caravan parks.

Touring caravans must be supplied from a 16 A industrial type socket outlet adjacent to the caravan park pitch, similar to that shown in Fig. 4.40. Each socket outlet must be supplied, either singly or in groups, with not more than three socket outlets, through a residual current circuit breaker with a rated tripping current of 30 mA, and which will operate within 40 ms with a 150 mA fault current flowing. Additionally, every socket outlet must be protected by an overcurrent device (Regulations 608–13–04 and 05). The distance between the caravan connector and the site socket outlet must be not more than 27 m. These requirements are shown in Fig. 4.42.

Information notices regarding type and voltage of the supply and maximum permitted load must be displayed at the source of supply. The supply cables must be installed outside the pitch area or be suitably armoured so that they cannot be damaged by caravan awning pegs (Regulation 608–12–02).

The caravan or motor caravan must be provided with a mains isolating switch and a residual current device to break all live conductors (Regulations 608–03–02 and 608–07–04). An adjacent notice detailing how to connect and disconnect the supply safely must also be provided, as shown in Regulation 608–07–05. Electrical equipment must not be installed in fuel storage compartments. Caravans flex when being towed, and therefore the installation must be wired in flexible or stranded conductors of at least 1.5 mm

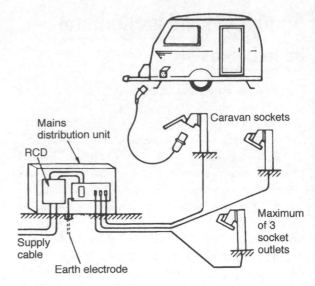

Fig. 4.42 Electrical supplies to caravans.

cross-section. The conductors must be supported on horizontal runs at least every 25 cm and the metalwork of the caravan and chassis must be bonded with 4.0 mm² cable.

The wiring of the extra low-voltage battery supply must be run in such a way that it does not come into contact with the 230 V wiring system (Regulation 608–06–04).

The caravan should be connected to the pitch socket outlet by means of a flexible cable, not longer than 27 m, and having a minimum cross-sectional area of 2.5 mm² or as detailed in Table 608A.

Because of the mobile nature of caravans it is recommended that the electrical installation be tested and inspected at intervals considered appropriate, between 1 and 3 years but not exceeding three years (Regulation 608–07–05).

Flammable and explosive installations

Most flammable liquids only form an explosive mixture between certain concentration limits. Above and below this level of concentration the mix will not explode. The lowest temperature at which sufficient vapour is given off from a flammable substance to form an explosive gas–air mixture is called the *flash-point*. A liquid which is safe at normal temperatures will require special consideration if heated to flash-point. An area in which an explosive gas–air mixture is present is called a *hazardous area*, as defined by BS 5345, and any electrical apparatus or equipment within a hazardous area must be classified as flameproof.

Flameproof electrical equipment is constructed so that it can withstand an internal explosion of the gas for which it is certified, and prevent any spark or flame resulting from that explosion leaking out and igniting the surrounding atmosphere. This is achieved by manufacturing flameproof equipment to a robust standard of construction. All access and connection points have wide machined flanges which damp the flame in its passage across the flange. Flanged surfaces are firmly bolted together with many recessed bolts, as shown in Fig. 4.43. Wiring systems within a hazardous area must be to flameproof fittings using an appropriate method, such as:

■ PVC cables encased in solid drawn heavy-gauge screwed steel conduit terminated at approved enclosures having wide flanges and bolted covers.

■ Mineral insulated cables terminated into accessories with approved flameproof glands. These have a longer gland thread than normal MICC glands of the type shown in Fig. 2.55 in Chapter 2 of this book. Where the cable is laid underground it must be protected by a PVC sheath and laid at a depth of not less than 500 mm.

■ PVC armoured cables terminated into accessories with approved flameproof glands or any other wiring system which is approved by BS 5345. All certified flameproof enclosures will be marked Ex, indicating that they are suitable for potentially explosive situations, or EEx, where equipment is certified to the harmonized European Standard. All the equipment used in a flameproof installation must carry the appropriate markings, as shown in Fig. 4.44, if the integrity of the wiring system is to be maintained.

Flammable and explosive installations are to be found in the petroleum and chemical industries, which are classified as group II industries. Mining is

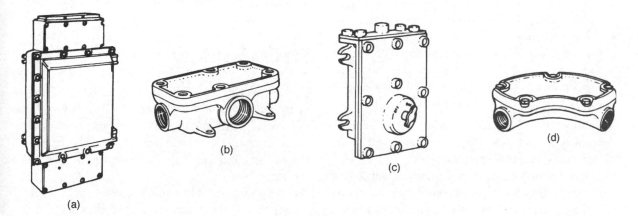

Fig. 4.43 Flameproof fittings: (a) flameproof distribution board; (b) flameproof rectangular junction box; (c) double-pole switch; (d) flameproof inspection bend.

Fig. 4.44 Flameproof equipment markings.

classified as group I and receives special consideration from the Mining Regulations because of the extreme hazards of working underground. Petrol filling pumps must be wired and controlled by flameproof equipment to BS 5345 and meet the requirements of the Petroleum Regulation Act 1928 and 1936 and any local licensing laws concerning the keeping and dispensing of petroleum spirit.

Hazardous area classification

British Standard 5345 divides the risk associated with inflammable gases and vapours into three classes or zones.

- Zone 0 is the most hazardous, and is defined as a zone or area in which an explosive gas–air mixture is *continuously present* or present for long periods. ('Long periods' is usually taken to mean that the gas–air mixture will be present for longer than 1000 hours per year.)
- Zone 1 is an area in which an explosive gas–air mixture is *likely to occur* in normal operation. (This is usually taken to mean that the gas–air mixture will be present for up to 1000 hours per year.)
- Zone 2 is an area in which an explosive gas–air mixture is *not likely* to occur in normal operation and if it does occur it will exist for a very short time. (This is usually taken to mean that the gas–air mixture will be present for less than 10 hours per year.)

If an area is not classified as zone 0, 1 or 2, then it is deemed to be non-hazardous, so that normal industrial electrical equipment may be used.

The electrical equipment used in zone 2 will contain a minimum amount of protection. For example, normal sockets and switches cannot be installed in a zone 2 area, but oil-filled radiators may be installed if they are directly connected and controlled from outside the area. Electrical equipment in this area should be marked Ex'o' for oil-immersed or Ex'p' for powder-filled.

In Zone 1 all electrical equipment must be flameproof, as shown in Fig. 4.43, and marked Ex'd' to indicate a flameproof enclosure.

Ordinary electrical equipment cannot be installed in Zone 0, even when it is flameproof protected. However, many chemical and oil-processing plants are entirely dependent upon instrumentation and data transmission for their safe operation. Therefore, very low-power instrumentation and data-transmission circuits can be used in special circumstances, but the equipment must be intrinsically safe, and used in conjunction with a 'safety barrier' installed outside the hazardous area. Intrinsically safe equipment must be marked Ex'ia' or Ex's', specially certified for use in zone 0.

Intrinsic safety

By definition, an intrinsically safe circuit is one in which no spark or thermal effect is capable of causing ignition of a given explosive atmosphere. The intrinsic safety of the equipment in a hazardous area is assured by incorporating a Zener diode safety barrier into the control circuit such as that shown in Fig. 4.45. In normal operation the voltage across a Zener diode is too low for it to conduct, but if a fault occurs the voltage across Z_1 and Z_2 will rise, switching them on and blowing the protective fuse. Z_2 is included in the circuit as a 'back-up' in case the first Zener diode fails.

An intrinsically safe system, suitable for use in Zone 0, is one in which *all* the equipment, apparatus and interconnecting wires and circuits are intrinsically safe.

Static electricity

Static electricity is a voltage charge which builds up to many thousands of volts between two surfaces when they rub together. A dangerous situation occurs when the static charge has built up to a potential capable of striking an arc through the airgap separating the two surfaces.

Static charges build up in a thunderstorm. A lightning strike is the discharge of the thunder cloud, which might have built up to a voltage of 100 MV, to the general mass of earth which is at zero volts. Lightning discharge currents are of the order of 20 kA, hence the need for lightning conductors on vulnerable buildings in order to discharge the energy safely.

Static charge builds up between any two insulating surfaces or between an insulating surface and a conducting surface, but it is not apparent between two conducting surfaces.

A motor car moving through the air builds up a static charge which sometimes gives the occupants a minor shock as they step out and touch the door handle.

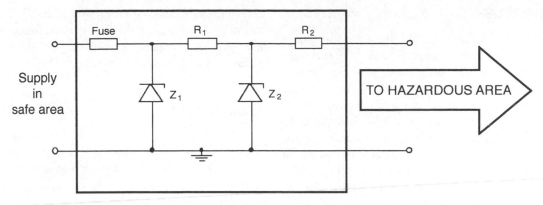

Fig. 4.45 Zener safety barrier.

A nylon overall and nylon bed sheets build up static charge which is the cause of the 'crackle' when you shake them. Many flammable liquids have the same properties as insulators, and therefore liquids, gases, powders and paints moving through pipes build up a static charge.

Petrol pumps, operating theatre oxygen masks and car spray booths are particularly at risk because a spark in these situations may ignite the flammable liquid, powder or gas.

So how do we protect ourselves against the risks associated with static electrical charges? I said earlier that a build-up of static charge is not apparent between two conducting surfaces, and this gives a clue to the solution. Bonding surfaces together with equipotential bonding conductors prevents a build-up of static electricity between the surfaces. If we use large-diameter pipes, we reduce the flow rates of liquids and powders and, therefore, we reduce the build-up of static charge. Hospitals use cotton sheets and uniforms, and use bonding extensively in operating theatres. Rubber, which contains a proportion of graphite, is used to manufacture antistatic trolley wheels and surgeons' boots. Rubber constructed in this manner enables any build-up of static charge to 'leak' away. Increasing humidity also reduces static charge because the water droplets carry away the static charge, thus removing the hazard.

Computer supplies

Every modern office now contains computers, and many systems are linked together or networked. Most computer systems are sensitive to variations or distortions in the mains supply and many computers incorporate filters which produce high protective conductor currents of around 2 or 3 milliamperes. This is clearly not a fault current, but is typical of the current which flows in the circuit protective conductor of IT equipment under normal operating conditions. Section 607 of the IEE Regulations deals with the earthing requirements for the installation of equipment having high protective conductor currents. IEE Guidance Note 7 recommends that IT equipment should be connected to double sockets as shown in Fig. 4.46.

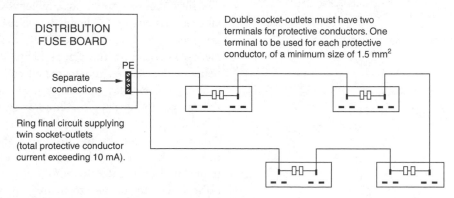

Fig. 4.46 Recommended method of connecting IT equipment to socket outlets.

CLEAN SUPPLIES

Supplies to computer circuits must be 'clean' and 'secure'. Mainframe computers and computer networks are sensitive to mains distortion or interference, which is referred to as 'noise'. Noise is mostly caused by switching an inductive circuit which causes a transient spike, or by brush gear making contact with the commutator segments of an electric motor. These distortions in the mains supply can cause computers to 'crash' or provoke errors and are shown in Fig. 4.47.

To avoid this, a 'clean' supply is required for the computer network. This can be provided by taking the ring or radial circuits for the computer supplies from a point as close as possible to the intake position of the electrical supply to the building. A clean earth can also be taken from this point, which is usually one

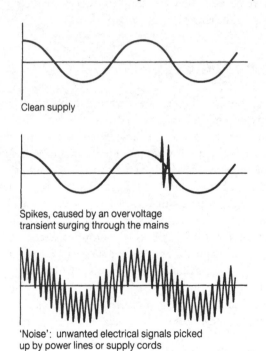

Clean supply

Spikes, caused by an overvoltage
transient surging through the mains

'Noise': unwanted electrical signals picked
up by power lines or supply cords

Fig. 4.47   Distortions in the a.c. mains supply.

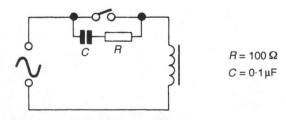

$R = 100\ \Omega$
$C = 0.1\ \mu F$

Fig. 4.48   A simple noise suppressor.

core of the cable and not the armour of an SWA cable, and distributed around the final wiring circuit. Alternatively, the computer supply can be cleaned by means of a filter such as that shown in Fig. 4. 48.

SECURE SUPPLIES

The mains electrical supply in the UK is extremely reliable and secure. However, the loss of supply to a mainframe computer or computer network for even a second can cause the system to 'crash', and hours or even days of work can be lost.

One solution to this problem is to protect 'precious' software systems with an uninterruptable power supply (UPS). A UPS is essentially a battery supply electronically modified to provide a clean and secure a.c. supply. The UPS is plugged into the mains supply and the computer systems are plugged into the UPS.

A UPS to protect a small network of, say, six PCs is physically about the size of one PC hard drive and is usually placed under or at the side of an operator's desk.

It is best to dedicate a ring or radial circuit to the UPS and either to connect the computer equipment permanently or to use non-standard outlets to discourage the unauthorized use and overloading of these special supplies by, for example, kettles.

Finally, remember that most premises these days contain some computer equipment and systems. Electricians intending to isolate supplies for testing or modification should first check and then check again before they finally isolate the supply in order to avoid loss or damage to computer systems.

OPTICAL FIBRE CABLES

The introduction of fibre-optic cable systems and digital transmissions will undoubtedly affect future cabling arrangements and the work of the electrician. Networks based on the digital technology currently being used so successfully by the telecommunications industry are very likely to become the long-term standard for computer systems. Fibre-optic systems dramatically reduce the number of cables required for control and communications systems, and this will in turn reduce the physical room required for these systems. Fibre-optic cables are also immune to electrical noise when run parallel to mains cables and, therefore, the present rules of segregation and screening may change in the future. There is no spark risk if the cable is accidentally cut and, therefore, such circuits are intrinsically safe.

Optical fibre cables are communication cables made from optical-quality plastic, the same material from which spectacle lenses are manufactured. The energy is transferred down the cable as digital pulses of laser light as against current flowing down a copper conductor in electrical installation terms. The light pulses stay within the fibre-optic cable because of a scientific principle known as 'total internal refraction' which means that the laser light bounces down the cable and when it strikes the outer wall it is always deflected inwards and, therefore, does not escape out of the cable, as shown in Fig. 4.49.

The cables are very small because the optical quality of the conductor is very high and signals can be transmitted over great distances. They are cheap to produce and lightweight because these new cables are made from high-quality plastic and not high-quality copper. Single-sheathed cables are often called 'simplex' cables and twin sheathed cables 'duplex', that is two simplex cables together in one sheath. Multicore cables are available containing up to 24 single fibres.

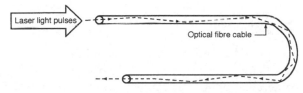

Fig. 4.49 Digital pulses of laser light down an optical fibre cable.

Fibre-optic cables look like steel wire armour cables (but of course are lighter) and should be installed in the same way and given the same level of protection as SWA cables. Avoid tight-radius bends if possible and kinks at all costs. Cables are terminated in special joint boxes which ensure cable ends are cleanly cut and butted together to ensure the continuity of the light pulses. Fibre-optic cables are Band I circuits when used for data transmission and must therefore be segregated from other mains cables to satisfy the IEE Regulations.

The testing of fibre-optic cables requires that special instruments be used to measure the light attenuation (that is, light loss) down the cable. Finally, when working with fibre-optic cables, electricians should avoid direct eye contact with the low-energy laser light transmitted down the conductors.

Fire alarm circuits (BS 5839)

Through one or more of the various statutory Acts, all public buildings are required to provide an effective means of giving a warning of fire so that life and property may be protected. An effective system is one which gives a warning of fire while sufficient time remains for the fire to be put out and any occupants to leave the building.

Fire alarm circuits are wired as either normally open or normally closed. In a *normally open circuit*, the alarm call points are connected in parallel with each other so that when any alarm point is initiated the circuit is completed and the sounder gives a warning of fire. The arrangement is shown in Fig. 4.50.

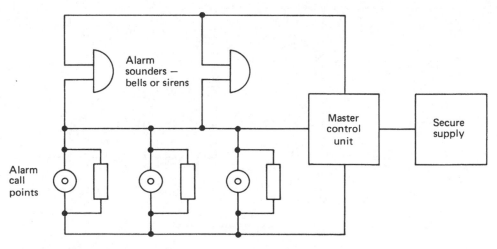

Fig. 4.50 A simple normally open fire alarm circuit.

It is essential for some parts of the wiring system to continue operating even when attacked by fire. For this reason the master control and sounders should be wired in MI or FP200 cable. The alarm call points of a normally open system must also be wired in MI or FP200 cable, unless a monitored system is used. In its simplest form this system requires a high-value resistor to be connected across the call-point contacts, which permits a small current to circulate and operate an indicator, declaring the circuit healthy. With a monitored system, PVC insulated cables may be used to wire the alarm call points.

In a *normally closed circuit*, the alarm call points are connected in series to normally closed contacts as shown in Fig. 4.51. When the alarm is initiated, or if a break occurs in the wiring, the alarm is activated. The sounders and master control unit must be wired in MI or FP200 cable, but the call points may be wired in PVC insulated cable since this circuit will always 'fail safe'.

ALARM CALL POINTS

Manually operated alarm call points should be provided in all parts of a building where people may be present, and should be located so that no one need to walk for more than 30 m from any position within the premises in order to give an alarm. A breakglass manual call point is shown in Fig. 4.52. They should be located on exit routes and, in particular, on the floor landings of staircases and exits to the street. They should be fixed at a height of 1.4 m above the

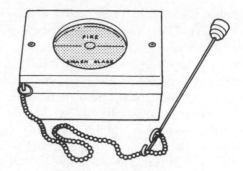

Fig. 4.52 Breakglass manual call point.

floor at easily accessible, well-illuminated and conspicuous positions.

Automatic detection of fire is possible with heat and smoke detectors. These are usually installed on the ceilings and at the top of stair wells of buildings because heat and smoke rise. Smoke detectors tend to give a faster response than heat detectors, but whether manual or automatic call points are used, should be determined by their suitability for the particular installation. They should be able to discriminate between a fire and the normal environment in which they are to be installed.

SOUNDERS

The positions and numbers of sounders should be such that the alarm can be distinctly heard above the background noise in every part of the premises. The sounders should produce a minimum of 65 decibels,

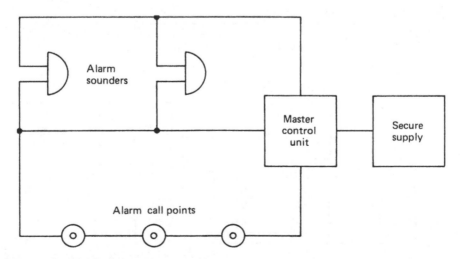

Fig. 4.51 A simple normally closed fire alarm circuit.

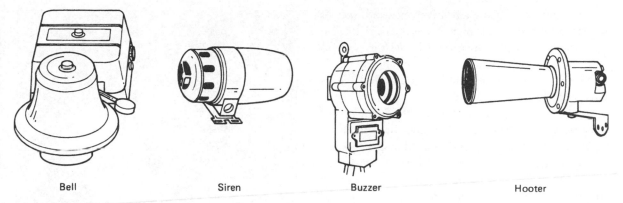

Bell   Siren   Buzzer   Hooter

Fig. 4.53 Typical fire alarm sounders.

or 5 decibels above any ambient sound which might persist, for more than 30 seconds. If the sounders are to arouse sleeping persons then the minimum sound level should be increased to 75 dB at the bedhead. Bells, hooters or sirens may be used but in any one installation they must all be of the same type. Examples of sounders are shown in Fig. 4.53.

FIRE ALARM DESIGN CONSIDERATIONS

Since all fire alarm installations must comply with the relevant statutory regulations, good practice recommends that contact be made with the local fire prevention officer at the design stage in order to identify any particular local regulations and obtain the necessary certification.

Larger buildings must be divided into zones so that the location of the fire can be quickly identified by the emergency services. The zones can be indicated on an indicator board situated in, for example, a supervisor's office or the main reception area.

In selecting the zones, the following rules must be considered:

1 Each zone should not have a floor area in excess of $2000 \, \text{m}^2$.
2 Each zone should be confined to one storey, except where the total floor area of the building does not exceed $300 \, \text{m}^2$.
3 Staircases and very small buildings should be treated as one zone.
4 Each zone should be a single fire compartment. This means that the walls, ceilings and floors are capable of containing the smoke and fire.

At least one fire alarm sounder will be required in each zone, but all sounders in the building must operate when the alarm is activated.

The main sounders may be silenced by an authorized person, once the general public have been evacuated from the building, but the current must be diverted to a supervisory buzzer which cannot be silenced until the system has been restored to its normal operational state.

A fire alarm installation may be linked to the local fire brigade's control room by the telecommunication network, if the permission of the fire authority and local telecommunication office is obtained.

The electricity supply to the fire alarm installation must be secure in the most serious conditions. In practice the most reliable supply is the mains supply, backed up by a 'standby' battery supply in case of mains failure. The supply should be exclusive to the fire alarm installation, fed from a separate switch fuse, painted red and labelled, 'Fire Alarm – Do Not Switch Off'. Standby battery supplies should be capable of maintaining the system in full normal operation for at least 24 hours and, at the end of that time, be capable of sounding the alarm for at least 30 minutes.

Fire alarm circuits are Band I circuits and consequently cables forming part of a fire alarm installation must be physically segregated from all Band II circuits unless they are insulated for the highest voltage (IEE Regulations 528–01–02).

Intruder alarms

The installation of security alarm systems in the United Kingdom is already a multi-million-pound business and yet it is also a relatively new industry. As society becomes increasingly aware of crime prevention, it is evident that the market for security systems will expand.

Not all homes are equally at risk, but all homes have something of value to a thief. Properties in cities are at highest risk, followed by homes in towns and villages, and at least risk are homes in rural areas. A nearby motorway junction can, however, greatly increase the risk factor. Flats and maisonettes are the most vulnerable, with other types of property at roughly equal risk. Most intruders are young, fit and foolhardy opportunists. They ideally want to get in and away quickly but, if they can work unseen, they may take a lot of trouble to gain access to a property by, for example, removing the putty from a window.

Most intruders are looking for portable and easily saleable items such as video recorders, television sets, home computers, jewellery, cameras, silverware, money, cheque books or credit cards. The Home Office has stated that only 7% of homes are sufficiently protected against intruders, although 75% of householders believe they are secure. Taking the simplest precautions will reduce the risk, while installing a security system can greatly reduce the risk of a successful burglary.

SECURITY LIGHTING

Security lighting is the first line of defence in the fight against crime. 'Bad men all hate the light and avoid it, for fear their practices should be shown up' (John 3:20). A recent study carried out by Middlesex University has shown that in two London boroughs the crime figures were reduced by improving the lighting levels. Police forces agree that homes which are externally well illuminated are a much less attractive target for the thief.

Security lighting installed on the outside of the home is activated by external detectors. These detectors sense the presence of a person outside the protected property and additional lighting is switched on. This will deter most potential intruders while also acting as courtesy lighting for visitors.

PASSIVE INFRA-RED DETECTORS

Passive infra-red (PIR) detector units allow a householder to switch on lighting units automatically

Fig. 4.54   Security lighting reduces crime.

whenever the area covered is approached by a moving body whose thermal radiation differs from the background. This type of detector is ideal for driveways or dark areas around the protected property. It also saves energy because the lamps are only switched on when someone approaches the protected area. The major contribution to security lighting comes from the 'unexpected' high-level illumination of an area when an intruder least expects it. This surprise factor often encourages the potential intruder to 'try next door'.

PIR detectors are designed to sense heat changes in the field of view dictated by the lens system. The field of view can be as wide as 180°, as shown by the diagram in Fig. 4.55. Many of the 'better' detectors use a split lens system so that a number of beams have to be broken before the detector switches on the security lighting. This capability overcomes the problem of false alarms, and a typical PIR is shown in Fig. 4.56.

PIR detectors are often used to switch tungsten halogen floodlights because, of all available luminaires, tungsten halogen offers instant high-level illumination. Light fittings must be installed out of reach of an intruder in order to prevent sabotage of the security lighting system.

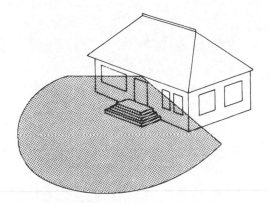

Fig. 4.55 PIR detector and field of detection.

Fig. 4.56 The Crabtree Minder, a typical PIR detector.

INTRUDER ALARM SYSTEMS

Alarm systems are now increasingly considered to be an essential feature of home security for all types of homes and not just property in high-risk areas. An intruder alarm system serves as a deterrent to a potential thief and often reduces home insurance premiums. In the event of a burglary they alert the occupants, neighbours and officials to a possible criminal act and generate fear and uncertainty in the mind of the intruder which encourages a more rapid departure. Intruder alarm systems can be broadly divided into three categories – those which give perimeter protection, space protection, or trap protection. A system can comprise one or a mixture of all three categories.

A perimeter protection system places alarm sensors on all external doors and windows so that an intruder can be detected as he or she attempts to gain access to the protected property. This involves fitting proximity switches to all external doors and windows.

A movement or heat detector placed in a room will detect the presence of anyone entering or leaving that room. PIR detectors and ultrasonic detectors give space protection. Space protection does have the disadvantage of being triggered by domestic pets but it is simpler and, therefore, cheaper to install. Perimeter protection involves a much more extensive and, therefore, expensive installation, but is easier to live with.

Trap protection places alarm sensors on internal doors and pressure pad switches under carpets on throughroutes between, for example, the main living area and the master bedroom. If an intruder gains access to one room he cannot move from it without triggering the alarm.

Proximity switches

These are designed for the discreet protection of doors and windows. They are made from moulded plastic and are about the size of a chewing-gum packet, as shown in Fig. 4.57. One moulding contains a reed switch, the other a magnet, and when they are placed close together the magnet maintains the contacts of the reed switch in either an open or closed position. Opening the door or window separates the two mouldings and the switch is activated, triggering the alarm.

Fig. 4.57 Proximity switches for perimeter protection.

PIR detectors

These are activated by a moving body which is warmer than the surroundings. The PIR shown in Fig. 4.58 has a range of 12 m and a detection zone of 110° when mounted between 1.8 m and 2 m high.

INTRUDER ALARM SOUNDERS

Alarm sounders give an audible warning of a possible criminal act. Bells or sirens enclosed in a waterproof enclosure, such as shown in Fig. 4.59, are suitable. It is usual to connect two sounders on an intruder alarm

Fig. 4.58 PIR intruder alarm detector.

Fig. 4.59 Intruder alarm sounder.

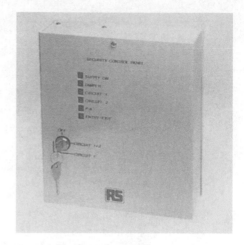

Fig. 4.60 Intruder alarm control panel.

supply and battery back-up. Nickel–cadmium recharge-able cells are usually mounted in the sounder housing box.

installation, one inside to make the intruder apprehensive and anxious, hopefully encouraging a rapid departure from the premises, and one outside. The outside sounder should be displayed prominently since the installation of an alarm system is thought to deter the casual intruder and a ringing alarm encourages neighbours and officials to investigate a possible criminal act.

Control panel

The control panel, such as that shown in Fig. 4.60, is at the centre of the intruder alarm system. All external sensors and warning devices radiate from the control panel. The system is switched on or off at the control panel using a switch or coded buttons. To avoid triggering the alarm as you enter or leave the premises, there are exit and entry delay times to allow movement between the control panel and the door.

Supply

The supply to the intruder alarm system must be secure and this is usually achieved by an a.c. mains

DESIGN CONSIDERATIONS

It is estimated that there is now a 5% chance of being burgled, but the installation of a security system does deter a potential intruder. Every home in Britain will almost certainly contain electrical goods, money or valuables of value to an intruder. Installing an intruder alarm system tells the potential intruder that you intend to make his job difficult, which in most cases encourages him to look for easier pickings.

The type and extent of the intruder alarm installation, and therefore the cost, will depend upon many factors including the type and position of the building, the contents of the building, the insurance risk involved and the peace of mind offered by an alarm system to the owner or occupier of the building.

The designer must ensure that an intruder cannot sabotage the alarm system by cutting the wires or pulling the alarm box from the wall. Most systems will trigger if the wires are cut and sounders should be mounted in any easy-to-see but difficult-to-reach position.

Intruder alarm circuits are Band I circuits and should, therefore, be segregated from mains supply cables which are designated as Band II circuits or insulated to the highest voltage present if run in a common enclosure with Band II cables (IEE Regulations).

Closed circuit television (CCTV)

Closed circuit television is now an integral part of many security systems. CCTV systems range from a single monitor with just one camera dedicated to monitoring perhaps a hotel car park, through to systems with many internal and external cameras connected to several locations for monitoring perhaps a shopping precinct.

CCTV cameras are often required to operate in total darkness when floodlighting is impractical. This is possible by using infra-red lighting which renders the scene under observation visible to the camera while to the human eye it appears to be in total darkness.

Cameras may be fixed or movable under remote control, such as those used for motorway traffic monitoring. Typically an external camera would be enclosed in a weatherproof housing such as those shown in Fig. 4.61. Using remote control, the camera can be panned, tilted or focused and have its viewing screen washed and wiped.

Pictures from several cameras can be multiplexed on to a single co-axial video cable, together with all the signals required for the remote control of the camera.

A permanent record of the CCTV pictures can be stored and replayed by incorporating a video tape recorder into the system as is the practice in most banks and building societies.

Security cameras should be robustly fixed and cable runs designed so that they cannot be sabotaged by a potential intruder.

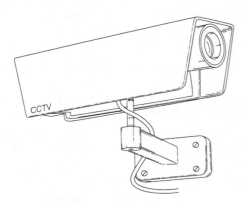

CCTV

Fig. 4.61 CCTV camera.

Emergency lighting (BS 5266)

Emergency lighting should be planned, installed and maintained to the highest standards of reliability and integrity, so that it will operate satisfactorily when called into action, no matter how infrequently this may be.

Emergency lighting is not required in private homes because the occupants are familiar with their surroundings, but in public buildings people are in unfamiliar surroundings. In an emergency people do not always act rationally, but well-illuminated and easily identified exit routes can help to reduce panic.

Emergency lighting is provided for two reasons; to illuminate escape routes, called 'escape' lighting; and to enable a process or activity to continue after a normal lights failure, called 'standby' lighting.

Escape lighting is usually required by local and national statutory authorities under legislative powers. The escape lighting scheme should be planned so that identifiable features and obstructions are visible in the lower levels of illumination which may prevail during an emergency. Exit routes should be clearly indicated by signs and illuminated to a uniform level, avoiding bright and dark areas.

Standby lighting is required in hospital operating theatres and in industry, where an operation or process once started must continue, even if the mains lighting fails. Standby lighting may also be required for security reasons. The cash points in public buildings may need to be illuminated at all times to discourage acts of theft occurring during a mains lighting failure.

EMERGENCY SUPPLIES

Since an emergency occurring in a building may cause the mains supply to fail, the emergency lighting should be supplied from a source which is independent from the main supply. In most premises the alternative power supply would be from batteries, but generators may also be used. Generators can have a large capacity and duration, but a major disadvantage is the delay of time while the generator runs up to speed and takes over the load. In some premises a delay of more than 5 seconds is considered unacceptable, and in these cases a battery supply is required to supply the load until the generator can take over.

The emergency lighting supply must have an adequate capacity and rating for the specified duration

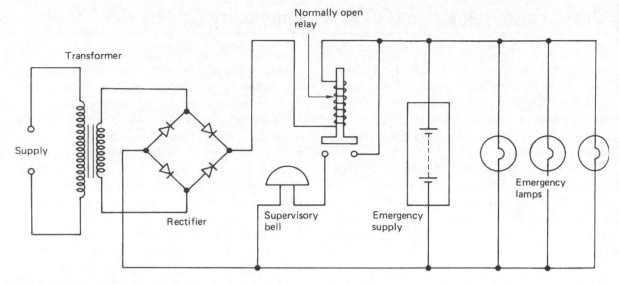

Fig. 4.62 Maintained emergency lighting.

of time (IEE Regulation 313–02). BS 5266 states that after a battery is discharged by being called into operation for its specified duration of time, it should be capable of once again operating for the specified duration of time following a recharge period of not longer than 24 hours. The duration of time for which the emergency lighting should operate will be specified by a statutory authority but is normally 1–3 hours. BS 5266 states that escape lighting should operate for a minimum of 1 hour. Standby lighting operation time will depend upon financial considerations and the importance of continuing the process or activity.

There are two possible modes of operation for emergency lighting installations: maintained and non-maintained.

Maintained emergency lighting

The emergency lamps are continuously lit using the normal supply when this is available, and change over to an alternative supply when the mains supply fails. The advantage of this system is that the lamps are continuously proven healthy and any failure is immediately obvious. It is a wise precaution to fit a supervisory buzzer in the emergency supply to prevent accidental discharge of the batteries, since it is not otherwise obvious which supply is being used.

Maintained emergency lighting is normally installed in theatres, cinemas, discotheques and places of entertainment where the normal lighting may be dimmed or extinguished while the building is occupied. The emergency supply for this type of installation is often supplied from a central battery, the emergency lamps being wired in parallel from the low-voltage supply as shown in Fig. 4.62.

Non-maintained emergency lighting

The emergency lamps are only illuminated if the normal mains supply fails. Failure of the main supply de-energizes a solenoid and a relay connects the emergency lamps to a battery supply, which is maintained in a state of readiness by a trickle charge from the normal mains supply. When the normal supply is restored, the relay solenoid is energized, breaking the relay contacts, which disconnects the emergency lamps, and the charger recharges the battery. Figure 4.63 illustrates this arrangement.

The disadvantage with this type of installation is that broken lamps are not detected until they are called into operation in an emergency, unless regularly maintained. The emergency supply is usually provided by a battery contained within the luminaire, together with the charger and relay, making the unit self-contained. Self-contained units are cheaper and easier to install than a central battery system, but the

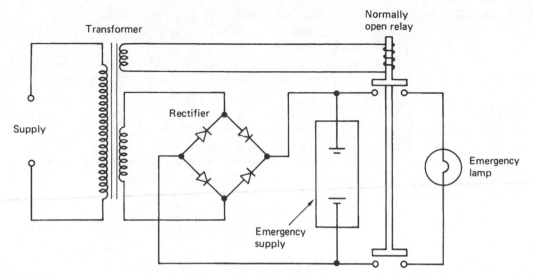

Fig. 4.63 Non-maintained emergency lighting.

central battery can have a greater capacity and duration, and permit a range of emergency lighting luminaires to be installed.

MAINTENANCE

The contractor installing the emergency lighting should provide a test facility which is simple to operate and secure against unauthorized interference. The emergency lighting installation must be segregated completely from any other wiring, so that a fault on the main electrical installation cannot damage the emergency lighting installation (IEE Regulation 528–01). Figure 4.30 shows a trunking which provides for segregation of circuits.

The batteries used for the emergency supply should be suitable for this purpose. Motor vehicle batteries are not suitable for emergency lighting applications, except in the starter system of motor-driven generators. The fuel supply to a motor-driven generator should be checked. The battery room of a central battery system must be well ventilated and, in the case of a motor-driven generator, adequately heated to ensure rapid starting in cold weather.

BS 5266 recommends that the full load should be carried by the emergency supply for at least 1 hour in every 6 months. After testing, the emergency system must be carefully restored to its normal operative state. A record should be kept of each item of equipment and the date of each test by a qualified or responsible person. It may be necessary to produce the record as evidence of satisfactory compliance with statutory legislation to a duly authorized person.

Self-contained units are suitable for small installations of up to about 12 units. The batteries contained within these units should be replaced about every 5 years, or as recommended by the manufacturer.

PRIMARY CELLS

A primary cell cannot be recharged. Once the active chemicals are exhausted, the cell must be discarded.

Primary cells, in the form of Leclanche cells, are used extensively as portable power sources for radios and torches and have an emf of 1.5 V. Larger voltages are achieved by connecting cells in series. Thus, a 6 V supply can be provided by connecting four cells in series.

Mercury primary cells have an emf of 1.35 V, and can have a very large capacity in a small physical size. They have a long shelf life and leakproof construction, and are used in watches and hearing aids.

SECONDARY CELLS

A secondary cell has the advantage of being rechargeable. If the cell is connected to a suitable electrical supply, electrical energy is stored on the plates of the cell as chemical energy. When the cell is connected to a load, the chemical energy is converted to electrical energy.

A lead-acid cell is a secondary cell. Each cell delivers about 2 V, and when six cells are connected in series a 12 V battery is formed of the type used on motor vehicles. Figure 4.64 shows the construction of a lead-acid battery.

A lead-acid battery is constructed of lead plates which are deeply ribbed to give maximum surface area for a given weight of plate. The plates are assembled in groups, with insulating separators between them. The separators are made of a porous insulating material, such as wood or ebonite, and the whole assembly is immersed in a dilute sulphuric acid solution in a plastic container.

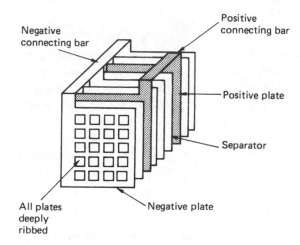

Fig. 4.64 The construction of a lead-acid battery.

BATTERY RATING

The capacity of a cell to store charge is a measure of the total quantity of electricity which it can cause to be displaced around a circuit after being fully charged. It is stated in ampere-hours, abbreviation Ah, and calculated at the 10 hour rate which is the steady load current which would completely discharge the battery in 10 hours. Therefore, a 50 Ah battery will provide a steady current of 5 A for ten hours.

MAINTENANCE OF LEAD-ACID BATTERIES

- The plates of the battery must always be covered by the dilute sulphuric acid. If the level falls, it must be topped up with distilled water.

- Battery connections must always be tight and should be covered with a thin coat of petroleum jelly.
- The specific gravity or relative density of the battery gives the best indication of its state of charge. A discharged cell will have a specific gravity of 1.150, which will rise to 1.280 when fully charged. The specific gravity of a cell can be tested with a hydrometer.
- To maintain a battery in good condition it should be regularly trickle-charged. A rapid charge or discharge encourages the plates to buckle, and may cause permanent damage.
- The room used to charge a battery must be well ventilated because the charged cell gives off hydrogen and oxygen, which are explosive in the correct proportions.

TELEPHONE SOCKET OUTLETS

The installation of telecommunications equipment could, for many years, only be undertaken by British Telecom engineers, but as a result of the recent liberalization of BT the electrical contractor may now supply and install telecommunications equipment.

On new premises the electrical contractor may install sockets and the associated wiring to the point of intended line entry, but the connection of the incoming line to the installed master socket must only be made by the telephone company's engineer.

On existing installations, additional secondary sockets may be installed to provide an extended plug-in facility as shown in Fig. 4.65. Any number of secondary sockets may be connected in parallel, but the number of telephones which may be connected at any one time is restricted.

Each telephone or extension bell is marked with a ringing equivalence number (REN) on the underside. Each exchange line has a maximum capacity of REN 4 and therefore, the total REN values of all the connected telephones must not exceed four if they are to work correctly.

An extension bell may be connected to the installation by connecting the two bell wires to terminals 3 and 5 of a telephone socket. The extension bell must be of the high impedance type having a REN rating. All equipment connected to a BT exchange line must display the green circle of approval.

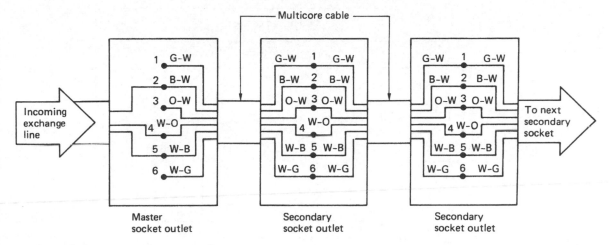

Fig. 4.65 Telephone circuit outlet connection diagram.

The multicore cable used for wiring extension socket outlets should be of a type intended for use with telephone circuits, which will normally be between 0.4 mm and 0.68 mm in cross-section. Telephone cable conductors are identified in Table 4.9 and the individual terminals in Table 4.10. The conductors should be connected as shown in Fig. 4.65. Telecommunications cables are Band I circuits and must be segregated from Band II circuits containing mains cables (IEE Regulations 528–01).

Table 4.9 Telephone cable identification

Code	Base colour	Stripe
G–W	Green	White
B–W	Blue	White
O–W	Orange	White
W–O	White	Orange
W–B	White	Blue
W–G	White	Green

Table 4.10 Telephone socket terminal identification. Terminals 1 and 6 are frequently unused, and therefore 4-core cable may normally be installed. Terminal 4, on the incoming exchange line, is only used on a PBX line for earth recall

Socket terminal	Circuit
1	Spare
2	Speech circuit
3	Bell circuit
4	Earth recall
5	Speech circuit
6	Spare

BATHROOM INSTALLATIONS

In rooms containing a fixed bath tub or shower basin, additional regulations are specified. This is to reduce the risk of electric shock to people in circumstances where body resistance is lowered because of contact with water. The regulations may be found in Section 601 and can be summarized as follows:

- Socket outlets must not be installed and no provision is made for connection of portable appliances.
- Only shaver sockets which comply with BS EN 61184 or BS EN 60238, that is, those which contain an isolating transformer, may be installed.
- Every switch must be inaccessible to anyone using the bath or shower unless it is of the cord-operated type.
- There are restrictions as to where appliances, switchgear and wiring accessories may be installed. See Zones for bath and shower rooms below.
- A supplementary bonding conductor must be provided, in addition to the main equipotential bonding discussed in Chapter 2 (Fig. 2.64)

Zones for bath and shower rooms

Locations that contain a bath or shower are divided in zones or separate areas as shown in Fig. 4.66.

Zone 0 – the bath tub or shower basin itself, which can contain water and is, therefore, the most dangerous zone

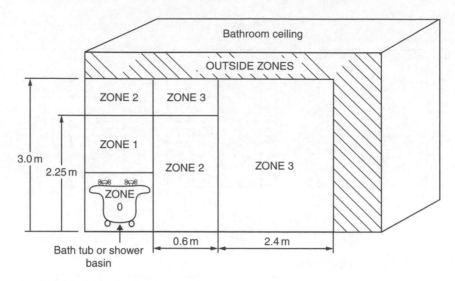

Fig 4.66 Cross section through bathroom showing zones.

Zone 1 – the next most dangerous zone in which people stand in water

Zone 2 – the next most dangerous zone in which people might be in contact with water

Zone 3 – people are least likely to be in contact with water but are still in a potentially dangerous environment.

Electrical equipment and accessories are restricted within the zones.

Zone 0 – being the most potentially dangerous zone for all practical purposes, no electrical equipment can be installed in this zone. However, the Regulations permit that where SELV fixed equipment cannot be located elsewhere, it may be installed in this zone.

Zone 1 – water heaters, showers and shower pumps and SELV fixed equipment

Zone 2 – luminaries, fans and heating appliances and equipment from zone 1 plus shaver units to BSEN 60742

Zone 3 – fixed appliances are allowed plus the equipment from zones 1 and 2

Outside Zones – appliances are allowed plus accessories except socket outlets.

If under-floor heating is installed in these areas it must have an overall earthed metallic grid or the heating cable must have an earthed metallic sheath, which must be supplementary bonded.

SUPPLEMENTARY BONDING

Modern plumbing methods make considerable use of non-metals (PTFE tape on joints, for example). Therefore, the metalwork of water and gas installations cannot be relied upon to be continuous throughout.

The IEE Regulations describe the need to consider additional or supplementary bonding in situations where there is a high risk of electric shock (for example, in kitchens and bathrooms).

In kitchens, supplementary bonding of hot and cold taps, sink tops and exposed water and gas pipes *is only required* if an earth continuity test proves that they are not already effectively and reliably connected to the main equipotential bonding, having negligible impedance, by the soldered pipe fittings of the installation. If the test proves unsatisfactory, the metalwork must be bonded using a single core copper conductor with PVC green/yellow insulation, which will normally be $4\,mm^2$ for domestic installations but must comply with Regulations 547–03–01 to 03.

In rooms containing a fixed bath or shower, supplementary bonding conductors *must* be installed to reduce to a minimum the risk of an electric shock (Regulation 601–04–02). Bonding conductors in domestic premises will normally be of $4\,mm^2$ copper with PVC insulation to comply with Regulations 547–03–01 to 03 and must be connected between all exposed metalwork (for example, between metal baths, bath and sink taps, shower fittings, metal waste pipes and radiators, as shown in Fig. 4.67.

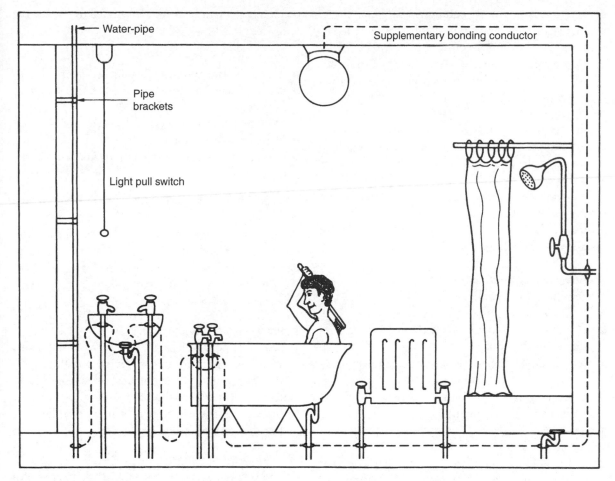

Fig. 4.67 Supplementary bonding in bathrooms to metal pipework.

The bonding connection must be made to a cleaned pipe, using a suitable bonding clip. Fixed at or near the connection must be a permanent label saying 'Safety electrical connection – do not remove' (Regulation 514–13–01) as shown in Fig. 4.68.

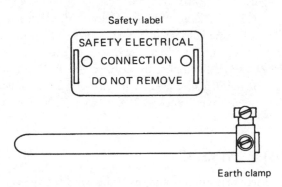

Fig. 4.68 Typical earth bonding clamp.

Inspection and testing techniques

The testing of an installation implies the use of instruments to obtain readings. However, a test is unlikely to identify a cracked socket outlet, a chipped or loose switch plate, a missing conduit-box lid or saddle, so it is also necessary to make a visual inspection of the installation.

All new installations must be inspected and tested before connection to the mains, and all existing installations should be periodically inspected and tested to ensure that they are safe and meet the regulations of the IEE (Regulations 711 to 744).

The method used to test an installation may inject a current into the system. This current must not cause danger to any person or equipment in contact with the installation, even if the circuit being tested is faulty. The test procedures must be followed carefully and in the correct sequence, as indicated by Regulation 713–01–01. This ensures that the protective conductors are correctly connected and secure before the circuit is energized.

The installation must be visually inspected before testing begins. The aim of the visual inspection is to confirm that all equipment and accessories are undamaged and comply with the relevant British and European Standards, and also that the installation has been securely and correctly erected. Regulation 712–01–03 gives a check-list for the initial visual inspection of an installation, including:

■ connection of conductors
■ identification of conductors
■ routing of cables in safe zones
■ selection of conductors for current carrying capacity and volt drop
■ connection of single-pole devices for protection or switching in phase conductors only
■ correct connection of socket outlets, lampholders, accessories and equipment
■ presence of fire barriers, suitable seals and protection against thermal effects
■ methods of protection against electric shock, including the insulation of live parts and placement of live parts out of reach by fitting appropriate barriers and enclosures
■ prevention of detrimental influences (e.g. corrosion)
■ presence of appropriate devices for isolation and switching
■ presence of undervoltage protection devices
■ choice and setting of protective devices
■ labelling of circuits, fuses, switches and terminals
■ selection of equipment and protective measures appropriate to external influences
■ adequate access to switchgear and equipment
■ presence of danger notices and other warning notices
■ presence of diagrams, instruction and similar information
■ appropriate erection method.

The check-list is a guide, it is not exhaustive or detailed, and should be used to identify relevant items for inspection, which can then be expanded upon. For example, the first item on the check-list, connection of conductors, might be further expanded to include the following:

■ Are connections secure?
■ Are connections correct? (conductor identification)
■ Is the cable adequately supported so that no strain is placed on the connections?
■ Does the outer sheath enter the accessory?
■ Is the insulation undamaged?
■ Does the insulation proceed up to but not *into* the connection?

This is repeated for each appropriate item on the checklist.

Those tests which are relevant to the installation must then be carried out in the sequence given in Regulation 713–01–01 and Sections 9 and 10 of the *On Site Guide* for reasons of safety and accuracy. These tests are as follows:

Before the supply is connected
1 Test for continuity of protective conductors, including main and supplementary bonding.
2 Test the continuity of all ring final circuit conductors.
3 Test for insulation resistance.
4 Test for polarity using the continuity method.
5 Test the earth electrode resistance.

With the supply connected
6 Recheck polarity using a voltmeter or approved test lamp.
7 Test the earth fault loop impedance.
8 Carry out functional testing (e.g. operation of RCDs).

If any test fails to comply with the Regulations, then *all* the preceding tests must be repeated after the fault has been rectified. This is because the earlier test results may have been influenced by the fault (Regulation 713–01–01).

There is an increased use of electronic devices in electrical installation work, for example, in dimmer switches and ignitor circuits of discharge lamps. These devices should temporarily be disconnected so that they are not damaged by the test voltage of, for example, the insulation resistance test (Regulation 713–04).

TEST INSTRUMENTS

The test instruments and test leads used by the electrician for testing an electrical installation must meet all the requirements of the relevant regulations. The Health and Safety Executive has published Guidance Notes GS 38 for test equipment used by electricians. The IEE Regulations (BS 7671) also specify the test voltage or current required to carry out particular tests satisfactorily. All testing must, therefore, be carried out using an 'approved' test instrument if the test results are to be valid. The test instrument must also carry a calibration certificate, otherwise the recorded results may be void. Calibration certificates usually last for a year. Test instruments must, therefore, be tested and recalibrated each year by an approved supplier. This will maintain the accuracy of the instrument to an acceptable level, usually within 2% of the true value.

Modern digital test instruments are reasonably robust, but to maintain them in good working order they must be treated with care. An approved test instrument costs equally as much as a good-quality camera; it should, therefore, receive the same care and consideration.

Continuity tester

To measure accurately the resistance of the conductors in an electrical installation we must use an instrument which is capable of producing an open circuit voltage of between 4 V and 24 V a.c. or d.c., and deliver a short-circuit current of not less than 200 mA (Regulation 713–02). The functions of continuity testing and insulation resistance testing are usually combined in one test instrument.

Insulation resistance tester

The test instrument must be capable of detecting insulation leakage between live conductors and between live conductors and earth. To do this and comply with Regulation 713–04 the test instrument must be capable of producing a test voltage of 250 V, 500 V or 1000 V and deliver an output current of not less than 1 mA at its normal voltage.

Earth fault loop impedance tester

The test instrument must be capable of delivering fault currents as high as 25 A for up to 40 ms using the supply voltage. During the test, the instrument does an Ohm's law calculation and displays the test result as a resistance reading.

RCD tester

Where circuits are protected by a residual current device we must carry out a test to ensure that the device will operate very quickly under fault conditions and within the time limits set by the IEE Regulations. The instrument must, therefore, simulate a fault and measure the time taken for the RCD to operate. The instrument is, therefore, calibrated to give a reading measured in milliseconds to an in-service accuracy of 10%.

If you purchase good-quality 'approved' test instruments and leads from specialist manufacturers they will meet all the Regulations and Standards and therefore give valid test results. However, to carry out all the tests required by the IEE Regulations will require a number of test instruments and this will represent a major capital investment in the region of £1000.

Let us now consider the individual tests.

1 CONTINUITY OF PROTECTIVE CONDUCTORS (713–02)

The object of the test is to ensure that the circuit protective conductor (CPC) is correctly connected, is electrically sound and has a total resistance which is low enough to permit the overcurrent protective device to operate within the disconnection time requirements of Regulation 413–02–08, should an earth fault occur. Every protective conductor must be separately tested from the consumer's earthing terminal to verify that it is electrically sound and correctly connected, including any main and supplementary bonding conductors.

A d.c. test using an ohmmeter continuity tester is suitable where the protective conductors are of copper or aluminium up to 35 mm². The test is made with the supply disconnected, measuring from the consumer's earthing terminal to the far end of each CPC, as shown in Fig. 4.69. The resistance of the long test lead is subtracted from these readings to give the resistance value of the CPC. The result is recorded on an installation schedule such as that given in Appendix 7 of the *On Site Guide*.

Where steel conduit or trunking forms the protective conductor, the standard test described above may

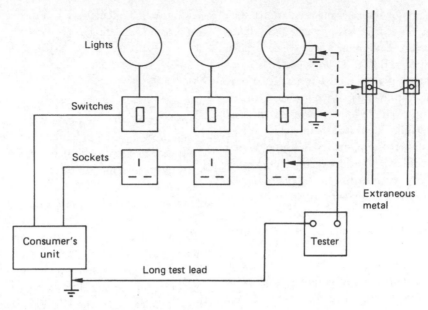

Fig. 4.69 Testing continuity of protective conductors.

be used, but additionally the enclosure must be visually checked along its length to verify the integrity of all the joints.

If the inspecting engineer has grounds to question the soundness and quality of these joints then the phase-earth loop impedance test described later in this chapter should be carried out.

If, after carrying out this further test, the inspecting engineer still questions the quality and soundness of the protective conductor formed by the metallic conduit or trunking then a further test can be done using an a.c. voltage not greater than 50 V at the frequency of the installation and a current approaching 1.5 times the design current of the circuit, but not greater than 25 A.

This test can be done using a low-voltage transformer and suitably connected ammeters and voltmeters, but a number of commercial instruments are available such as the Clare tester, which give a direct reading in ohms.

Because fault currents will flow around the earth fault loop path, the measured resistance values must be low enough to allow the overcurrent protective device to operate quickly. For a satisfactory test result, the resistance of the protective conductor should be consistent with those values calculated for a phase conductor of similar length and cross-sectional area. Values of resistance per metre for copper and aluminium conductors are given in Table 9A of the *On*

Table 4.11 Resistance values of some metallic containers

Metallic sheath	Size (mm)	Resistance at 20°C (mΩ/m)
Conduit	20	1.25
	25	1.14
	32	0.85
Trunking	50 × 50	0.949
	75 × 75	0.526
	100 × 100	0.337

Site Guide and shown in Table 3.1 of Chapter 3 in this book. The resistances of some other metallic containers are given in Table 4.11.

EXAMPLE

The CPC for a ring final circuit is formed by a 1.5 mm² copper conductor of 50 m approximate length. Determine a satisfactory continuity test value for the CPC using the value given in Table 9A of the *On Site Guide*. From Table 9A (shown in Table 3.1 of Chapter 3.)

Resistance/metre for a
1.5 mm² copper conductor $= 12.10 \, \text{m}\Omega/\text{m}$
Therefore,
the resistance of 50 m $\quad = 50 \times 12.10 \times 10^{-3}$
$\quad = 0.605 \, \Omega$

The protective conductor resistance values calculated by this method can only be an approximation since the length of the CPC can only be estimated. Therefore, in this case, a satisfactory test result would be obtained if the resistance of the protective conductor was about $0.6\,\Omega$. A more precise result is indicated by the earth fault loop impedance test which is carried out later in the sequence of tests.

2 TESTING FOR CONTINUITY OF RING FINAL CIRCUIT CONDUCTORS (713–03)

The object of the test is to ensure that all ring circuit cables are continuous around the ring, that is, that there are no breaks and no interconnections in the ring and that all connections are electrically and mechanically sound. This test also verifies the polarity of each socket outlet.

The test is made with the supply disconnected, using an ohmmeter as follows:

Disconnect and separate the conductors of both legs of the ring at the main fuse. There are three steps to this test.

Step 1

Measure the resistance of the phase conductors (L_1 and L_2), the neutral conductors (N_1 and N_2) and the protective conductors (E_1 and E_2) at the mains position as shown in Fig. 4.70. End-to-end live and neutral conductor readings should be approximately the same (i.e. within $0.05\,\Omega$) if the ring is continuous. The protective conductor reading will be 1.67 times as great as these readings if 2.5/1.5 mm cable is used. Record the results on a table such as that shown in Table 4.12.

Step 2

The live and neutral conductors should now be temporarily joined together as shown in Fig. 4.71. An ohmmeter reading should then be taken between live and neutral at *every* socket outlet on the ring circuit. The readings obtained should be substantially the same, provided that there are no breaks or multiple loops in the ring. Each reading should have a value of approximately half the live and neutral ohmmeter readings measured in step 1 of this test. Sockets connected as a spur will have a slightly higher value of resistance because they are fed by only one cable, while each socket on the ring is fed by two cables. Record the results on a table such as that shown in Table 4.12.

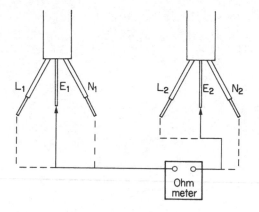

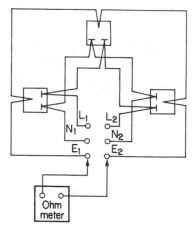

Fig. 4.70 Step 1 test: measuring the resistance of phase, neutral and protective conductors.

Table 4.12 Table which may be used to record the readings taken when carrying out the continuity of ring final circuit conductors tests according to IEE Regulation 713–02

Test	Ohmmeter connected to	Ohmmeter readings	This gives a value for
Step 1	L_1 and L_2 N_1 and N_2		r_1
	E_1 and E_2		r_2
Step 2	Live and neutral at each socket		
Step 3	Live and earth at each socket		$R_1 + R_2$
As a check ($R_1 + R_2$) value should equal			$\dfrac{r_1 + r_2}{4}$

Step 3

Where the circuit protective conductor is wired as a ring, for example where twin and earth cables or plastic conduit is used to wire the ring, temporarily join

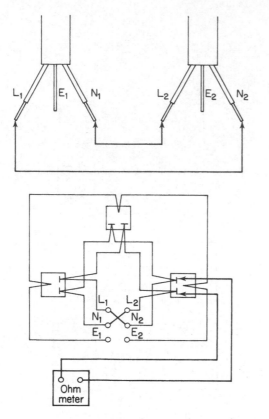

Fig. 4.71   Step 2 test: connection of mains conductors and test circuit conditions.

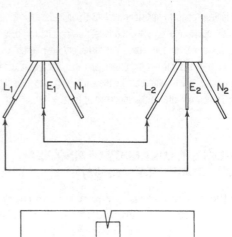

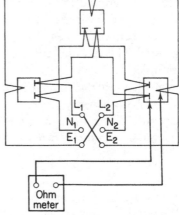

Fig. 4.72   Step 3 test: connection of mains conductors and test circuit conditions.

the live and circuit protective conductors together as shown in Fig. 4.72. An ohmmeter reading should then be taken between live and earth at *every* socket outlet on the ring. The readings obtained should be substantially the same provided that there are no breaks or multiple loops in the ring. This value is equal to $R_1 + R_2$ for the circuit. Record the results on an installation schedule such as that given in Appendix 7 of the *On Site Guide* or a table such as that shown in Table 4.12. The step 3 value of $R_1 + R_2$ should be equal to $(r_1 + r_2)/4$, where r_1 and r_2 are the ohmmeter readings from step 1 of this test (see Table 4.12).

3 INSULATION RESISTANCE (713–04)

The object of the test is to verify that the quality of the insulation is satisfactory and has not deteriorated or short-circuited. The test should be made at the consumer's unit with the mains switch off, all fuses in place and all switches closed. Neon lamps, capacitors and electronic circuits should be disconnected, since they will respectively glow, charge up or be damaged by the test.

There are two tests to be carried out using an insulation resistance tester which must supply a voltage of 500 V d.c. for 230 V and 400 V installations. These are phase and neutral conductors to earth and between phase conductors. The procedures are:

Phase and neutral conductors to earth
1 Remove all lamps.
2 Close all switches and circuit breakers.
3 Disconnect appliances.
4 Test separately between the phase conductor and earth, *and* between the neutral conductor and earth, for *every* distribution circuit at the consumer's unit as shown in Fig. 4.73(a). Record the results on an installation schedule such as that given in Appendix 7 of the *On Site Guide*.

Between phase conductors
1 Remove all lamps.
2 Close all switches and circuit breakers.

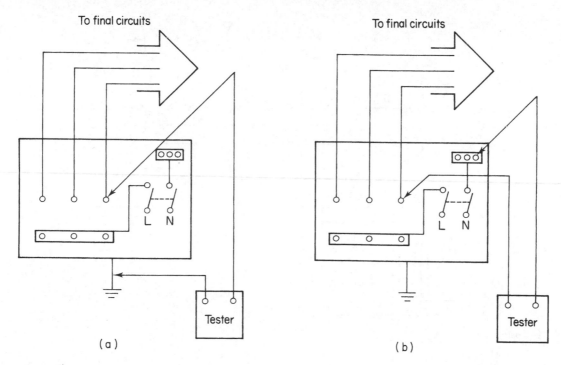

Fig. 4.73   Insulation resistance test.

3 Disconnect appliances.

4 Test between phase and neutral conductors of *every* distribution circuit at the consumer's unit as shown in Fig. 4.73(b) and record the result.

The insulation resistance readings for each test must be not less than 0.5 MΩ for a satisfactory result (IEE Regulation 713–04–02).

Where equipment is disconnected for the purpose of the insulation resistance test, the equipment itself must be insulation resistance tested between all live parts (that is, live and neutral conductors connected together) and the exposed conductive parts. The insulation resistance of these tests should be not less than 0.5 MΩ (IEE Regulation 713–04–04).

Although an insulation resistance reading of 0.5 MΩ complies with the Regulations, the IEE Guidance Notes tell us that a reading of less than 2 MΩ might indicate a latent but not yet visible fault in the installation. In these cases each circuit should be separately tested to obtain a reading greater than 2 MΩ.

4 POLARITY (713–09)

The object of this test is to verify that all fuses, circuit breakers and switches are connected in the phase or live conductor only, and that all socket outlets are correctly wired and Edison screw-type lampholders have the centre contact connected to the live conductor. It is important to make a polarity test on the installation since a visual inspection will only indicate conductor identification.

The test is done with the supply disconnected using an ohmmeter or continuity tester as follows:

1 Switch off the supply at the main switch.

2 Remove all lamps and appliances.

3 Fix a temporary link between the phase and earth connections on the consumer's side of the main switch.

4 Test between the 'common' terminal and earth at each switch position.

5 Test between the centre pin of any Edison screw lampholders and any convenient earth connection.

6 Test between the live pin (that is, the pin to the right of earth) and earth at each socket outlet as shown in Fig. 4.74.

For a satisfactory test result the ohmmeter or continuity meter should read approximately zero.

Remove the test link and record the results on an installation schedule such as that given in Appendix 7 of the *On Site Guide*.

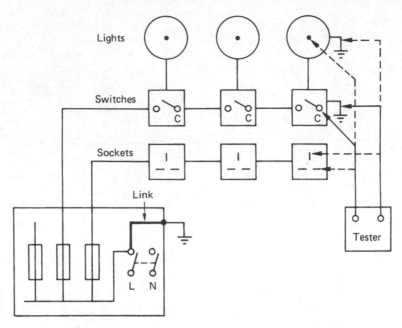

Fig. 4.74 Polarity test.

5 EARTH ELECTRODE RESISTANCE (713–10)

Low-voltage supplies having earthing arrangements which are independent of the supply cable are classified as TT systems. For this type of supply it is necessary to sink an earth electrode into the general mass of earth, which then forms a part of the earth return in conjunction with a residual current device. To verify the resistance of an electrode used in this way, the following test method may be applied:

1 Disconnect the installation equipotential bonding from the earth electrode to ensure that the test current passes only through the earth electrode.
2 Switch off the consumer's unit to isolate the installation.
3 Using a phase earth loop impedance tester, test between the incoming phase conductor and the earth electrode.

Record the result on an installation schedule such as that given in Appendix 7 of the *On Site Guide*.

Section 10.3.5 of the *On Site Guide* tells us that the recommended maximum value of the earth fault loop impedance for a TT installation is $220\,\Omega$. Since most of the circuit impedance will be made up of the earth electrode resistance, we can say that an acceptable value for the measurement of the earth electrode resistance will be about $200\,\Omega$ or less.

Providing the first five tests were satisfactory, the supply may now be switched on and the final tests completed with the supply connected.

6 POLARITY – SUPPLY CONNECTED

Using an approved voltmeter or test lamp and probes which comply with the HSE Guidance Note GS38, again carry out a polarity test to verify that all fuses, circuit breakers and switches are connected in the live conductor. Test from the common terminal of switches to earth, the live pin of each socket outlet to earth and the centre pin of any Edison screw lampholders to earth. In each case the voltmeter or test lamp should indicate the supply voltage for a satisfactory result.

7 EARTH FAULT LOOP IMPEDANCE (SUPPLY CONNECTED) (713–11)

The object of this test is to verify that the impedance of the whole earth fault current loop phase to earth is low enough to allow the overcurrent protective device to operate within the disconnection time requirements of Regulations 413–02–08 and 09, should an earth fault occur.

The whole earth fault current loop examined by this test is comprised of all the installation protective conductors, the earthing terminal and earth conductors, the earthed neutral point and the secondary winding of the supply transformer and the phase conductor from the transformer to the point of the fault in the installation.

The test will, in most cases, be done with a purpose-made phase earth loop impedance tester which circulates a current in excess of 10 A around the loop for a very short time, so reducing the danger of a faulty circuit. The test is made with the supply switched on, from the furthest point of *every* final circuit, including lighting, socket outlets and any fixed appliances. Record the results on an installation schedule.

Purpose-built testers give a readout in ohms and a satisfactory result is obtained when the loop impedance does not exceed the appropriate values given in Tables 2A, 2B and 2C of Appendix 2 of the *On Site Guide* or Table 41B1 and 41B2 or Table 604B2, 605B1 and 605B2 of the IEE Regulations. See Table 3.2 in Chapter 3 of this book.

8 FUNCTIONAL TESTING OF RCD – SUPPLY CONNECTED (713–12)

The object of the test is to verify the effectiveness of the residual current device, that it is operating with the correct sensitivity and proving the integrity of the electrical and mechanical elements. The test must simulate an appropriate fault condition and be independent of any test facility incorporated in the device.

When carrying out the test, all loads normally supplied through the device are disconnected.

Functional testing of a ring circuit protected by a general-purpose RCD to BS 4293 in a split-board consumer unit is carried out as follows:

1 Using the standard lead supplied with the test instrument, disconnect all other loads and plug in the test lead to the socket at the centre of the ring (that is, the socket at the furthest point from the source of supply).
2 Set the test instrument to the tripping current of the device and at a phase angle of 0°.
3 Press the test button – the RCD should trip and disconnect the supply within 200 ms.
4 Change the phase angle from 0° to 180° and press the test button once again. The RCD should again

trip within 200 ms. Record the highest value of these two results on an installation schedule such as that given in Appendix 7 of the *On Site Guide*.
5 Now set the test instrument to 50% of the rated tripping current of the RCD and press the test button. The RCD should *not trip* within 2 seconds. This test is testing the RCD for inconvenience *or* nuisance tripping.
6 Finally, the effective operation of the test button incorporated within the RCD should be tested to prove the integrity of the mechanical elements in the tripping device. This test should be repeated every 3 months.

If the RCD fails any of the above tests it should be changed for a new one.

Where the residual current device has a rated tripping current not exceeding 30 mA and has been installed to reduce the risk associated with direct contact, as indicated in Regulation 412–06–02, a residual current of 150 mA should cause the circuit breaker to open within 40 ms.

Certification and reporting

Following the inspection and testing of an installation, a certificate should be given by the electrical contractor or responsible person to the person ordering the work (Chapter 74 of the IEE Regulations).

The certificate should be in the form set out in Appendix 6 of the IEE Regulations and Appendix 7 of the *On Site Guide*. It should include the test values which verify that the installation complies with the regulations for electrical installations at the time of testing.

An 'Electrical Installation certificate' should be used for the initial certification of a new electrical installation or for an alteration or addition to an existing installation.

All installations should be tested and inspected periodically and a 'periodic inspection' certificate issued. Suggested periodic inspection intervals are given below:

- domestic installations – 10 years
- industrial installations – 3 years
- agricultural installations – 3 years
- caravan site installations – 1 year

■ caravans – 3 years
■ temporary installations on
   construction sites – 3 months

Exercises

1 The temporary electrical installation on a construction site must be inspected and tested:
   (a) every 3 weeks
   (b) every month
   (c) every 3 months
   (d) at least once each year.
2 Portable hand-tools on construction sites should be supplied at:
   (a) 50 V
   (b) 110 V
   (c) 230 V
   (d) 400 V.
3 Industrial socket outlets and plugs are colour-coded for easy identification. 400 V, 230 V and 110 V plugs are respectively colour-coded
   (a) red, blue and yellow
   (b) white, blue and green
   (c) yellow, blue and white
   (d) blue, yellow and red.
4 Agricultural and horticultural electrical installations must be tested and inspected every:
   (a) 3 months
   (b) year
   (c) 3 years
   (d) 5 years.
5 Mobile caravan electrical installations must be tested and inspected:
   (a) before every road journey
   (b) at least once each year
   (c) every 3 months
   (d) at least every 3 years.
6 Caravan site electrical installations must be tested and inspected at least once every:
   (a) 3 months
   (b) year
   (c) 3 years
   (d) 5 years.
7 'An area in which an explosive gas–air mixture is present' is one definition of:
   (a) a flashpoint
   (b) an intrinsically safe area

   (c) a hazardous area
   (d) a safe area.
8 In a zone 0 hazardous area:
   (a) no electrical installation equipment may be installed
   (b) any electrical installation equipment may be installed
   (c) only flameproof electrical equipment may be installed
   (d) only oil-filled or powder-filled equipment may be installed.
9 In a zone 1 hazardous area:
   (a) no electrical installation equipment may be installed
   (b) any electrical installation equipment may be installed
   (c) only flameproof electrical equipment may be installed
   (d) only oil-filled or powder-filled equipment may be installed.
10 'A circuit in which no spark or thermal effect is capable of causing ignition of a given explosive atmosphere' is one definition of:
   (a) a flameproof circuit
   (b) an intrinsic circuit
   (c) a special installation
   (d) an explosive circuit.
11 A static charge cannot exist between two pieces of equipment which are:
   (a) made of insulating material
   (b) made of a mixture of insulating and conducting material
   (c) separated by a gap greater than 100 mm
   (d) bonded together.
12 Overload or overcurrent protection is offered by a:
   (a) transistor
   (b) transformer
   (c) functional switch
   (d) circuit breaker.
13 The current rating of a protective device is the current which:
   (a) it will carry continuously without deterioration
   (b) will cause the device to operate
   (c) will cause the device to operate within 4 hours
   (d) is equal to the fusing factor.
14 On a domestic installation the lighting circuit comprises 15 outlets. Good practice would suggest:
   (a) one final circuit
   (b) two final circuits

(c) five final circuits

(d) 15 final circuits.

15 Domestic lighting circuits are usually wired in:

(a) 1.0 mm cable

(b) 2.5 mm cable

(c) 4.0 mm cable

(d) 6.0 mm cable.

16 Domestic ring final circuits are usually wired in:

(a) 1.0 mm cable

(b) 2.5 mm cable

(c) 4.0 mm cable

(d) 6.0 mm cable.

17 The regulations recommend that the flexible cable connecting an appliance to a socket outlet should not exceed:

(a) 1 m

(b) 2 m

(c) 3 m

(d) 4 m.

18 A 13 A plug top always:

(a) has round pins

(b) has copper pins

(c) contains a thermal overload

(d) contains a cartridge fuse.

19 A radial circuit, wired in 2.5 mm PVC insulated and sheathed cable and protected by a 20 A fuse, may feed any number of socket outlets, provided that the total floor area does not exceed:

(a) $10 \, m^2$

(b) $20 \, m^2$

(c) $50 \, m^2$

(d) $100 \, m^2$.

20 A ring circuit wired in 2.5 mm PVC insulated and sheathed cable and protected by a 30 A fuse, may feed any number of socket outlets, provided that the total floor area does not exceed:

(a) $10 \, m^2$

(b) $20 \, m^2$

(c) $50 \, m^2$

(d) $100 \, m^2$.

21 A non-fused spur to a ring circuit may feed:

(a) any number of socket outlets

(b) any number of fixed appliances

(c) only one single- or twin-socket outlet

(d) only two fixed appliances.

22 The capacity of a battery to store charge is measured in:

(a) watts

(b) volt-amperes

(c) ampere-ohms

(d) ampere-hours.

23 Block storage radiators connected to an off-peak supply have the advantage of:

(a) a higher thermal output than other radiators

(b) using half-price electricity

(c) a slim, lightweight and portable construction

(d) controllable radiant heat.

24 In a room containing a fixed bath or shower:

(a) no electrical equipment may be installed

(b) only switched socket outlets may be installed

(c) every socket outlet must be inaccessible to anyone using the bath or shower

(d) every switch must be inaccessible to anyone using the bath or shower.

25 A cooking appliance must be controlled by a switch which:

(a) is incorporated in the appliance

(b) is separate from the cooker but within 2 m of it

(c) is cord-operated

(d) incorporates a pilot light.

26 Equipment which displays the BSI kite mark:

(a) is guaranteed to perform efficiently

(b) has been produced under a system of supervision and control by a manufacturer holding a licence

(c) will reduce the risk of an electric shock, under fault conditions, to anyone using the product

(d) carries a guarantee of the product's electrical, mechanical and thermal safety.

27 Equipment which carries the BSI safety mark:

(a) is guaranteed to perform efficiently

(b) has been produced under a system of supervision and control by a manufacturer holding a licence

(c) will reduce the risk of an electric shock, under fault conditions, to anyone using the product

(d) carries a guarantee of the product's electrical, mechanical and thermal safety.

28 The CE mark:

(a) is a quality symbol

(b) is an indication that the product meets the legal safety requirements of the European Commission

(c) guarantees the product's efficiency

(d) will reduce the risk of electric shock under fault conditions to anyone using the product.

29 The Health and Safety at Work Act places the responsibility for safety at work on:
(a) the employer
(b) the employee
(c) both the employer and employee
(d) the main contractor.

30 One advantage of a steel conduit installation, compared with a PVC conduit installation, is that it:
(a) may be easily rewired
(b) may be installed more quickly
(c) offers greater mechanical protection
(d) may hold more conductors for a given conduit size.

31 The earth continuity of a metallic conduit installation will be improved if:
(a) black enamel conduit is replaced by galvanized conduit
(b) the installation is painted with galvanized paint
(c) the installation is painted with bright orange paint
(d) all connections are made tight and secure during installation.

32 The earth continuity of a metallic trunking installation may be improved if:
(a) copper earth straps are fitted across all joints
(b) galvanized trunking is used
(c) all joints are painted with galvanized paint
(d) a space factor of 45% is not exceeded.

33 Circuits of Band I and II can
(a) never be installed in the same trunking
(b) only be installed in the same trunking if they are segregated by metal enclosures
(c) only be installed in the same trunking if a space factor of 45% is not exceeded
(d) only be wired in MI cables.

34 The test required by the Regulations to ascertain that the circuit protective conductor is correctly connected is:
(a) continuity of ring final circuit conductors
(b) continuity of protective conductors
(c) earth electrode resistance
(d) protection by electrical separation.

35 A visual inspection of a new installation must be carried out:
(a) during the erection period
(b) during testing upon completion
(c) after testing upon completion
(d) before testing upon completion.

36 One objective of the polarity test is to verify that:
(a) lampholders are correctly earthed
(b) final circuits are correctly fused
(c) the CPC is continuous throughout the installation
(d) the protective devices are connected in the live conductor.

37 When testing a 230 V installation an insulation resistance tester must supply a voltage of:
(a) less than 50 V
(b) 500 V
(c) less than 500 V
(d) greater than twice the supply voltage but less than 1000 V.

38 The value of a satisfactory insulation resistance test on each final circuit of a 230 V installation must be:
(a) less than $1\,\Omega$
(b) less than $0.5\,M\Omega$
(c) not less than $0.5\,M\Omega$
(d) not less than $1\,M\Omega$.

39 The value of a satisfactory insulation resistance test on a disconnected piece of equipment is:
(a) less than $1\,\Omega$
(b) less than $0.5\,M\Omega$
(c) not less than $0.5\,M\Omega$
(d) not less than $1\,M\Omega$.

40 The maximum inspection and retest period for a general electrical installation is:
(a) 3 months
(b) 3 years
(c) 5 years
(d) 10 years.

41 An industrial installation of PVC/SWA cables laid on cable tray offers the advantage over other types of installation of:
(a) greater mechanical protection
(b) greater flexibility in response to changing requirements
(c) higher resistance to corrosion in an industrial atmosphere
(d) flameproof installation suitable for hazardous areas.

42 The cables which can best withstand high temperatures are:
(a) MI cables
(b) PVC cables with asbestos oversleeves
(c) PVC/SWA cables
(d) PVC cables in galvanized conduit.

43 The CPC of a lighting final circuit is formed by approximately 70 m of 1.0 mm copper conductor. Calculate a satisfactory value for a continuity test on the CPC given that the resistance per metre of 1.0 mm copper is 18.1 mΩ/m.

44 The CPC of an installation is formed by approximately 200 m of 50 mm $\times$ 50 mm trunking. Determine a satisfactory test result for this CPC, using the information given in Table 4.11. Describe briefly a suitable instrument to carry out this test.

45 Four 1 mm cables and four 2.5 mm cables are to be run in a metal conduit which contains one right-angle bend and one double set. The distance between the boxes is 8 m. Find the size of conduit required to enclose these cables.

46 Determine the minimum size of trunking required to contain the following stranded cables:
(a) 20 $\times$ 1.5 mm cables
(b) 16 $\times$ 2.5 mm cables
(c) 10 $\times$ 4.0 mm cables
(d) 20 $\times$ 6.0 mm cables.

47 Calculate the number of 1.0 mm cables which may be drawn into a 5 m straight run of 20 mm conduit.

48 Calculate the number of 2.5 mm cables which may be drawn into a 20 mm plastic conduit along with a 4.0 mm CPC if the distance between the boxes is 10 m and contains one right-angled bend.

49 Determine the size of galvanized steel conduit required to contain PVC insulated conductors if the distance between two boxes is 5 m and the conduit has two bends of 90°. The conduit must contain ten 1.5 mm cables and four 2.5 mm cables.

50 Calculate the number of PVC insulated 4.0 mm cables which may be installed in a 75 $\times$ 75 mm trunking.

51 (a) Calculate the minimum size of vertical trunking required to contain 20 $\times$ 10 mm PVC insulated cables.
(b) Explain why fire barriers are fitted in vertical trunking.
(c) Explain how and why cables are supported in vertical trunking.

52 List four things which an employer must do in order to comply with the Health and Safety at Work Act.

53 List three things which an electrician must do at work in order to comply with the Health and Safety at Work Act.

54 What are the IEE Regulations and how do they influence the work of an electrician?

55 What is the NICEIC and how does it seek to maintain high standards in the electrical contracting industry?

56 Describe what is meant when a piece of equipment is protected to IP2X. What does the 'X' mean in IP2X?

57 Describe how a polarity test should be carried out on a domestic installation comprising eight light positions and ten socket outlets. The final circuits are to be supplied by a consumer unit.

58 Describe how to carry out a continuity test of ring final circuit conductors. State the values to be obtained for a satisfactory test.

59 Describe how to carry out an earth fault loop impedance test. Sketch a circuit diagram and indicate the test circuit path.

60 Describe how to carry out an insulation resistance test on a domestic installation. State the type of instrument to be used and the values of a satisfactory test.

61 Sketch the circuit diagram for a non-maintained emergency lighting installation. State the advantages and disadvantages of a maintained system and a non-maintained system.

62 Sketch the circuit diagram for a normally open fire alarm circuit, with four call points and three sounders. Describe the appearance and installation position of a breakglass call point.

63 Sketch the circuit diagram for a normally closed fire alarm circuit, with six call points and three sounders. Describe two typical fire alarm sounders.

64 Describe a simple intruder alarm installation suitable for a domestic dwelling. Indicate the type and position of detectors, sounders, supply and wiring.

65 Sketch the circuit diagram for a maintained emergency lighting installation. Describe some of the factors to be considered when choosing the position of the luminaires.

66 A large construction site is provided with a 400 V/230 V supply. Describe, with simple sketches, how the electrical contractor would supply the following loads from the mains supply:
(a) a 400 V crane
(b) robustly installed 230 V perimeter lighting
(c) 110 V sockets for portable tools

(d) SELV to special hand lamps

(e) 230 V to site offices.

State the type of cable to be used for each supply.

67 State the precautions to be considered by an electrical contractor asked to wire sockets and lights in a farm building which will be used to accommodate animals.

68 Describe the installation and operation of an electric fence controller.

69 An electrical contractor was asked to design an electrical installation which would be suitable for a caravan site with space for ten vans. Each van was to be supplied with a 230 V socket outlet adjacent to the van and the main electrical supply was to be contained in a central services brick building. Sketch the arrangements and describe the type of cable, socket outlets and control and protection equipment to be used to wire the caravan socket outlets.

70 Describe the type of cable and equipment to be used to wire a petrol pump on a garage forecourt.

71 Sketch the circuit diagram for a normally open fire alarm circuit, with four call points and three sounders. Describe the appearance and installation position of a breakglass call point.

72 Sketch the circuit diagram for a normally closed fire alarm circuit, with six call points and three sounders. Describe two typical fire alarm sounders.

73 Describe a simple intruder alarm installation suitable for a domestic dwelling. Indicate the type and position of detectors, sounders, supply and wiring.

74 Describe what is meant by a static charge. Explain why it can be dangerous and how the risk is reduced or eliminated in a hospital operating theatre and a paint spray booth.

75 Describe with sketches the construction of a 230 V, 5 A flameproof light switch. Why is it safe to use this type of equipment in a zone 1 hazardous area?

76 Describe what is meant by a clean supply for electronic equipment.

77 Describe with sketches how you would provide a secure supply to a network of four personal computers.

78 Sketch and label a 'duplex' fibre-optic cable.

79 Describe, with a sketch, how digital transmission occurs in a fibre-optic cable. State three advantages of optical fibre-optic cable, compared with co-axial copper cable, for data transmissions.

LIGHTING

—

In ancient times, much of the indoor work done by humans depended upon daylight being available to light the interior. Today almost all buildings have electric lighting installed and we automatically assume that we can work indoors or out of doors at any time of the day or night, and that light will always be available.

Good lighting is important in all building interiors, helping work to be done efficiently and safely and also playing an important part in creating pleasant and comfortable surroundings.

Lighting schemes are designed using many different types of light fitting or luminaire. 'Luminaire' is the modern term given to the equipment which supports and surrounds the lamp and may control the distribution of the light. Modern lamps use the very latest technology to provide illumination cheaply and efficiently. To begin to understand the lamps and lighting technology used today, we must first define some of the terms we will be using.

Common lighting terms

LUMINOUS INTENSITY – SYMBOL I

This is the illuminating power of the light source to radiate luminous flux in a particular direction. The earliest term used for the unit of luminous intensity was the candle power because the early standard was the wax candle. The SI unit is the candela (pronounced candeela and abbreviated to cd).

LUMINOUS FLUX – SYMBOL F

This is the flow of light which is radiated from a source. The SI unit is the lumen, one lumen being the light flux which is emitted within a unit solid angle (volume of a cone) from a point source of 1 candela.

ILLUMINANCE – SYMBOL E

This is a measure of the light falling on a surface, which is also called the incident radiation. The SI unit is the lux (lx) and is the illumination produced by 1 lumen over an area of $1\ m^2$.

LUMINANCE – SYMBOL L

Since this is a measure of the brightness of a surface it is also a measure of the light which is reflected from a surface. The objects we see vary in appearance according to the light which they emit or reflect towards the eye.

The SI units of luminance vary with the type of surface being considered. For a diffusing surface such as blotting paper or a matt white painted surface the unit of luminance is the lumen per square metre. With polished surfaces such as silvered glass reflector, the brightness is specified in terms of the light intensity and the unit is the candela per square metre.

Illumination laws

Rays of light falling upon a surface from some distance d will illuminate that surface with an illuminance of say 1 lux. If the distance d is doubled as shown in Fig. 5.1, the illuminance of 1 lux will fall over four square units of area. Thus the illumination of a surface follows the *inverse square law*, where

$$E = \frac{I}{d^2} \text{ (lx)}$$

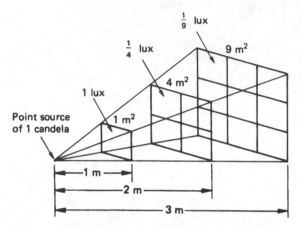

Fig. 5.1   The inverse square law.

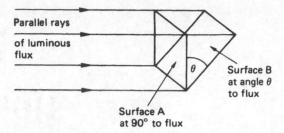

Fig. 5.2   The cosine law.

EXAMPLE 2

A street lantern suspends a 2000 cd light source 4 m above the ground. Determine the illuminance directly below the lamp and 3 m to one side of the lamp base.

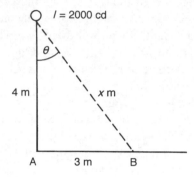

The illuminance below the lamp, E_A, is

$$E_A = \frac{I}{d^2} \text{ (lx)}$$

$$\therefore E_A = \frac{2000 \text{ cd}}{(4 \text{ m})^2} = 125 \text{ lx}$$

To work out the illuminance at 3 m to one side of the lantern, E_B, we need the distance between the light source and the position on the ground at B; this can be found by Pythagoras' theorem:

$$x \text{ (m)} = \sqrt{(4 \text{ m})^2 + (3 \text{ m})^2} = \sqrt{25} \text{ m}$$
$$x = 5 \text{ m}$$

$$\therefore E_B = \frac{I \cos \theta}{d^2} \text{ (lx) and } \cos \theta = \frac{4}{5}$$

$$\therefore E_B = \frac{2000 \text{ cd} \times 4}{(5 \text{ m})^2 \times 5} = 64 \text{ lx}$$

EXAMPLE 1

A lamp of luminous intensity 1000 cd is suspended 2 metres above a laboratory bench. Calculate the illuminance directly below the lamp

$$E = \frac{I}{d^2} \text{ (lx)}$$

$$\therefore E = \frac{1000 \text{ cd}}{(2 \text{ m})^2} = 250 \text{ lx}$$

The illumination of surface A in Fig. 5.2 will follow the inverse square law described above. If this surface were removed, the same luminous flux would then fall on surface B. Since the parallel rays of light falling on the inclined surface B are spread over a larger surface area, the illuminance will be reduced by a factor θ, and therefore

$$E = \frac{I \cos \theta}{d^2} \text{ (lx)}$$

Since the two surfaces are joined together by the trigonometry of the cosine rules this equation is known as the *cosine law*.

EXAMPLE 3

A discharge lamp is suspended from a ceiling 4 m above a bench. The illuminance on the bench below the lamp was 300 lux. Find:
(a) the luminous intensity of the lamp
(b) the distance along the bench where the illuminance falls to 153.6 lux.

For (a),

$$E_A = \frac{I}{d^2} \text{ (lx)} \qquad \therefore I = E_A \, d^2 \text{ (cd)}$$

$$I = 300 \text{ lx} \times 16 \text{ m} = 4800 \text{ cd}$$

For (b),

$$E_B = \frac{I}{d^2} \cos\theta \text{ (lx)}$$

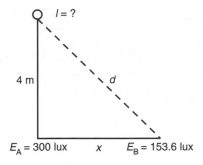

$$\therefore d^2 = \frac{I \cos\theta}{E_B} \text{ (m}^2)$$

$$d^2 = \frac{4800 \text{ cd}}{153.6 \text{ lx}} \times \frac{4 \text{ m}}{d \text{ m}}$$

$$d^3 = 125$$

$$\therefore d = \sqrt[3]{125} = 5 \text{ m}$$

By Pythagoras,

$$x = \sqrt{5^2 - 4^2} = 3 \text{ m}$$

Measurement of illuminance

To take a reading, place a suitable illuminance meter, calibrated in lux, on to the surface whose illumination level is to be measured. For general light levels hold the instrument 85 cm above the floor in a horizontal plane.

Take readings from the appropriate scale but take care not to obscure the photocell in any way when taking measurements, for example, by casting a shadow with the body or the hand. The recommended levels of illuminance for various types of installation are given by the IES (Illumination Engineers Society) code which is usually printed on the back of the meter, as shown in Fig. 5.3. Some examples are given in Table 5.1.

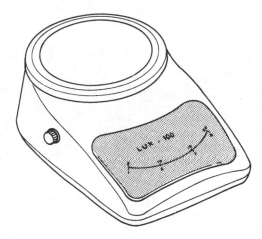

Fig. 5.3 A typical lightmeter.

Table 5.1 Illuminance values

Task	Working Situation	Illuminance (lux)
Casual vision	Storage rooms, stairs and washrooms	100
Rough assembly	Workshops and garages	300
Reading, writing and drawing	Classrooms and offices	500
Fine assembly	Electronic component assembly	1000
Minute assembly	Watchmaking	3000

The activities being carried out in a room will determine the levels of illuminance required since different levels of illumination are required for the successful operation or completion of different tasks. The assembly of electronic components in a factory will require a higher level of illumination than, say, the assembly of engine components in a garage because the electronic components are much smaller and finer detail is required for their successful assembly.

Table 5.2　Characteristics of a Thorn Lighting 1500 mm 65 W bi-pin tube

Tube colour	Initial lamp lumens*	Lighting design lumens†	Colour rendering quality	Colour appearance
Artifical daylight	2600	2100	Excellent	Cool
De Luxe Natural	2900	2500	Very Good	Intermediate
De Luxe Warm white	3500	3200	Good	Warm
Natural	3700	3400	Good	Intermediate
Daylight	4800	4450	Fair	Cool
Warm white	4950	4600	Fair	Warm
White	5100	4750	Fair	Warm
Red	250*	250	Poor	Deep red

*The initial lumens are the measured lumens after 100 hours of life.
†The lighting design lumens are the output lumens after 2000 hours.
Coloured tubes are intended for decorative purposes only.

Burning position	Lamp may be operated in any position
Rated life	7500 hours
Efficacy	30 to 70 lm/W depending upon the tube colour

The inverse square law calculations considered earlier are only suitable for designing lighting schemes where there are no reflecting surfaces producing secondary additional illumination. This method could be used to design an outdoor lighting scheme for a cathedral, bridge or public building.

Interior luminaires produce light directly on to the working surface but additionally there is a secondary source of illumination from light reflected from the walls and ceilings. When designing interior lighting schemes the method most frequently used depends upon a determination of the total flux required to provide a given value of illuminance at the working place. This method is generally known as the *lumen method*.

To determine the total number of luminaires required to produce a given illuminance by the lumen method we apply the following formula:

$$\begin{array}{l}\text{Total number of} \\ \text{luminaires required} \\ \text{to provide a chosen} \\ \text{level of illumination} \\ \text{at a surface}\end{array} = \frac{\begin{array}{l}\text{illuminance level (lx)} \\ \times \text{ area (m}^2) \end{array}}{\begin{array}{l}\text{lumen output} \\ \text{of each} \\ \text{luminaire (lm)} \times \text{UF} \times \text{LLF}\end{array}}$$

where:

■ the illuminance level is chosen after consideration of the IES code,
■ the area is the working area to be illuminated,
■ the lumen output of each luminaire is that given in the manufacturer's specification and may be found by reference to tables such as Table 5.2.
■ UF is the utilization factor,
■ LLF is the light loss factor.

Utilization factor (UF)

The light flux reaching the working plane is always less than the lumen output of the lamp since some of the light is absorbed by the various surface textures. The method of calculating the utilization factor is detailed in Chartered Institution of Building Services Engineers (CIBSE) Technical Memorandum No 5, although lighting manufacturers' catalogues give factors for standard conditions. The UF is expressed as a number which is always less than unity; a typical value might be 0.9 for a modern office building.

Light Loss Factor (LLF)

The light output of a luminaire is reduced during its life because of an accumulation of dust and dirt on the lamp and fitting. Decorations also deteriorate with time, and this results in more light flux being absorbed by the walls and ceiling.

You can see from Table 5.2 that the output lumens of the lamp decrease with time – for example, a warm white tube gives out 4950 lumens after the first 100 hours of its life but this falls to 4600 lumens after 2000 hours.

The total light loss can be considered under four headings:

1 light loss due to luminaire dirt depreciation (LDD),
2 light loss due to room dirt depreciation (RDD),
3 light loss due to a lamp failure factor (LFF),
4 light loss due to lamp lumen depreciation (LLD).

The LLF is the total loss due to these four separate factors and typically has a value between 0.8 and 0.9.

When using the LLF in lumen method calculations we always use the manufacturer's initial lamp lumens for the particular lamp because the LLF takes account of the depreciation in lumen output with time. Let us now consider a calculation using the lumen method.

EXAMPLE

It is proposed to illuminate an electronic workshop of dimensions 9 m × 8 m × 3 m to an illuminance of 550 lux at the bench level. The specification calls for luminaires having one 1500 mm 65 W natural tube with an initial output of 3700 lumens (see Table 5.2). Determine the number of luminaires required for this installation when the UF and LLF are 0.9 and 0.8 respectively.

$$\frac{\text{The number of}}{\text{luminaires required}} = \frac{E \text{ (lx)} \times \text{area (m}^2\text{)}}{\text{lumens from each luminaire} \times \text{UF} \times \text{LLF}}$$

$$\frac{\text{The number of}}{\text{luminaires}} = \frac{550 \text{ lx} \times 9 \text{ m} \times 8 \text{ m}}{3700 \times 0.9 \times 0.8} = 14.86$$

Therefore 15 luminaires will be required to illuminate this workshop to a level of 550 lux.

Space to mounting height ratio

The correct mounting height of luminaires is important since glare may result if fittings are placed in the line of vision. Excessive height will result in a rapid reduction of illuminance, as demonstrated by the inverse square law, and make lamp replacement and maintenance difficult. The correct spacing of luminaires is important since large spaces between the fittings may result in a fall-off of illuminance at the working

plane midway between adjacent fittings. The illuminance between the luminaires must not be allowed to fall below 70% of the value directly below the fitting. For most installations a spacing to mounting height ratio of 1 : 1 to 2 : 1 above the working surface is usually considered adequate and the working surface is normally taken as 0.85 m above the floor level, as shown in Fig. 5.4.

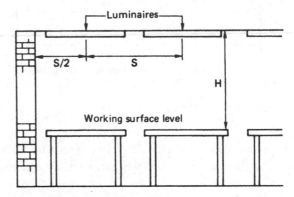

Fig. 5.4 Space to height ratio.

Layout of luminaires

To maintain an even distribution of illuminance from the luminaires, those adjacent to the walls of the room should be fixed at half the spacing distance. This is because a point in the middle of the room receives luminous flux from two adjacent luminaires, while a point close to the wall is illuminated mainly from only one luminaire.

Considering the previous example of an electronic workshop requiring 15 luminaires to provide the required illuminance, if we assume a space to height ratio of 1 : 1, the best layout may be four rows, with four luminaires in each row. This would necessitate using one extra luminaire than the calculations suggested. This is quite acceptable since the overall illuminance will be raised by only about 6% and the resultant layout will be more symmetrical while complying very closely with the space to height ratio. The layout is shown in Fig. 5.5.

Luminaires used for the general illumination of a room are normally arranged parallel to the longest wall, as shown in Fig. 5.5 or they might be recessed modular fluorescent light fittings installed in a suspended ceiling. Fittings are then switched in rows parallel to the wall

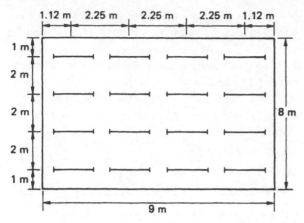

The mounting height in this case is the ceiling height minus the height of the working surface $\therefore$ H = 3.0 − 0.85 = 2.15 m

Fig. 5.5 Layout of luminaires.

containing the daylight windows. In this way the back of the room, furthest away from the windows, can be switched on earlier than those closest to the natural daylight, thus saving electrical energy.

In corridors, luminaires are normally fixed parallel to the length of the corridor to avoid glare from the tubes when walking down the length of the corridor.

Comparison of light sources

When comparing one light source with another we are interested in the colour reproducing qualities of the lamp and the efficiency with which the lamp converts electricity into illumination. These qualities are expressed by the lamp's efficacy and colour rendering qualities.

Lamp efficacy

The performance of a lamp is quoted as a ratio of the number of lumens of light flux which it emits to the electrical energy input which it consumes. Thus *efficacy* is measured in lumens per watt; the greater the efficacy the better is the lamp's performance in converting electrical energy into light energy.

A general lighting service (GLS) lamp, for example, has an efficacy of 14 lumens per watt, while a fluorescent tube, which is much more efficient at converting electricity into light, has an efficacy of about 50 lumens per watt.

Colour rendering

We recognize various materials and surfaces as having a particular colour because luminous flux of a frequency corresponding to that colour is reflected from the surface to our eye which is then processed by our brain. White light is made up of the combined frequencies of the colours red, orange, yellow, green, blue, indigo and violet. Colours can only be seen if the lamp supplying the illuminance is emitting light of that particular frequency. The ability to show colours faithfully as they would appear in daylight is a measure of the colour rendering property of the light source.

GLS LAMPS

General lighting service lamps produce light as a result of the heating effect of an electrical current. A fine tungsten wire is first coiled and coiled again to form the incandescent filament of the GLS lamp. The coiled coil arrangement reduces filament cooling and increases the light output by allowing the filament to operate at a higher temperature. The light output covers the visible spectrum, giving a warm white to yellow light with a colour rendering quality classified as fairly good. The efficacy of the GLS lamp is 14 lumens per watt over its intended life-span of 1000 hours.

The filament lamp in its simplest form is a purely functional light source which is unchallenged on the domestic market despite the manufacture of more efficient lamps. One factor which may have contributed to its popularity is that lamp designers have been able to modify the glass envelope of the lamp to give a very pleasing decorative appearance, as shown by Fig. 5.6.

TUNGSTEN HALOGEN LAMPS

In the GLS lamp, the high operating temperature of the filament causes some evaporation of the tungsten which is carried by convection currents on to the bulb wall. When the lamp has been in service for some time, evaporated tungsten darkens the bulb wall, with the result that the light output is reduced and the filament becomes thinner and eventually fails.

To overcome these problems the envelope of the tungsten halogen lamp contains a trace of one of the halogen gases; iodine, chlorine, bromine or fluorine. This allows a reversible chemical reaction to occur between the tungsten filament and the halogen gas. When tungsten is evaporated from the incandescent

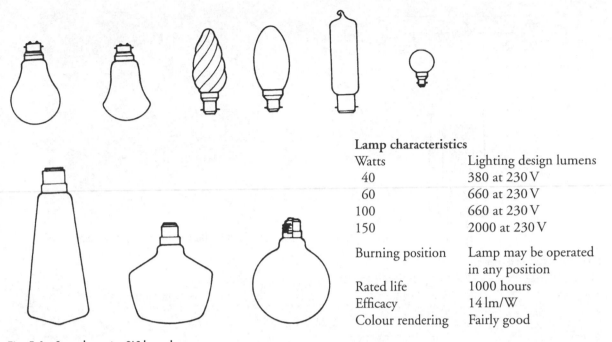

Lamp characteristics

Watts	Lighting design lumens
40	380 at 230 V
60	660 at 230 V
100	660 at 230 V
150	2000 at 230 V
Burning position	Lamp may be operated in any position
Rated life	1000 hours
Efficacy	14 lm/W
Colour rendering	Fairly good

Fig. 5.6 Some decorative GLS lamp shapes.

filament, some part of it spreads out towards the bulb wall, but at a point close to the wall where the temperature conditions are favourable, the tungsten combines with the halogen. This tungsten halide molecule then drifts back towards the filament where it once more separates, depositing the tungsten back on to the filament, leaving the halogen available for a further reaction cycle.

Since all the evaporated tungsten is returned to the filament, the bulb blackening normally associated with tungsten lamps is completely eliminated and a high efficacy is maintained throughout the life of the lamp.

A minimum bulb wall temperature of 250°C is required to maintain the halogen cycle and consequently a small glass envelope is required. This also permits a much higher gas pressure to be used which increases the lamp life to 2000 hours and allows the filament to be operated at a higher temperature, giving more light. The lamp is very small and produces a very white intense light giving it a colour rendering classification of good and an efficacy of 20 lumens per watt.

The tungsten halogen lamp, as shown in Fig. 5.7 was a major development in lamp design and resulted in Thorn Lighting gaining the Queen's Award for Technical Innovation in Industry in 1972.

Tungsten Halogen Dichroic Reflector Miniature Spot Lamps such as the one shown in Fig. 5.8 are extremely popular in the lighting schemes of the new millennium. Their small size and bright white illumination makes them very popular in both commercial and domestic installations. They are available as a 12 volt bi-pin package in 20, 35 and 50 watts and as a 230 volt bayonet type cap (called a GU10 or GZ10 cap) in 20, 35 and 50 watts.

DISCHARGE LAMPS

Discharge lamps do not produce light by means of an incandescent filament but by the excitation of a gas or metallic vapour contained within a glass envelope. A voltage applied to two terminals or electrodes sealed into the end of a glass tube containing a gas or metallic vapour will excite the contents and produce light directly.

The colour of the light produced depends upon the type of gas or metallic vapour contained within the tube. Some examples are given below:

Gases	neon	red
	argon	green/blue
	hydrogen	pink
	helium	ivory
	mercury	blue
Metallic vapours	sodium	yellow
	magnesium	grass green

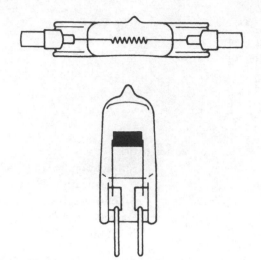

Lamp characteristics

Watts	Lighting design lumens
300	5000 at 230 V
500	9500 at 230 V

Burning position	Linear lamps must be operated horizontally or within 4° of the horizontal
Rated life	2000 hours
Efficacy	20 lm/W
Colour rendering	Good

When installing the lamp, grease contamination of the glass envelope by touching must be avoided. Any grease present on the outer surface will cause cracking and premature failure of the lamp because of the high operating temperatures. A paper sleeve should be used when handling the lamp, and if accidentally touched with bare hands the lamp should be cleaned with methylated spirit.

Fig. 5.7 A tungsten halogen lamp.

Let us now consider some of the more frequently used discharge lamps.

Fluorescent tube

A fluorescent lamp is a linear arc tube, internally coated with a fluorescent powder, containing a low-pressure mercury vapour discharge and given the designation MCF by lamp manufacturers. The lamp construction is shown in Fig. 5.9 and the characteristics of the variously coloured tubes are given in Table 5.2.

Passing a current through the cathodes of the tube causes them to become hot and produce a cloud of

Fig. 5.8 Tungsten Halogen Dichroic Reflector Lamp.

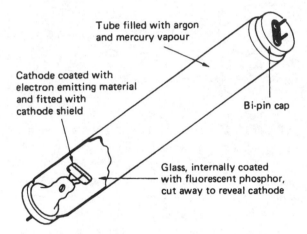

Tube filled with argon and mercury vapour

Cathode coated with electron emitting material and fitted with cathode shield

Bi-pin cap

Glass, internally coated with fluorescent phosphor, cut away to reveal cathode

The arc radiates much more UV than visible light: almost all the light from a fluorescent tube comes from the phosphors

Fig. 5.9 Fluorescent lamp construction.

electrons which ionize the gas in the immediate vicinity of the cathodes. This ionization then spreads to the whole length of the tube, producing invisible ultraviolet rays and some blue light. The fluorescent powder on the inside of the tube is sensitive to ultra-violet rays and converts this radiation into visible light. The fluorescent powder on the inside of the tube can be mixed to give light of almost any desired colour or grade of white light. Some mixes have their maximum light output in the yellow-green region of the spectrum giving maximum efficacy but poor colour rendering. Other mixes give better colour rendering at the cost of reduced lumen output as can be seen from

Table 5.2. The lamp has many domestic, industrial and commercial applications. Its efficacy varies between 30 and 70 lumens per watt depending upon the colour rendering qualities of the tube.

Energy-efficient lamps

Energy-efficient lamps are miniature fluorescent lamps designed to replace ordinary GLS lamps. They are available in a variety of shapes and sizes so that they can be fitted into existing light fittings. Figure 5.10 shows three typical shapes. The 'stick' type give most of their light output radially while the flat 'double D' type give most of their light output above and below.

Energy-efficient lamps use electricity much more efficiently than an equivalent GLS lamp. For example, a 20 watt energy efficient lamp will give the same light output as a 100 watt GLS lamp. An 11 watt energy efficient lamp is equivalent to a 60 watt GLS lamp. Energy-efficient lamps also have a lifespan of about eight times longer than a GLS lamp and so, they do use energy very efficiently.

However, energy-efficient lamps are very expensive to purchase and they do take a few minutes to attain full brilliance after switching on. They cannot be controlled by a dimmer switch and are unsuitable for incorporating in an automatic presence detector because they are usually not switched on long enough to be worthwhile, but energy efficient lamps are excellent for outside security lighting which is left on for several hours each night.

The electrical contractor, in discussion with a customer, must balance the advantages and disadvantages of energy-efficient lamps compared to other sources of illumination for each individual installation.

High-pressure mercury vapour lamp

The high-pressure mercury discharge takes place in a quartz glass arc tube contained within an outer bulb which, in the case of the lamp classified as MBF, is internally coated with fluorescent powder. The lamp's construction and characteristics are shown in Fig. 5.11.

The inner discharge tube contains the mercury vapour and a small amount of argon gas to assist starting. The main electrodes are positioned at either end of the tube and a starting electrode is positioned close to one main electrode.

Fig. 5.10 Energy-efficient lamps.

When the supply is switched on the current is insufficient to initiate a discharge between the main electrodes, but ionization does occur between the starting electrode and one main electrode in the argon gas.

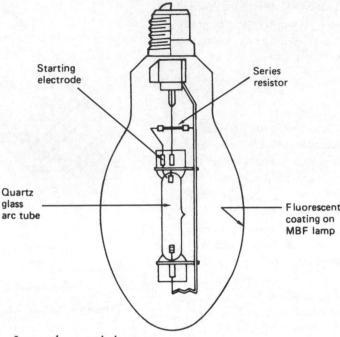

Lamp characteristics

Watts	Lighting design lumens
50	1 800
80	3 350
125	5 550
250	12 000
400	21 500
700	38 000
1000	54 000

Burning position	Lamp may be operated in any position
Rated life	7500 hours
Efficacy	38 to 56 lm/W
Colour rendering	Fairly good

Fig. 5.11 High-pressure mercury vapour lamp.

This spreads through the arc tube to the other main electrode. As the lamp warms the mercury is vaporized, the pressure builds up and the lamp achieves full brilliance after about 5 to 7 minutes.

If the supply is switched off the lamp cannot be relit until the pressure in the arc tube has reduced. It may take a further 5 minutes to restrike the lamp.

The lamp is used for commercial and industrial installations, street lighting, shopping centre illumination and area floodlighting.

Low-pressure sodium lamps

The low-pressure sodium discharge takes place in a U-shaped arc tube made of special glass which is resistant to sodium attack. This U-tube is encased in a tubular outer bulb of clear glass as shown in Fig. 5.12. Lamps classified as type SOX have a BC lampholder while the SLI/H lamp has a bi-pin lampholder at each end.

Since at room temperature the pressure of sodium is very low, a discharge cannot be initiated in sodium vapour alone. Therefore, the arc tube also contains neon gas to start the lamp. The arc path of the low-pressure sodium lamp is much longer than that of mercury lamps and starting is achieved by imposing a high voltage equal to about twice the main voltage across the electrodes by means of a leakage transformer. This voltage initiates a discharge in the neon gas which heats up the sodium. The sodium vaporizes and over a period of 6 to 11 minutes the lamp reaches full brilliance, changing colour from red to bright yellow.

The lamp must be operated horizontally so that when the lamp is switched off the condensing sodium is evenly distributed around the U-tube.

The light output is yellow and has poor colour rendering properties but this is compensated by the fact that the wavelength of the light is close to that at which the human eye has its maximum sensitivity, giving the lamp a high efficacy. The main application for this lamp is street lighting where the light output meets the requirements of the Ministry of Transport.

High-pressure sodium lamp

The high-pressure sodium discharge takes place in a sintered aluminium oxide arc tube contained within a hard glass outer bulb. Until recently no suitable material was available which would withstand the extreme chemical activity of sodium at high pressure. The construction and characteristics of the high-pressure sodium lamp classified as type SON are given in Fig. 5.13.

The arc tube contains sodium and a small amount of argon or xenon to assist starting. When the lamp is switched on an electronic pulse igniter of 2 kV or more initiates a discharge in the starter gas. This heats up the sodium and in about 5 to 7 minutes the sodium vaporizes and the lamp achieves full brilliance. Both colour and efficacy improve as the pressure of the sodium rises giving a pleasant golden white colour to the light which is classified as having a fair colour rendering quality.

The SON lamp is suitable for many applications. Because of the warming glow of the illuminance it is used in food halls and hotel reception areas. Also, because of the high efficacy and long lamp life it is used for high bay lighting in factories and warehouses and for area floodlighting at airports, car parks and dockyards.

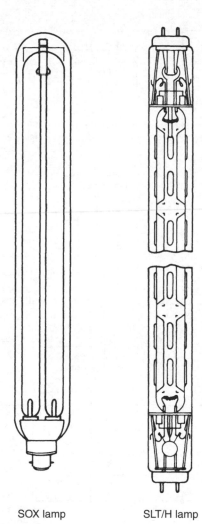

SOX lamp   SLT/H lamp

Lamp characteristics

Watts	Lighting design lumens
Type SOX	
35	4 300
55	7 500
90	12 500
135	21 500
Type SLI/H	
140	20 000
200	25 000
200 HO	27 500
Burning position	Horizontal or within 20° of the horizontal
Rated life	6000 hours
Guaranteed life	4000 hours
Efficacy	61 to 160 lm/W
Colour rendering	Very poor

Fig. 5.12 Low-pressure sodium lamp.

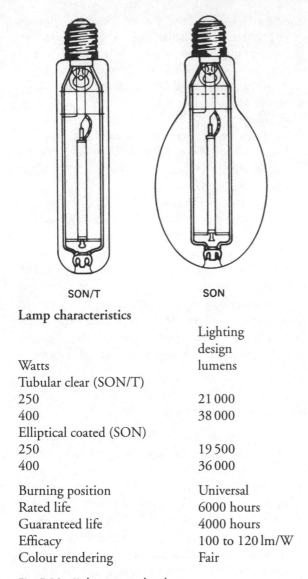

SON/T   SON

Lamp characteristics

Watts	Lighting design lumens
Tubular clear (SON/T)	
250	21 000
400	38 000
Elliptical coated (SON)	
250	19 500
400	36 000
Burning position	Universal
Rated life	6000 hours
Guaranteed life	4000 hours
Efficacy	100 to 120 lm/W
Colour rendering	Fair

Fig. 5.13 High-pressure sodium lamp.

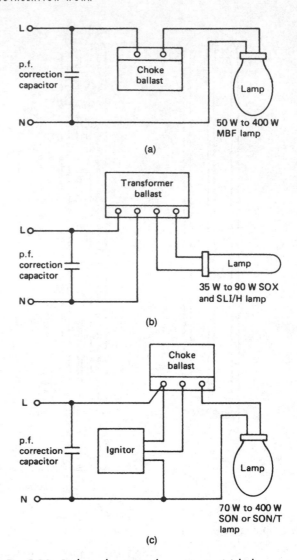

Fig. 5.14 Discharge lamp control gear circuits: (a) high-pressure mercury vapour lamp; (b) low-pressure sodium lamp; (c) high-pressure sodium lamp.

Control gear for lamps

Luminaires are wired using the standard lighting circuits described in Chapter 6 of *Basic Electrical Installation Work*, but discharge lamps require additional control gear and circuitry for their efficient and safe operation. The circuit diagrams for high-pressure mercury vapour and high- and low-pressure sodium lamps are given in Fig. 5.14. Each of these circuits requires the inclusion of a choke or transformer, creating a lagging power factor which must be corrected.

This is usually achieved by connecting a capacitor across the supply to the luminaire as shown.

Fluorescent lamp control circuits

A fluorescent lamp requires some means of initiating the discharge in the tube, and a device to control the current once the arc is struck. Since the lamps are usually operated on a.c. supplies, these functions are usually achieved by means of a choke ballast. Three basic circuits are commonly used to achieve starting – switch-start, quick-start and semi-resonant.

SWITCH-START FLUORESCENT LAMP CIRCUIT

Figure 5.15 shows a switch-start fluorescent lamp circuit in which a glow-type starter switch is now standard. A glow-type starter switch consists of two bimetallic strip electrodes encased in a glass bulb containing an inert gas. The starter switch is initially open-circuit. When the supply is switched on the full mains voltage is developed across these contacts and a glow discharge takes place between them. This warms the switch electrodes and they bend towards each other until the switch makes contact. This allows current to flow through the lamp electrodes, which become heated so that a cloud of electrons is formed at each end of the tube, which in turn glows.

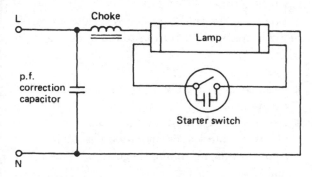

Fig. 5.15 Switch-start fluorescent lamp circuit.

When the contacts in the starter switch are made the glow discharge between the contacts is extinguished since no voltage is developed across the switch. The starter switch contacts cool and after a few seconds spring apart. Since there is a choke in series with the lamp, the breaking of this inductive circuit causes a voltage surge across the lamp electrodes which is sufficient to strike the main arc in the tube. If the lamp does not strike first time the process is repeated.

When the main arc has been struck in the low-pressure mercury vapour, the current is limited by the choke. The capacitor across the mains supply provides power-factor correction and the capacitor across the starter switch contact is for radio interference suppression.

QUICK-START FLUORESCENT LAMP CIRCUIT

When the circuit is switched on the tube cathodes are heated by a small auto-transformer. After a short pre-heating period the mercury vapour is ionized and spreads rapidly through the tube to strike the arc. The luminaire or some earthed metal must be in close proximity to the lamp to assist in the striking of the main arc. In some cases a metal strip may be bonded along the tube length to assist starting.

When the main arc has been struck the current flowing in the circuit is limited by the choke. A capacitor connected across the supply provides power-factor correction. The circuit is shown in Fig. 5.16.

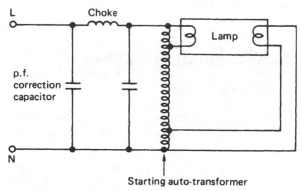

Fig. 5.16 Quick-start fluorescent lamp circuit.

SEMI-RESONANT START FLUORESCENT LAMP CIRCUIT

In this circuit a specially wound transformer takes the place of the choke. When the circuit is switched on, a current flows through the primary winding to one cathode of the lamp, through the secondary winding and a large capacitor to the other cathode. The secondary winding is wound in opposition to the primary winding. Therefore, the voltage developed across the transformer windings is 180° out of phase.

The current flowing through the electrodes causes an electron cloud to form around each cathode. This cloud spreads rapidly through the tube due to the voltage across the tube being increased by winding the transformer windings in opposition. When the main arc has been struck the current is limited by the primary winding of the transformer which behaves as a choke. A power-factor correction capacitor is not necessary since the circuit is predominantly capacitive and has a high power factor. With the luminaire earthed to assist starting this circuit will start very easily at temperatures as low as −5°C. The circuit is shown in Fig. 5.17.

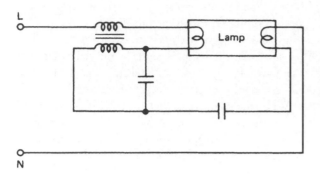

Fig. 5.17 Semi-resonant start fluorescent lamp circuit.

Installation of luminaires

OPERATING POSITION

Some lamps, particularly discharge lamps, have limitations placed upon their operating position. Since the luminaire is designed to support the lamp, any restrictions upon the operating position of the lamp will affect the position of the luminaire. Some indications of the operating position of lamps were given earlier under the individual lamp characteristics. The luminaire must be suitable for the environment in which it is to operate. This may be a corrosive atmosphere, an outdoor situation or a low-temperature zone. On the other hand, the luminaire may be required to look attractive in a commercial environment. It must satisfy all these requirements while at the same time providing adequate illumination without glare.

Many lamps contain a wire filament and delicate supports enclosed in a glass envelope. The luminaire is designed to give adequate support to the lamp under normal conditions, but a luminaire subjected to excessive vibration will encourage the lamp to fail prematurely by either breaking the filament, cracking the glass envelope or breaking the lampholder seal.

CONTROL GEAR

Chokes and ballasts for discharge lamps have laminated sheet steel cores in which a constant reversal of the magnetic field due to the a.c. supply sets up vibrations.

In most standard chokes the noise level is extremely low, and those manufactured to BS 2818 have a maximum permitted noise level of 30 dB. This noise level is about equal to the sound produced by a Swatch watch at a distance of 1 metre in a very quiet room.

Chokes must be rigidly fixed, otherwise metal fittings can amplify choke noise. Plasterboard, hardboard or wooden panels can also act as a sounding board for control gear or luminaires mounted upon them, thereby amplifying choke noise. The background noise will obviously affect people's ability to detect choke noise, and so control gear and luminaires which would be considered noisy in a library or church may be unnoticeable in a busy shop or office.

The cable, accessories and fixing box must be suitable for the mass suspended and the ambient temperature in accordance with Regulations 553–03 and 554–01. Self-contained luminaires must have an adjacent means of isolation provided in addition to the functional switch to facilitate safe maintenance and repair (Regulation 476–02–04).

Control gear should be mounted as close as possible to the lamp. Where it is liable to cause overheating it must be either.

- enclosed in a suitably designed non-combustible enclosure or
- mounted so as to allow heat to dissipate or
- placed at a sufficient distance from adjacent materials to prevent the risk of fire (Regulations 422–01–01 and 02).

Discharge lighting may also cause a stroboscopic effect where rotating or reciprocating machinery is being used. This effect causes rotating machinery to appear stationary, and we will consider the elimination of this dangerous effect next.

STROBOSCOPIC EFFECT

The flicker effect of any light source can lead to the risk of a stroboscopic effect. This causes rotating or reciprocating machinery to appear to be running at speeds other than their actual speed, and in extreme cases a circular saw or lathe chuck may appear stationary when rotating. A stroboscopic light is used to good effect when electronically 'timing' a car, by making the crank shaft appear stationary when the engine is running so that the top dead centre position (TDC) may be found.

All discharge lamps used on a.c. circuits flicker, often unobtrusively, due to the arc being extinguished every half cycle as the lamp current passes through zero. This variation in current and light output is illustrated in Fig. 5.18.

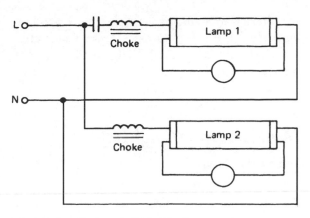

Fig. 5.19 Circuit diagrams for lead-lag fluorescent lamp circuits.

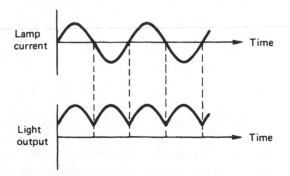

Fig. 5.18 Variation of current and light output for a discharge lamp.

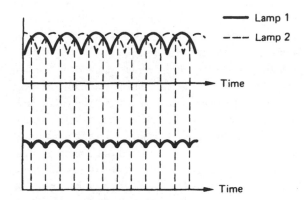

Fig. 5.20 Elimination of stroboscopic effect.

The elimination of this flicker is desirable in all commercial installations and particularly those which use rotating machinery. Fluorescent phosphors with a long afterglow help to eliminate the brightness peaks. Where a three-phase supply is available, discharge lamps may be connected to a different phase, so that they reach their brightness peaks at different times. The combined effect is to produce a much more uniform overall level of illumination, which eliminates the flicker effect.

When only a single-phase supply is available, adjacent fluorescent tubes in twin fittings may be connected to a lead-lag circuit, as illustrated in Fig. 5.19. This is one lamp connected in series with a choke and the other in series with a choke and capacitor, which causes the currents in the two lamps to differ by between 120 and 180 degrees. The lamps flicker out of phase, producing a more uniform level of illumination which eliminates the stroboscopic effect, as shown in Fig. 5.20.

GLS lamps do not flicker because the incandescent filament produces a carrying-over of the light output as the current reverses. Thus the light remains almost constant and, therefore, the stroboscopic flicker effect is not apparent.

Loading and switching of discharge circuits

Discharge circuits must be capable of carrying the total steady current (the current required by the lamp plus the current required by any control gear). Appendix 1 of the *On Site Guide* states that where more exact information is not available, the rating of the final circuits for discharge lamps may be taken as the rated lamp wattage multiplied by 1.8. Therefore, an 80 W fluorescent lamp luminaire will have an assumed demand of $80 \times 1.8 = 144\,\text{W}$.

All discharge lighting circuits are inductive and will cause excessive wear of the functional switch contacts. Where discharge lighting circuits are to be switched, the rating of the functional switch must be suitable for breaking an inductive load. To comply with this requirement we usually assume that the rating of the functional switch should be *twice* the total steady current of the inductive circuit from information previously given in the 15th edition of the IEE Regulations.

Maintenance of lighting installations

The extent of the maintenance required will depend upon the size and type of lighting installation and the installed conditions – for example, whether the environment is dusty or clean. However, lamps and luminaires will need cleaning, and lamps will need to be replaced. They can be replaced either by 'spot replacement' or 'group replacement'.

Spot replacement is the replacement of individual lamps as and when they fail. This is probably the most suitable method of maintaining small lighting installations in shops, offices and nursing homes, and is certainly the preferred method in domestic property. But each time a lamp is replaced or cleaned, a small disturbance occurs to the normal environment. Access equipment must be set up, furniture must be moved and the electrician chats to the people around him. In some environments this small disturbance and potential hazard from someone working on steps or a mobile tower is not acceptable and, therefore, group replacement of lamps must be considered.

Group replacement is the replacement of all the lamps at the same time. The time interval for lamp replacement is determined by taking into account the manufacturer's rated life for the particular lamp and the number of hours each week that the lamp is illuminated.

Group replacement and cleaning can reduce labour costs and inconvenience to customers and is the preferred method for big stores and major retail outlets. Group replacement in these commercial installations is often carried out at night or at the weekend to avoid a disturbance of the normal working environment.

High-voltage discharge lighting

The popularity of high-voltage discharge lighting has arisen almost entirely from its use in advertising.

Piccadilly Circus in London attracts visitors from around the world to admire the high-voltage discharge signs as well as the statue of Eros.

A gas sealed into a glass tube with electrodes at each end will ionize when a high voltage is applied to the electrodes. The ionized gas will glow with a colour which is characteristic of the gas contained in the tube. The luminous flux radiated by hydrogen is pink, argon green/blue and neon red. Since many advertisement signs incorporate neon gas, the term 'neon sign' has become synonymous with high-voltage discharge signs.

The glass letters of the advertisement sign may be built up individually and joined together in series with nickel wire enclosed in a thin glass tube. Alternatively, the words may be formed from one piece of glass tube, the glass links between the letters being painted over so that only the letters illuminate (see Fig. 5.21).

Individual letters are formed by heating the glass tube with a blowtorch and bending it into the required shape. An electrode is then sealed to the end of the tube and injected with the desired gas.

The diameter of the tube will to some extent be influenced by the size of the letters to be formed, but common sizes are 10, 15, 20 and 30 mm. These tubes typically carry currents of 25, 35, 60 and 150 mA respectively.

In order to calculate what sort of transformer we need to feed our discharge lighting, as well as the power output, we need to calculate the total volt drop. This is given by:

$$\frac{\text{Total volt}}{\text{drop}} = \frac{\text{Volt drop}}{\text{in tube}} + \frac{\text{Volt drop}}{\text{across electrodes}}$$

For neon gas, the volt drop per pair of electrodes is 300 V, and the volt drop per metre of 15 mm tube is 400 V.

EXAMPLE

A high-voltage neon-filled discharge sign spells the word 'Victoria' in separate letters made from 19 m of 15 mm glass tube. Calculate the secondary voltage of the transformer and the output power if the p.f. is 0.8.

Since the word 'Victoria' is made up of eight individual letters, eight pairs of electrodes will be required. Now the volt drop (VD) is given by

$VD = VD$ in tube $+ \, VD$ across electrodes
$VD = (19\,\text{m} \times 400\,\text{V/m}) + (8 \times 300\,\text{V})$
$VD = 7600\,\text{V} + 2400\,\text{V} = 10\,\text{kV}$

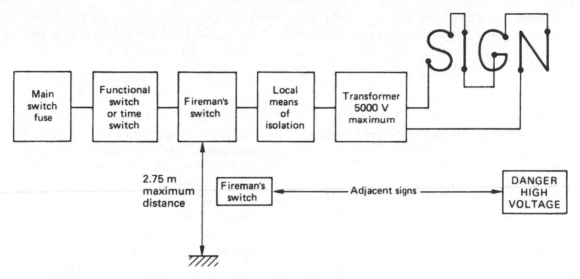

Fig. 5.21 Block diagram showing the layout of a high-voltage discharge lighting circuit.

The sign could be fed from a 10 kV centre tapped transformer since this would give 5 kV to earth and satisfy the requirements of BS 559.

$$\text{Power} = V/\cos\theta\,(\text{W})$$
$$\text{Power} = 10\,000 \times 35 \times 10^{-3} \times 0.8 = 280\,\text{W}$$

When installing high-voltage discharge lighting special attention should be paid to Regulations 554–02. A means of isolation complying with Regulation 476–02–04 must be provided at the point of supply and also adjacent to the control equipment for safe maintenance. The means of isolation should preferably be provided with a lock or removable handle so that the supply cannot accidentally be switched on when maintenance is being carried out.

The secondary voltage of the transformer must not exceed 5 kV to earth. Any transformer with an output of more than 500 W must be protected by a circuit breaker. The auxillary equipment such as transformers and capacitors must be contained in substantial metal enclosures and a 'Danger – High Voltage' notice placed on each container.

A fireman's emergency switch must be provided adjacent to the discharge lamps for all exterior installations and for interior installations which run unattended. The switch must be red with the off position at the top and mounted not more than 2.75 m from the ground (Regulations 476–03–05 and 537–04–06).

On interior installations the switch should be installed in the main entrance and a warning sign should be fixed alongside, stating 'Fireman's Switch' in letters not less than 13 mm high.

For the purpose of these regulations a high-voltage discharge sign installation in a closed market or arcade is considered to be an exterior installation. Only small portable discharge lighting, luminaires or signs with a rating not exceeding 100 W, fed from a readily accessible socket outlet, are exempt from these regulations.

Exercises

1 The illuminance directly below a light source of 1000 cd suspended 5 m above the surface will be equal to:
   (a) 20 lx
   (b) 40 lx
   (c) 100 lx
   (d) 200 lx.

2 A surface is illuminated to 125 lux by a light source suspended 4 m directly above the surface. The luminous intensity of the light source will be equal to:
   (a) 7.8128 cd
   (b) 31.25 cd
   (c) 2000 cd
   (d) 3906.25 cd.

3 The efficacy of a light source is measured in terms of the rate at which electricity can be converted to light. The units used are:
(a) amperes per candela
(b) lux per watt
(c) volts per candela
(d) lumens per watt.

4 The light output of a luminaire is reduced during its life because of the accumulation of dust and dirt on the luminaire and the interior decorations. The factor which takes this into account is called the:
(a) illuminance factor
(b) lamp lumens factor
(c) light loss factor
(d) depreciation factor.

5 A light source which emits the whole range of wavelengths corresponding to visible radiation can be said to have:
(a) good colour rendering properties
(b) poor colour rendering properties
(c) good efficacy
(d) poor efficacy.

6 Lamps which produce light as a result of the excitation of a gas or metallic vapour are known as:
(a) general lighting service lamps
(b) discharge lamps
(c) filament lamps
(d) incandescent lamps.

7 A fluorescent tube can be more accurately described as:
(a) an incandescent lamp
(b) a low-pressure sodium lamp
(c) a high-pressure mercury vapour lamp
(d) a low-pressure mercury vapour lamp.

8 The rating of the final circuits for discharge lamps may be taken as the rated lamp watts multiplied by a factor of:
(a) 0.75
(b) 1.5
(c) 1.8
(d) 2.0.

9 If eight 60 W fluorescent luminaires were connected to one final circuit, the assumed rating of the circuit would be:
(a) 7.5 W
(b) 48 W
(c) 480 W
(d) 864 W.

10 The secondary voltage of the transformer used for high-voltage discharge lighting must not exceed a voltage to earth of:
(a) 55 V
(b) 110 V
(c) 5 kV
(d) 10 kV.

11 Use a sketch to describe the construction and operation of a tungsten halogen lamp. State the lamp's efficacy, rated life, burning position and colour rendering properties.

12 Use a sketch to describe the construction and operation of a high-pressure mercury vapour lamp. State the lamp's efficacy, rated life, burning position and colour rendering properties.

13 Use a sketch to describe the construction and operation of a low-pressure sodium lamp. State the lamp's efficacy, rated life, burning position and colour rendering properties.

14 Use a sketch to describe the construction and operation of a high-pressure sodium lamp. State the efficacy, rated life, burning position and colour rendering properties of these lamps.

15 Sketch the circuit diagram and describe the operation of:
(a) a switch-start fluorescent lamp circuit
(b) a quick-start fluorescent lamp circuit
(c) a semi-resonant start fluorescent lamp circuit.

16 Use a block diagram to describe the control isolation and protection equipment to be considered when installing a high-voltage discharge lighting circuit.

17 State one typical application for each of the following lamps:
(a) GLS lamps
(b) tungsten halogen lamps
(c) low-pressure mercury vapour lamps
(d) high-pressure mercury vapour lamps
(e) low-pressure sodium lamps
(f) high-pressure sodium lamps.

18 Explain why the light output from a high-pressure sodium lamp does not reach a maximum value until some time after the lamp is switched on.

19 A high-pressure mercury vapour lamp takes some minutes to restrike when it is switched off and immediately switched on again. Explain the reason for this and the limitations which this places on such a lamp.

20 A large machine shop is to be illuminated by dis-
    charge lamps. Explain what is meant by the
    stroboscopic effect and how it can be eliminated
    in a machine shop.

21 A street lantern suspends a 3000 cd high-pressure
    sodium lamp 5 m above the ground. Determine
    the illuminance directly below the lamp and at a
    distance of 5 m from the base of the lantern.

22 A workshop measuring 12 m × 8 m is to be illumi-
    nated to an illuminance of 300 lux. The electrical
    specification requires that 1500 mm fluorescent
    luminaires be used, each having an output of
    3900 lumens. Calculate the number of luminaires
    required for this installation when the utilization
    factor and light loss factor are respectively 0.9
    and 0.7.

    Draw to scale a plan of the workshop and
    show a suitable layout for the luminaires.

SOLUTIONS TO EXERCISES

—

CHAPTER 1

1: D, 2: B, 3: C, 4: D, 5: A, 6: B, 7: B, 8: C, 9: B, 10: C, 11: D, 12: B, 13: D, 14: A, 15 to 30 Answers in text.

CHAPTER 2

1: C, 2: D, 3: D, 4: A, 5: C, 6: D, 7: C, 8: C, 9: D, 10: C, 11: B, 12: C, 13: D, 14: B, 15: A, 16: B, 17: C, 18: A, 19: D, 20: D, 21: A, 22: D, 23: A, 24: C, 25: C, 26: A, 27: B, 28: B, 29: B, 30: B, 31: A, 32: A, 33: D, 34: D, 35: C, 36: B, 37: B, 38: B, 39: A, 40: B, 41: C, 42: D, 43: A, 44: D, 45: C, 46: B, 47: B, 48: D, 49: C, 50: A, 51: C, 52: D, 53: D, 54: C, 55: D, 56: C, 57: C, 58: B, 59: D, 60: B, 61: A, 62: A, 63: D, 64: C, 65: D, 66: C, 67: D, 68: B, 69: A, 70: B, 71: A, 72: A, 73: B, 74: C, 75: D, 76: A, 77 to 92 Answers in text.

CHAPTER 3

1: A, 2: B, 3: A, 4: A, 5: D, 6: C, 7: C, 8: C, 9: C, 10: B, 11: A, 12: B, 13: C, 14: D, 15: A, 16: B, 17: C, 18: B, 19: C, 20: A, 21: B, 22: D, 23: B, 24: C, 25: B, 26: A, 27: C, 28: B, 29: D, 30: B, 31: D, 32: C, 33: A, 34: B, 35: D, 36: A, 37: C, 38: D, 39: C, 40: B, 41: D, 42: B, 43: B, 44: B, 45: C, 46: C, 47: B, 48: D, 49: C, 50: D, 51: D, 52: C, 53: C, 54: A, 55: A, 56: C, 57: C, 58: B, 59: B, 60: C, 61 to 100 Answers in text.

CHAPTER 4

1: C, 2: B, 3: A, 4: C, 5: D, 6: C, 7: C, 8: A, 9: C, 10: B, 11: D, 12: D, 13: A, 14: B, 15: A, 16: B, 17: B, 18: D, 19: B, 20: D, 21: C, 22: D, 23: B, 24: D, 25: B, 26: B, 27: D, 28: B, 29: C, 30: C, 31: D, 32: A, 33: B, 34: B, 35: D, 36: D, 37: B, 38: C, 39: C, 40: C, 41: B, 42: A, 43: $1.27\,\Omega$ 44: $190\,m\Omega$ 45: 25 mm 46: 100×25 47: 17 48: 5 49: 25 mm 50: 155 51: 75×25, 52 to 79 Answers in text.

CHAPTER 5

1: B, 2: C, 3: D, 4: C, 5: A, 6: B, 7: D, 8: C, 9: D, 10: C, 11 to 20 Answers in text 21: 120 lx and 42.43 lx 22: 11.72 or 12 luminaires.

APPENDICES

Appendix A: Abbreviations, symbols and codes

Abbreviations used in electronics for multiples and submultiples

T	tera	10^{12}
G	giga	10^{9}
M	mega or meg	10^{6}
k	kilo	10^{3}
d	deci	10^{-1}
c	centi	10^{-2}
m	milli	10^{-3}
μ	micro	10^{-6}
n	nano	10^{-9}
p	pico	10^{-12}

Terms and symbols used in electronics

Term	Symbol
Approximately equal to	$\simeq$
Proportional to	$\propto$
Infinity	∞
Sum of	$\sum$
Greater than	$>$
Less than	$<$
Much greater than	$\gg$
Much less than	$\ll$
Base of natural logarithms	e
Common logarithms of x	log x
Temperature	θ
Time constant	T
Efficiency	η
Per unit	p.u.

Appendix B: Health and Safety Executive (HSE) Publications and Information

HSE Books, Information Leaflets and Guides may be obtained from
HSE Books, P.O. Box 1999, Sudbury, Suffolk CO10 6FS

HSE Infoline – Telephone No. 01541 545500 or write to
HSE Information Centre, Broad Lane, Sheffield S3 7HO

HSE home page on the World Wide Web
http:/www.open.gov.uk/hse/hsehome.htm

The Health and Safety Poster (Figure 1.2) and other HSE publications are available from
www.hsebooks.com

Environmental Health Department of the Local Authority
Look in the local telephone directory under the name of the authority

HSE AREA OFFICES

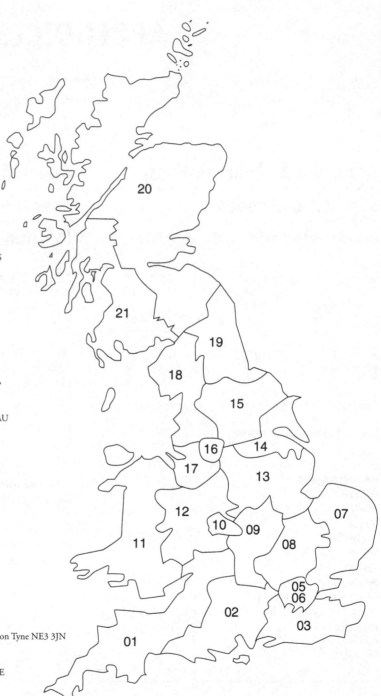

01 South West
Inter City House, Mitchell Lane, Victoria Street, Bristol BS1 6AN
Telephone: 01171 290681

02 South
Priestley House, Priestley Road,
Basingstoke RG24 9NW Telephone: 01256 473181

03 South East
3 East Grinstead House, London Road,
East Grinstead, West Sussex RH19 1RR
Telephone: 01342 326922

05 London North
Maritime House, 1 Linton Road, Barking,
Essex IG11 8HF Telephone: 0208 594 5522

06 London South
1 Long Lane London SE1 4PG
Telephone: 0207 407 8911

07 East Anglia
39 Baddow Road, Chelmsford, Essex CM2 OHL
Telephone: 0207 407 8911

08 Northern Home Counties
14 Cardiff Road, Luton, Beds LU1 1PP
Telephone: 01582 34121

09 East Midlands
Belgrave House, 1 Greyfriars, Northampton NN1 2BS
Telephone: 01604 21233

10 West Midlands
McLaren Building, 2 Masshouse Circus, Queensway
Birmingham B4 7NP
Telephone: 0121 200 2299

11 Wales
Brunel House, Nizalan Road, Cardiff CF2 1SH
Telephone: 02920 473777

12 Marches
The Marches House, Midway, Newcastle-under-Lyme,
Staffs ST5 1DT Telephone: 01782 717181

13 North Midlands
Brikbeck House, Trinity Square, Nottingham NG1 4AU
Telephone: 0115 470712

14 South Yorkshire
Sovereign House, 40 Silver Street, Sheffield S1 2ES
Telephone: 0114 739081

15 West and North Yorkshire
8 St Paul's Street, Leeds LS1 2LE
Telephone: 0113 446191

16 Greater Manchester
Quay House, Quay Street, Manchester M3 3JB
Telephone: 0161 831 7111

17 Merseyside
The Triad, Stanley Road, Bootle L20 3PG
Telephone: 01229 922 7211

18 North West
Victoria House, Ormskirk Road, Preston PR1 1HH
Telephone: 01772 59321

19 North East
Arden House, Regent Centre, Gosforth, Newcastle upon Tyne NE3 3JN
Telephone: 0191 284 8448

20 Scotland East
Belford House, 59 Belford Road, Edinburgh EH4 3UE
Telephone: 0181 225 1313

21 Scotland West
314 St Vincent Street, Glasgow G3 8XG
Telephone: 0141 204 2646

INDEX

—